Performance and Physics of Bicycles

Manu Konchady

Mustru Publishing,
Frisco, TX

Performance and Physics of Bicycles
by Manu Konchady

Mustru Publishing,
11608 Lenox Lane,
Frisco, TX 75033.

ISBN 978-1-79476-615-0

The author has taken every precaution to verify the contents of the book, but assumes no responsibility for errors or omissions in the book and any damages resulting from the use of the information contained herein.

All brand names and product names mentioned in this book are trademarks or service marks of their respective companies.

Any omission or misuse (of any kind) of services or trademarks should not be regarded as intent to infringe on property of others.

The author recognizes and respects all marks used by companies, manufacturers, and developers as a means to distinguish their products.

Contents

Preface

A bicycle is an uniquely efficient machine, but elements such as the wheels, tyres, frame, and brakes will affect the performance of your ride. This book answers questions like –

How much faster will you ride with a light bicycle? Bicycle manufacturers go to great lengths to reduce the weight of a frame and wheels. In Chapter 3 you can estimate your gain in speed from a lighter bicycle and evaluate the drag force due to wind and slope.

Should you reduce air drag or rolling resistance? Air drag becomes significant at higher speeds and it will take more power to sustain the same speed. Chapter 3 explains how much power you will need to maintain speed and the affects of wind on speed. A uphill slope will also slow you down. Chapter 4 describes rolling resistance and how it is measured. Although not as significant as air drag, rolling resistance is also a constant drag force that is relatively low at high speeds.

Will wide tyres slow you down?, What is the "right" tyre pressure? Road bicycles usually have narrower tyres than mountain or hybrid bicycles. A common assumption was that narrow tyres are more aerodynamic and therefore more efficient. The explanation in Chapter 4 reveals that there is more to efficiency than just the width of the tyre. Also what is a suitable tyre pressure given the width and diameter of a tyre and the load due to the rider.

How fast can you turn? Does a lighter wheel help? Taking a turn at a safe speed is one of the thrills of riding a bicycle. Chapter 5 describes the physics of turning on a flat and banked road. A lighter wheel does help to accelerate quickly and in chapter 5 estimates of the power needed to accelerate at different speeds are computed.

Why are front brakes more effective than rear brakes? What is a safe braking distance? In chapter 6 the reason why front brakes are far more effective than rear brakes is described. Also included are the safest braking forces that you can apply on a slope, wet road, and dry road before you will fall over. Several methods to calculate the safe braking distance are evaluated given road conditions.

Are you safer riding in a bicycle lane? A frequently cited reason for not cycling is that it is a risky mode of transportation. In Chapter 8 the reasons for traffic accidents, the types of accidents, and some of the ways of avoid an accident are elaborated. Some of the public domain statistics on accidents are evaluated to identify the most common accidents and locations on the roadway.

Android App to track rides: GPS Trip Analyzer is an ad-free app to track your rides (Chapter 7). You can view your speed, altitude, latitude, longitude, gradient, satellite signal strength, power, wind magnitude and direction, temperature, and bearing at locations on your route. The app uses the Google Maps and Openweather APIs to collect altitude, temperature, and wind data (requires API keys). A calculator to compute power or velocity given bicycle and rider parameters is also included.

Benefits of cycling: The first chapter describes the benefits of cycling for an individual and society. Several studies have shown that cycling has proven benefits. However, some of these benefits are offset by pollution. There are thresholds for the duration of a ride beyond which cycling loses any benefits. The degree of pollution varies across cities and a bicycle ride in a highly polluted city does have a higher health risk than a ride in a less polluted city. A common reason for not riding bicycles is the perceived risk of riding on a congested road with fast moving motorized vehicles. While the risk is undeniably greater for a cyclist than other motorists, riding predictably and following traffic rules can mitigate the risks.

Sources: Data for the analysis in this book was collected from government websites as well as public domain sources. The book includes over one hundred plots and one hundred illustrations to describe the performance and physics of bicycle components. Citations for the public domain images used in the book are included. Equations explaining how components perform and work are included for the interested reader.

Some of the Python scripts that were used in the book are available at `https://github.com/mkonchady/cycle`. The source code for the Android GPS trip analyzer app that accompanies the book is maintained at `https://github.com/mkonchady/MyTripoo`. Please report bugs, errors, and questions to mkonchady@gmail.com. Your feedback is valuable and will help to correct errors in the book. I have attempted to make the contents of the book comprehensible and correct. Any errors or omissions in the book are mine alone. The accompanying code is for demonstrative purposes and comes with no warranty.

1 Introduction

A bicycle is one of the most efficient and fun modes of transport. Among human powered vehicles, none can match the efficiency of a bicycle. The bicycle is one of the few human powered machines that uses muscles in a near optimal way. Since leg muscles are among the most powerful muscles in the human body, attaching a crank and pedals to a wheel was an obvious way to power a bicycle. The use of pedals to harvest human energy was common in other machines such as lathes, saws, cutting machines, and power generators.

The circular pedalling motion uses the powerful thigh muscles or quadriceps to power a bicycle. Most of us can generate sufficient power to ride a bicycle - other factors to consider when riding a bicycle are matching individual power with the load. Each of us is unique and has the ability to generate varying amounts of power for some duration. A professional cyclist can generate 300+ watts [1] for several hours and is roughly twice as powerful as an amateur cyclist. If we assume that you can generate 125 Watts, then in an ideal situation you would match your generated power to a smaller load than the load of a professional cyclist.

A geared bicycle makes it possible to adjust the load to match your generated power. At a high gear for a faster speed, the load is proportionately higher compared to the load for a low gear. If you attempt to generate more power than your body is capable of for a sustained period, then you are certain to tire soon and end your ride. However, with a geared bicycle you can adjust your load such that it matches your generated power. The addition of gears and derailleurs (see Glossary) to a bicycle made it possible for the average rider to travel long distances without becoming exhausted.

1.1 Reasons to Ride a Bicycle

You need some mode of transportation to get to work or shop and a bicycle is a great alternative to the common motorized vehicle options. You may even find that you sometimes reach your destination sooner on a bicycle than with other vehicles. While spending time riding to work or on an errand, you are actually burning calories as well and possibly saving time and money visiting a gym. At first, you might find riding a bicycle is slow and tiring. It may also be discouraging to note passengers in motorized vehicles overtake you effortlessly. However, in a congested city, the differences in commute times between a bicycle and other vehicles is not as much as you would expect.

1.1.1 Health Benefits

One of the benefits of riding a bicycle besides saving money on fuel is improved health. A study [1] of about 14, 000 people has shown that riding a bicycle for just 30 minutes a day reduces the risk of type 2 diabetes by about 40%. Another study [2] of 2400 individuals showed that people

---

[1] A watt is the product of force and velocity. One horsepower (HP) is 746 watts

who rode about $2\frac{1}{2}$ hours a week were nine years "biologically younger" than their sedentary counterparts.

Manage your Weight

Among the health benefits are a stronger heart, close to ideal weight, and improved joint mobility. An average male and average female consume about 2500 and 2000 calories per day respectively. If you burn more calories than what you consume, you are more likely to lose weight. You could also consume less calories (go on a diet) and similarly lose weight. However, riding a bicycle and burning calories is definitely a better option than going on a diet – not only is it more fun, but you can eat at least some of the food that you would have avoided on a diet. If you ride a bicycle for an hour at a moderate pace ($20-24$ kmph) you can burn roughly $500-600$ calories. This of course depends on the gradient of your ride - a climb would mean an even higher burn rate while a downhill ride will do the opposite.

The amount of energy (calories) that you need to function in a day while resting, varies from person to person. This rate, also called the "Basal Metabolic Rate" (BMR) is the number of calories required to support your normal body functions. BMR is roughly calculated using the Mifflin St. Jeor Equation – $10 \times w + 6.25 \times h + x$ where w is your weight in kgs., h is your height in cms. and x is $+5$ for males and -161 for females. Over time as you continue to ride and your fitness increases, your BMR rises and therefore you burn more calories even when sedentary.

In theory, you should be able to control your weight by simply managing calorie intake and calories consumed through exercise and BMR. However, all calories are not equal and how the body processes calories depends on the source. All the calories from fibrous foods may not be absorbed while calories from proteins may require more energy to digest and there is some margin of error in food labels. In general, processed foods require less energy to digest than natural foods.

The benefits of a ride do not end when the ride ends. Following a ride, the body continues to consume calories at a higher rate than the normal rate without any exercise. Cyclists who rode vigorously for 45 minutes burned an additional 190 calories in the 14 hours following a ride compared to days when they did not ride [3].

There is no universal agreement on an ideal weight per person. The Devine formula for a male who is 67 inches tall in kgs. is $50 + 2.3 \times (67-60) = 66.1$ kgs., similarly for a female of the same height it is $45.5 + 2.3 \times (67-60) = 61.6$ kgs. The ideal weights from other weight formulas are in a range of $\pm 5\%$ of the values shown. The Body Mass Index (BMI) is another value to check if your weight is within a healthy range. It is the ratio of your weight in kilos to your height in meters squared. So for a person who is 67 inches (1.7 meters) tall and weighs 66 kgs., BMI is $\frac{66.0}{1.7^2} = 22.83$. A BMI in the range of $18-25$ is generally considered healthy.

Generally, gaining weight is considered easier than loosing weight. Further once weight is lost, it takes some effort to keep the weight loss permanent and prevent a gradual increase to the original weight. Cycling like other physical activities consumes calories and the intensity of your ride is proportionate to the number of calories consumed. A ride on unpaved or pot holed roads will burn more calories than a ride on a smooth road. Finally, the best way to burn calories is to ride uphill at a moderate pace – not only will your heart rate and power output increase, the number of calories consumed over a short period will rise sharply. So, if you do not have the time for long rides, then riding up and down hills is a great option to get a quick workout that is as beneficial as a longer ride on flat terrain.

Caveat: Since bicycling is a form of exercise, you will have a stronger appetite following a ride. Loosing weight by consuming fewer calories and burning more calories through exercise, is hard to accomplish. Eventually you may end up consuming as much or more calories necessary to regain the lost energy. Therefore, bicycling will help you maintain your weight, but not necessarily lose weight.

Although you may have a larger appetite after exercise, your diet maybe healthier compared to those who don't exercise. A study [4] of 2680 adults revealed that following a 15 week exercise training program, the participants were more likely to choose healthy foods in their diets.

Strengthen your Heart

Cycling can also improve the health of your heart, lungs, and circulation. Since your heart rate increases as you cycle, you will strengthen your heart muscle and also lower your resting pulse rate. A Harvard study of $29,000$ people for 10 years [5] found that the participants whose pulse rate (beats per minute) increased from under 70 to over 85 were about 90% more likely to have died during the study. In general aging tends to speed up the pulse rate while regular exercise tends to slow it down. Some of the best cyclists have pulse rates of 50 or less. However, a lower pulse rate is just one of the factors in evaluating your health.

Since cycling is a cardiovascular activity, it reduces the risk of heart disease. A study by the British Medical Association of $10,000$ participants found that those cycling about 32 kms. a week were 50% less likely to develop heart disease [6]. You can also reverse heart damage (stiffening of the heart muscle) provided you start riding before your mid-60s after which the heart loses some of its flexibility. However, the activity should be intense enough to raise your heart rate to about 70% the maximum heart rate (MHR). The MHR is roughly calculated as $220 - age$, so the MHR for a 40 year old would be 180 beats per minute. Exceeding 80% of MHR is not advised and benefits from physical activity will be lost due to the higher risks and consequences of over training. If you don't have a heart monitor to track your current pulse rate, you will know that you approaching the limit when you begin to feel light headed and short of breath. The risks of cycling near your MHR will outweigh the benefits of the exercise and the time to become healthier does not reduce with workouts at or near your MHR. An alternate formula for MHR has been suggested for healthy adults, $208 - 0.7 \times age$, which gives a slightly higher MHR for adults over the age of 40 than the prior formula.

What is your "Real" Age?   Even though two individuals may share the same birth date, they can age differently depending on their diet, environment, and exercise. Researchers [7] have found that exercise capacity is a good indicator of life expectancy. The energy that you expend will depend on the type of activity. If you are at rest, then the energy you generate will be much less than the energy you need to ride a bicycle at 20 kmph. A metabolic equivalent of task or MET is one method of quantifying energy expended depending on the type of task.

The Compendium of Physical Activities (`https://bit.ly/1oAWE2t`) publishes a table of METs associated with a number of activities including bicycling. Riding a non-racing bicycle at 15 kmph. requires 5.8 METs, while riding at over 32 kmph. will need 15.8 METs. As you increase your riding speed, you will generate more power and your heart rate will also increase. The heart rate of a more fit individual will increase less rapidly than the heart rate of a sedentary person. Therefore, the peak METs that a fit individual can achieve at the maximum heart rate will be higher, even though both the fit and sedentary person may share the same chronological age.

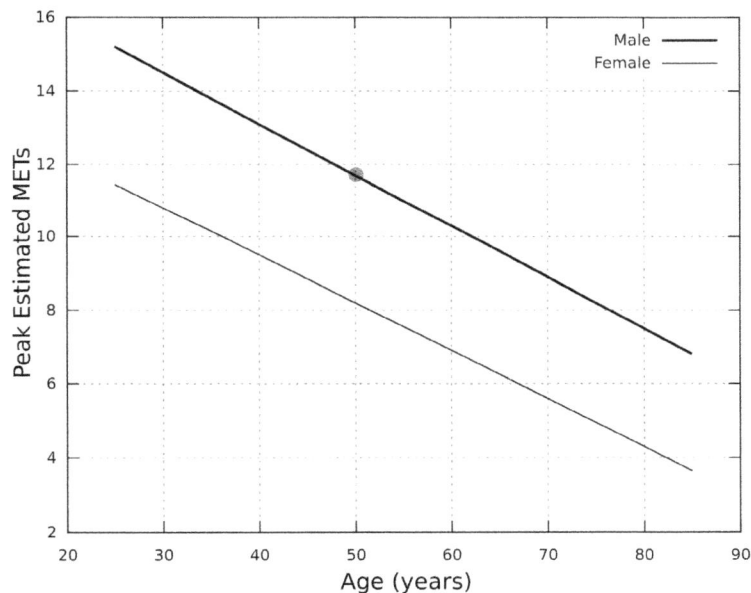

Figure 1.1: Metabolic Equivalent of Task vs. Age in years using the St James Take Heart Project formula

In Figure 1.1, the peak estimated METs is calculated using St James Take Heart Project formula [7]. For men and women, peak $METs = 18.7 - (0.15 \times age)$ and $14.7 - (0.13 \times age)$ respectively. As you age, you can expect to lose an average of 0.14 METs per year. If you are male and your chronological age is 60, but your peak estimated METs is 11.7, then your "real" age is about 50 years. Unfortunately, computing the peak METs for a person is usually performed in a laboratory with a treadmill test. Although the Compendium of Physical Activities publishes the required METs for various bicycle speeds, there are many variables that can affect your speed including gradient, wind, type and conditions of bicycle and road surface,

Other Health Benefits

High blood pressure (BP), also called a "silent killer" because it usually has no warning signs, can cause a heart attack or stroke. The guidelines for normal BP are < 80 for diastolic (expansion) and < 120 for systolic (contraction). The numbers are measured in millimeters of mercury and indicate pressure on the walls of arteries. For people with high BP, there are several methods to lower BP. Among them is cycling three times a week for an hour, that can lower your BP by five points or more [8]. The other methods include using medication, diet, and meditation.

Other health benefits from cycling and other forms of exercise include a reduced risk of cancer, a stable blood sugar level, a lower risk of depression, improved digestion, lower risk of impotence, protection from arthritis, better self esteem, improved productivity, lower stress levels, fewer illnesses, higher bone density, improved sleep, and happiness [9].

Figure 1.2 lists some of these benefits. These benefits are not necessarily unique to cycling and may also be found in other forms of exercise. Cycling does boost your mental health and can lower anxiety and depression. Further riding at a high intensity will elevate your heart rate leading to better cardiovascular fitness. Training regimen such as the High Intensity Interval Training (HIIT) can over a short daily period of about 30 minutes, lower body fat (see Chapter 9).

A fairly more obvious benefit is higher muscular strength. At first, riding uphill may seem to take a great deal of effort, but over time as you build muscular strength from repeated efforts, the ride uphill becomes easier and more manageable. A short ride of three kilometers in a medium gear (3 meters per pedal revolution) will require you to flex your knees 1000 times. This repeated flexing of the knee will maintain joint mobility. Longer rides of 100 kilometers or more will improve your stamina and confidence. For those concerned that cycling may damage their sexual or urinary health, a study in the Journal of Urology of over 2500 cyclists has shown no significant difference in health between those who cycle, swim, or run [10].

After the age of 35, you could lose 1% of muscle mass per year, without some form of physical activity. This process of aging cannot be stopped, but can be limited. A study tested one of the largest muscles in the human body that is attached to the top of the femur and the kneecap at the lower end, of "highly active" male and female cyclists between the ages of 55-79. A "highly active" male was defined as someone who could cycle 100 kms. at 15+ kmph. and for a female the required distance was 60 kms. at 12+ kmph. This muscle was chosen since any age related deterioration should be more visible in such a large muscle. The study found that there was less muscle deterioration and loss of immunity in the group of cyclists compared to their sedentary counterparts.

Do the Health Benefits of Cycling Outweigh the Risks?

This is a logical question for those concerned about riding in city traffic with no designated bike lanes. In Chapter 8, ways to reduce risk while riding a bicycle will be described. The argument in favour of cycling is that it reduces air pollution, lowers greenhouse gas emissions, and increases

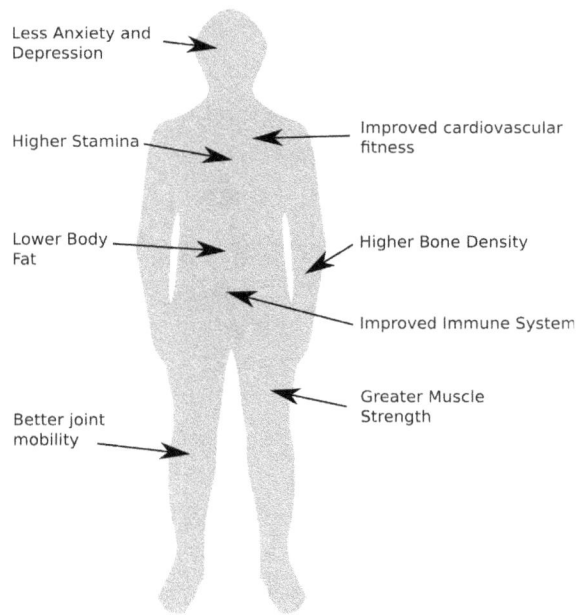

Figure 1.2: Some of the Health Benefits of Cycling

physical activity. Governments in some European capitals have developed policies to encourage cycling as a mode of transportation for short trips. The goals behind these policies were to reduce traffic congestion and improve public health. The bicycle sharing system has become popular in many cities - citizens do not need to own a bicycle and instead rent a bicycle when they need transportation. A bicycle can be rented at point A and returned at point B making a round trip optional. For someone who uses a bicycle on a regular basis, the ownership costs are nominal (see 1.1.2) compared to the cost of a motorized vehicle.

While it is clear that commuters will benefit from more physical activity and society benefits from lower pollution, there is the risk of inhaling more pollutants, since breathing rates increase while cycling and also the chances of a severe or fatal injury to a cyclist is higher than an injury to a person shielded in a car or bus. Pollution rates are also the highest in the most congested parts of a city where a move to use cycles is encouraged. A Dutch study [11] concluded that the estimated health benefits of cycling were substantially larger than the risks relative to car driving. The study evaluated the effects on a population of half a million between the ages of 18-64, if individuals chose to cycle instead of driving for short round trips of between 7.5 and 15 kms.

1.1.2 Monetary Benefits

While the health benefits alone should be sufficient to persuade you to ride a bicycle, the monetary benefits are an additional bonus. Estimates of your savings will vary widely depending on how much you commute and the cost of your current vehicle plus fuel.

One way to compute costs is to use a government estimated cost to run a motor vehicle. In 2018, the U.S. Internal Revenue Service code allowed a deduction of 0.34 USD per kilometer, for business travel expenses by car. This operational cost can be compared with a similar cost for maintaining a bicycle. Since the annual distances covered by riders varies widely, we can broadly categorize bicycle riders by their mileage (in kilometers) per month in Table 1.1.

Category	Monthly Distance (kms.)	Car Costs	Bicycle Costs	Savings
A	0-100	34.00	8.00	26.00
B	100-300	102.00	10.00	92.00
C	300-500	170.00	12.00	158.00

Table 1.1: Car vs. Bicycle Monthly Costs

The bicycle maintenance costs are based on a rough estimate of 100 USD per year. If you consider the annual maintenance costs, then a category B rider saves over 1000 USD per year. The purchase cost of the average bicycle and car are vastly different, but are not included in this comparison since a bicycle cannot replace a car if you need to transport 4 people and luggage. Only the operational costs are shown since a bicycle can replace a car when the purpose is to commute to work.

While these are U.S. cost estimates, the estimates in other countries will vary depending on the cost of fuel and maintenance of a car. The cost of maintaining a bicycle is minimal compared to the cost of maintaining a car. To begin, all the components of a bicycle are easy to access and you can replace parts with a small set of tools. Secondly, a half hour to an hour a week is enough time to maintain a bicycle. Finally, you may have to visit your local bike store just once a year for a job such as replacing a cassette, changing cables, or truing a wheel.

1.1.3 Societal Benefits

Does the ownership and usage of bicycles have any benefits for society? The ownership of bicycles worldwide was summarized in a study [12] at the Johns Hopkins University (JHU) from a collection of surveys conducted in over 140 countries. In the 20th century, there was a rapid change in the mode of transportation from bicycles to motorized vehicles. With greater prosperity and affordability of cars, motor cycles, and scooters, there was little incentive to ride bicycles. The perceived benefits of a motorized vehicle were hard to resist. The speed of even an athletic bicyclist could not match the speed of a rider on a scooter. This was specially true if a route had an uphill gradient and besides the ride was practically effortless compared to a bicycle ride.

GDP and Bicycles

However, the transition from bicycles came at a cost to public health - increased air pollution, more traffic accidents, and less physical activity. In the $70s$ with soaring energy costs, a few countries began looking at alternatives to the prevailing modes of transportation and the bicycle was appealing for its simplicity, affordability, and health benefits. Despite the obvious benefits, cycling did not regain popularity in all countries. In fact, cycling was embraced in some countries and ignored in others. In the study [12] from JHU, the percentage of ownership of cycles ranged from a low 4% in Armenia to a high 84.2% in Burkina Faso.

Group	Avg. Ownership%	No. of Countries	Avg. GDP/Capita
1	80	9	41,127
2	60	34	24,224
3	40	45	8,896
4	20	56	2,878

Table 1.2: Bicycle Ownership and GDP (USD) / Capita in 144 countries

The list of 140 countries in the survey were divided into four categories based on the mean ownership percentage (see Table 1.2). Clearly, very few countries have a high ownership percentage, while the majority of countries have less than 50% ownership. There is a positive correlation (0.6) between groups and the mean GDP (USD) per capita per group [13]. Groups with a higher bicycle ownership percentage tend to have a higher GDP per capita and the difference between the first and last group is substantial (see Figure 1.3). However, simply having a high bicycle ownership percentage does not guarantee prosperity. Group 1 included countries such as Denmark, Netherlands, and Sweden and an outlier Burkina Faso.

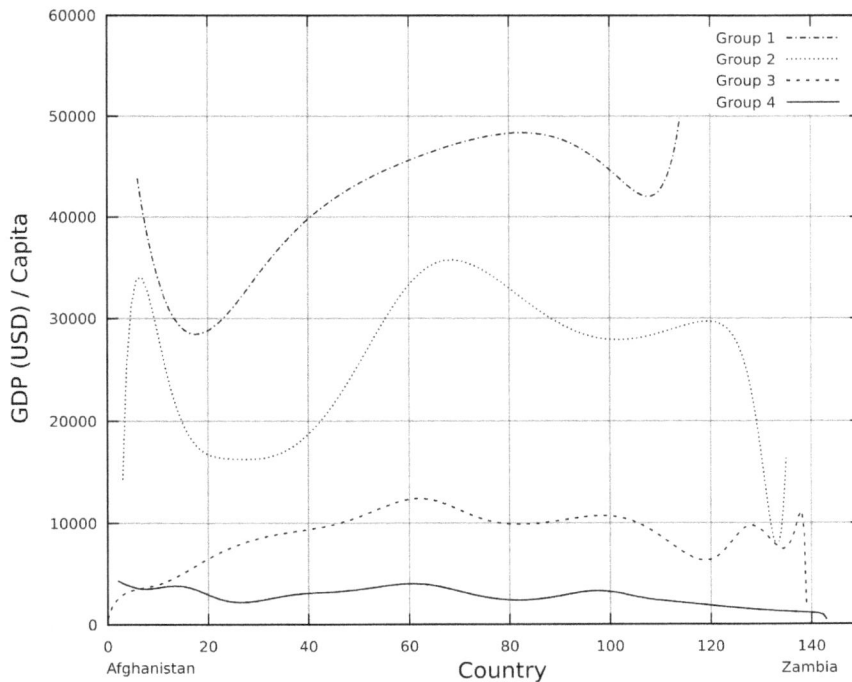

Figure 1.3: GDP (USD) per Capita vs. Country by Group

Membership in the groups of countries was not based on geographic location. Two neighbouring countries such as Ethiopia and Uganda had a large difference in ownership percentage. Differences in ownership percentage also changed over time. China had an ownership percentage of 97% in 1992, which declined by almost half to 48% in 2007. Dumping the humble bicycle in favour of a motorized vehicle was seen as a step up the social ladder and an indicator of greater individual wealth. This decrease coincided with a rapid increase in GDP per capita which would appear to counter the argument that bicycle ownership is associated with greater property.

The move away from bicycles to other modes of transportation did have consequences that were not foreseen or ignored by the governments in countries like China and India. Meanwhile, other nations like the US and European countries that had adopted motor vehicles earlier, decided to combat the accompanying air pollution through regulations for auto manufacturers and unleaded fuel. Still in 2013, the Union of Concerned Scientists concluded that the single largest source of air pollution in the US was transportation [14].

Pollution and Bicycles

Air pollution that is visible in the atmosphere is as harmful as the smaller invisible particles to our health and environment. The visible part of air pollution is seen as a haze in the atmosphere and is made up of sulphur dioxide, ozone, and other particulate matter (PM). Although pollutants such as carbon monoxide, hydrocarbons, and nitrogen oxides are invisible, they are also equally responsible for poor air quality. Some of the pollutants are emitted by vehicles, while others are formed by chemical reactions between pollutants and sunlight.

Particulate Matter

One of the most commonly measured pollutants is $PM_{2.5}$ or particulate matter that is less than $2.5\mu m$ in diameter. Since the size of these pollutants is measured in millionths of a meter, they have low weight and therefore can stay suspended in the air that we breathe, for many days or even weeks. The focus here is on $PM_{2.5}$ alone, since the number of deaths due to traffic accidents is comparable to number of deaths from exposure to $PM_{2.5}$ [15]. The World Bank publishes a list [16] of the average annual mean $PM_{2.5}$ for over 140 countries and in Figure 1.4 the group 4 countries tend to have higher pollution levels than countries in other groups. In a list [17] of the top 500 polluted cities ranked by $PM_{2.5}$ between 2008 and 2015, over half of the cities were located in China (179) and India (97). Countries with a higher bicycle ownership percentage also tend to have lower air pollution measured by $PM_{2.5}$ (see Figure 1.4).

The $PM_{2.5}$ levels also increased over time - in China the average levels increased from $48\mu g/m^3$ to $58\mu g/m^3$ between 1990 and 2015, with the cities having much higher pollution levels. While this period coincided with a fall in bicycle ownership percentage by almost half, there were other reasons for the rise in pollution levels. Yet, higher usage of alternate modes of transportation did contribute to the higher levels of pollution. There are other pollutants such as sulphur dioxide

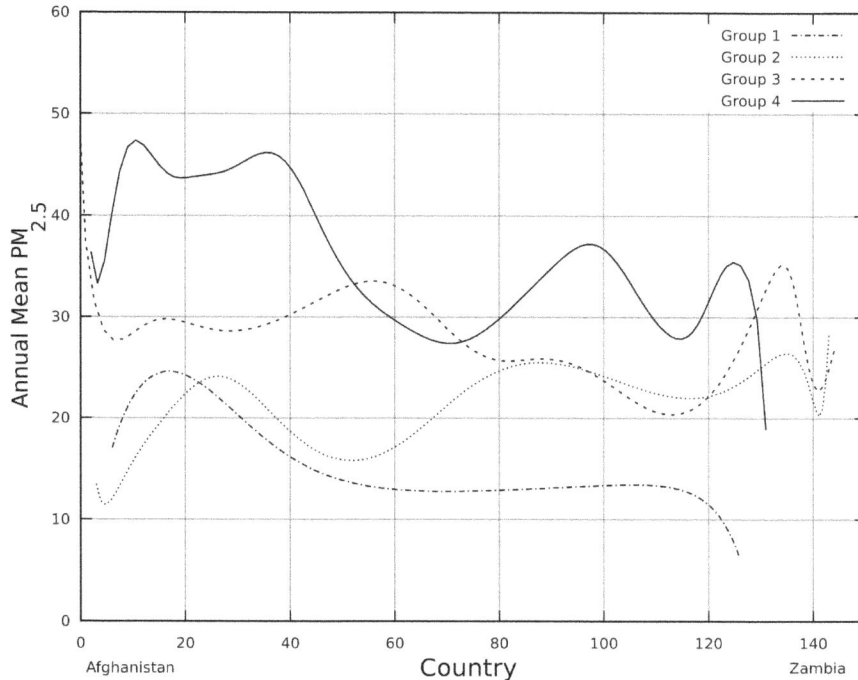

Figure 1.4: $PM_{2.5}$ vs. Country by Group

and carbon monoxide generated by motorized vehicles not considered here, but which are also equally harmful to health.

Carbon Dioxide

While carbon dioxide is not considered as harmful as particulate matter, it has steadily increased and accumulated in the atmosphere leading to climate change worldwide. Transportation based CO_2 emissions are one of the leading causes of this accumulation in the atmosphere. On a per capita basis, countries like India and China have lower CO_2 emission rates compared to other countries in the West. However, the large population of India leads to a high CO_2 country emission rate that in turn means it is one of the top five CO_2 generating countries with about six percent of worldwide emissions. With the growing evidence of the harmful effects on climate from unrestrained emissions of CO_2, there is a need to encourage non motorized and public transport [18].

Cars and Bicycles   A car with an internal combustion engine runs at about 15% efficiency [19], a diesel car at about 35% efficiency and an electric car at about 80% efficiency. A well maintained bicycle runs at over 95% efficiency. This means that if you apply a force of 100 N on your pedals, then 95 N will be transferred to the wheel. While the bicycle runs at high efficiency, the human body is not as efficient in generating forces. Chemical energy stored in the body is converted to mechanical energy to apply pressure on the pedals at about 25% efficiency.

17

The car itself may weigh 10-15 times the weight of a single passenger. The typical city bicycle may weigh 14 kgs. which is 20% the weight of an average 70 kg. cyclist. According to an estimate of the usage of fuel in a car [19], just 1% is used to actually transport a passenger with the remaining used to the transport the car or converted to waste heat energy.

In 1991, the city of Münster, Germany commissioned a photo to compare the space occupied on roads by public transportation vs. bicycles vs. cars. The experiment showed that in the space it takes to accommodate 60 cars, cities can accommodate around sixteen buses or more than 600 bikes (see Chapter 8). Further bicycles take much less space compared to a car, to park at home.

1.1.4 Mental Benefits

Besides saving money and improved physical health, studies have shown that exercise can reduce the rate at which muscles including the brain shrink with age. Maintaining a bigger and more connected brain implies better memory, higher concentration levels, and greater problem solving skills. Riding a bicycle is exercise that can be easily combined with other activities such as commuting or shopping.

Build a Healthier Brain

While cycling does increase your heart rate, it also produces chemicals used to form new brain cells. The results are better memory, stronger concentration, and improved reasoning and problem solving ability. However, not all studies showed an improvement in cognition with increased physical activity. It is not necessarily true that the students who exercise the most will perform the best in academics.

To understand the mental benefits from physical activity, a study [20] in India evaluated a small sample of 10 medical students in their 20s. In the study, the subjects were given similar cognitive tests on two separate days. On the first day, they were evaluated on a standard test without any recent physical activity and on the second day they were given similar tests following a 30 minute cycle ride at about 70% of MHR. There was a consistent increase in scores for questions that required memory, reasoning, and planning concurrent with a decrease in the time to take the test. The conclusion was that even a single bout of exercise can improve an individual's cognitive abilities.

While a short cycle ride before an exam may improve the results of a student taking an exam, there are reasons for adults to consider cycling as well. The size of the human brain shrinks at a rate of about 0.4% per year for healthy adults and this rate increases later in life. While brain shrinkage with age is unavoidable, it can be limited with physical activity like cycling. In a study [21] of 59 participants aged between 60 and 79 for six months who were divided into two groups - an aerobic and a control group, it was found that the participants in the aerobic group not only increased their oxygen uptake, but also increased their volume in areas of the brain that typically shrink with age. The shrinking in areas of the brain that is associated with ailments such as Dementia or Alzheimer's disease can be limited through aerobic exercise like cycling.

If a short period of exercise improves mental acuity, then it would seem logical to expect a long workout to generate even greater benefits. However, this is not true. Long endurance rides will not turn you into a scholar, you need some time to recover before using your mind for anything challenging. A ride of between 30-60 minutes at about 75% of maximum heart rate is sufficient to stimulate the brain for a few hours following.

Happiness

Exercise including cycling boosts the production of feel-good chemicals like dopamine and serotonin. These chemicals reduce the risk of depression and can at least temporarily elevate your mood. However, duration is not proportional to happiness. Taking a very long ride will not necessarily make you more happier than if you rode a half hour.

Cycling allows you to explore further than you would have while running or walking and finding new places or scenic routes is an added incentive to ride. Being outside you neighbourhood implies that you have to be prepared to handle breakdowns and you will need a sense of self sufficiency to tackle simple mechanical problems away from home. Once you start riding without a great deal of effort, you can also think, reflect, and come up with new ideas on a long ride.

1.2   Riding in Netherlands vs. New Delhi

Average pollution levels ($PM_{2.5}$) vary across the globe from $< 50 \mu g/m^3$ in the Netherlands to $> 180 \mu g/m^3$ in New Delhi. The harmful effects of pollution on the cyclist will be correspondingly different depending on the locale. There is a sharp contrast between cycling in a polluted vs. non-polluted environment.

Netherlands

A study[11] evaluated the outcome if 500,000 people rode bicycles instead of driving a car for short trips ranging from 7.5 to 15 kms. Some of the rides were round trip rides to work or shop. The risks and benefits of the two modes of transportation, cycling and driving, were compared in the study. A rigorous comparison is hard, since there are many factors to consider such as the route, speed, open window / close window, trip duration, and weather conditions.

Air Pollution: The study specifically looked the impact on mortality, i.e. life years gained or lost from the switch to cycling. When 500,000 people stop driving and start cycling, there is an obvious reduction in air pollution. Transportation is one of the leading causes of air pollution and may contribute upto 25% or more of total air pollution. Given that some percentage of the population will continue to drive, a cyclist will continue to be exposed to air pollution.

A cyclist exerts more effort than a driver during a commute and therefore will need to inhale more often during a ride. While the inhaled doses of $PM_{2.5}$ for a cyclist maybe twice as much as that of a car driver during the same period, the difference in the total inhaled dosage per day between the two modes of transportation was much less. This is due to the fact that during the rest of the day, both populations are exposed to the identical pollution levels. When the ratio of total exposure levels between a cyclist and driver was considered, it was about 1.1 or lower. Therefore, the additional health risk due to increased air pollution exposure is not as substantial as one would imagine, provided the commute duration is not excessive.

Accidents: The second risk for a cyclist on a road that is shared with other vehicles is the higher probablity of an accident. Clearly, since a cyclist has no steel or glass shield for protection, the risks from an accident are higher. The number of reported traffic deaths due to cycling was about 5.5 times higher than a driver per kilometer. However, this difference is probably exaggerated since motor vehicles are often used for longer journeys on highways and therefore the fatalities per kilometer will be lower. When highways were excluded, the mortality rates for were similar for cyclists and drivers (see Chapter 8).

These two risks faced by cyclists - a higher risk of a traffic accident and increased inhaled pollution - were offset by the benefits of greater physical activity. The estimated gain in life expectancy due to cycling was about 8.5 months, while loss from the risks was much smaller at 20 days. In a country like Netherlands, where air pollution is lower and bicycling lanes are widespread these conclusions are not unexpected. But, how does a polluted city like New Delhi with little bicycle infrastructure compare?

New Delhi

A study [15] in New Delhi evaluated the levels of $PM_{2.5}$ pollution for 11 transport modes including cycling. The pollution levels in the city were strongly dependent on the season - in winter (November - February), the pollution levels ($225\mu g/m^3$) were about 2.5 times that of levels during the monsoon (July - September). Pollution due to transportation is relatively constant throughout the year and the seasonal change in pollution levels is due to other added pollutants, particular to winter.

Air Pollution: The study evaluated air pollution levels on a busy street in New Delhi from numerous trips back and forth on the same street at different times of the year. The ambient concentration is the pollution level outdoors and is the base level against which the increase / decrease in exposure for various transportation modes can be evaluated.

The ratio of the in-vehicle to ambient concentration is referred to as γ. One would expect the γ value for an air conditioned vehicle to be less than one, while γ for a cyclist would be greater than one. In New Delhi, the value for γ for a cyclist was 1.1 and actually lower than the γ value in some European cities. This is due to a higher concentration of pollution along the roadways in European cities, leading to a lower ambient concentration. Transport is one of the leading causes of pollution in a developed economy, while in New Delhi there are numerous other sources of pollution besides transportation.

A typical cyclist is possibly the slowest vehicle on the road and therefore rides on the left extreme (in left hand traffic). Unlike a bicyclist, riders of motorized two wheelers can accelerate and manoeuver within a stream of traffic with ease. However, this ability to move swiftly between vehicles across the road leads to greater exposure to pollution. The γ level for a motorized two wheeler was the highest compared to all other mode of tranportation, at 1.7. While a cyclist cannot ride within traffic like a motor cycle rider, the benefits are exposure to a slightly lower pollution levels.

When the surrounding pollution levels rise beyond $150 \mu g/m^3$ (in winter), the differences between exposure levels for all modes of transportation practically vanishes. Everyone faces similar health risks due to very high pollution levels, except for occupants in air conditioned cars or passengers in underground trains.

Inhaled Dose: Unfortunately, a cyclist has to inhale a higher volume of air to get sufficient oxygen to power a bicycle. The inhaled dose of $PM_{2.5}$ for the same distance is roughly four times higher for a cyclist compared to a motor cyclist. This is due to the longer duration spent by a cyclist on the road for the same distance and the greater volume of air inhaled. One solution for some bicyclists is to ride fast. The benefits of lower exposure to pollution from a shorter duration ride outweighs the harm caused by increased respiration to support a higher speed.

An individual cycling for a hour in New Delhi will have a higher dose of $PM_{2.5}$ compared to the inhaled dose of a resident of Netherlands for the entire day. The study did not consider the positive health benefits of cycling that could offset the greater exposure to pollution. Clearly, the benefits cannot compensate for harmful effects of pollution from a long ride (> 1 hour) ride. Since the inhaled doses for a cyclist are much higher compared to drivers of other vehicles, the lower γ level for cyclists cannot compensate for the additional exposure to pollution.

Yet, the cumulative benefits when large numbers of a population use cycles for short trips will lead to a corresponding reduction in air pollution making it a better option for transportation [11]. While air conditioned cars do offer a bubble of relatively clean air in a mass of polluted air, it is not practical for the larger population who either cannot afford air conditioned transport or need to use open vehicles.

Accidents: The risks of a traffic accident were not considered in the study, but are clearly present. Bicycling in India is considered a high risk activity. Media reports of accidents and the dangers of riding in traffic that is not segregated, on pot holed roads, and with rash drivers who are impatient, give the impression that being a bicyclist in India is either foolhardy or brave.

There is no denying that when you ride a vehicle that can easily topple over and when you do not have a steel / glass cage to protect you, you are at a higher risk of an injury or worse. Yet, suprisingly the reported number of deaths of bicyclists in India has fallen from 6600 in 2012 to 2585 in 2016, a drop of almost 60% [22]. This reduction occured while cities showed a slow but increasing interest in bicycling as an alternate mode of transportation, implies that the risk of riding a bicycle can be mitigated by increasing numbers of riders on the road. The Ministry of Road Transport reported that in 2016, the largest number of fatalities (33%) occurred due to accidents involving motorized two wheelers. By comparison, just 1.7% of all fatal accidents involved bicycles. However, an observer on the road in any urban Indian city can easily notice that the number of motorized two wheelers far exceeds the number of bicycles. In 2015, the number of motorized two wheelers in New Delhi was about 5.5 million and a significant contributor to air pollution. While the risk of riding a bicycle on a busy road with other types of vehicles will always be present, it is not as high as perceived and depends to a large extent on a rider taking precautions, giving way to faster traffic and following traffic rules.

One of the claims for the popularity of cycling in countries like Denmark and the Netherlands is that the terrain is relatively flat. However, a study [23] found that regions with rolling terrain did not have fewer numbers of cyclist than regions with flat terrain. Most cyclists adapted to the local conditions. Besides terrain, weather is another factor that can limit the number of cyclists. Few riders will venture out in heavy rain or snow.

1.2.1 Cost Benefit Analysis

Cycling is a form of exercise and most doctors would agree that the benefits of physical activity are significant. With sufficient exercise, you will live longer and have a healthier life. However, air pollution due to motorized traffic is universal in any city and a cyclist will inevitably be exposed to effects of breathing polluted air. At some point, the cost of breathing polluted air will exceed the benefits of cycling. However, this threshold depends on the level of pollution in a city.

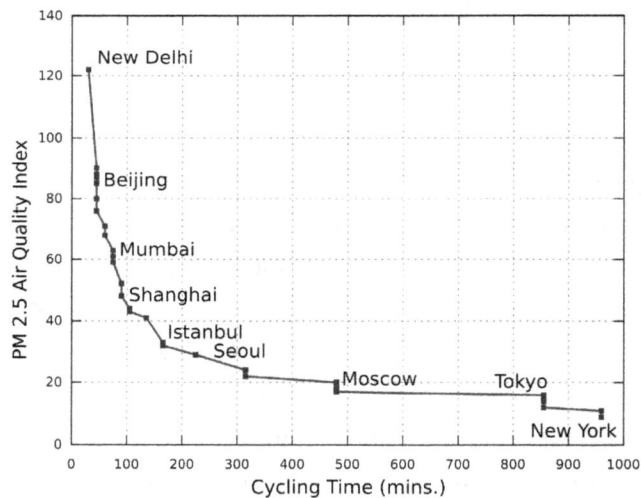

Figure 1.5: $PM_{2.5}$ Air Quality Index of some Cities vs. Cycling Time in Minutes for Health Benefits

Researchers [24] estimated the threshold for health benefits from cycling in different cities. As you would expect, the threshold time is lower in cities with higher pollution levels. The study differentiated between a tipping time threshold (shown in Figure 1.5) beyond which additional time spent cycling would not be of any benefit and a breaking time threshold beyond which cycling would be more harmful than beneficial. In New Delhi, the tipping time and breaking time thresholds were 30 minutes and 60 minutes respectively. Therefore, riding upto half an hour would be beneficial, but beyond 30 minutes there would be no additional health benefits and beyond 60 minutes, cycling would actually be harmful to health. Figure 1.5 shows the vast difference in pollution levels across cities and the corresponding cycling tipping time thresholds. New York with a $PM_{2.5}$ level of 9 and Istanbul with a $PM_{2.5}$ level of 33 have a tipping time of 960 minutes and 165 minutes respectively.

The researchers used a model based on $PM_{2.5}$ pollution levels alone, even though there are a number of other air pollutants as well. Yet, including other pollutants would not susbstantially change the results of the study. The pollution levels in a large city can also vary from a central location with a lot of traffic to a suburb which is likely to have lower than average $PM_{2.5}$ levels.

1.3 Urban Cycling

For some of us, a bicycle was the first means of transport after walking to travel in a neighbourhood. It still is a common vehicle for students attending school, but quickly loses popularity beyond high school. The main reason is the introduction to motorized vehicles such as a scooter or motorcycle. Even though such vehicles cost many times the price of a basic bicycle, the common sight of motorized two wheelers on the road along with the speed and effortless journey are attractions that most cannot resist.

The Low Income Riders

In India, bicycling has long been associated with the urban poor and anyone who wanted to escape being labelled poor would buy a motorized two wheeler, even if it meant taking a loan to buy a vehicle. While 3% of middle or higher income citizens use a bicycle, it is the primary means of transportation for about 40% of low income citizens [25]. For the urban poor, a bicycle provides access to a livelihood and can be less expensive than public transport. The cumulative cost of daily fares on public transport exceeds the cost of a bicycle [26].

A cheap used bicycle can be purchased for a few thousand rupees. However, daily usage of such a bicycle would have high maintenance costs, depending on the distance travelled. In New Delhi, 85% of daily travel is within 10 kilometers [25]. This appears to be a low estimate for cyclists, since a bicycle is more often used to travel and transport goods within a city than to commute to an office. Therefore, the maintenance costs for a bicycle that is heavily used will be relatively high and yet it is a favoured mode of transportation among the urban poor. The most favoured bicycle is the sturdy roadster in Figure 1.6 below. The high volume, low taxes, and mass production of such bicycles has kept their price low enough, so that it is afforable to a large majority of the population.

These bicycles were not necessarily built for comfort, although the saddle with large springs may indicate otherwise. Some of the main features of these types of bicycles include

Figure 1.6: A Roadster Bicycle

- Roughly 32 steel spokes per wheel with a thickness of $\frac{3}{32}$ inch or 13 gauge to support a relatively substantial weight.

- A single gear drivetrain with about 32 and 12 teeth in the front and rear sprockets respectively.

- Wheels between 26-28 inches in diameter and of 1-2 inches wide with nylon tires

- Since weight is not a primary consideration, this bicycle comes with accessories such as a stand, chain guard, mud guards or fenders, and a carrier.

The total weight of such a bicycle would be near 20 kgs. or more. By comparison, a road bicycle built for speed, weighs about seven kgs. or less and may cost 20 times as much. Riders of these sturdy bicycles move along the road in a slow but steady manner towards their destination with no hurry, since it is hard to accelerate on such bicycles. A slight incline with a load on such a bicycle can make it taxing on the rider who may have to simply get off the bicycle and wheel it up, till the next flat or downhill road.

Common users of such bicycles are delivery persons, vendors of goods, and service providers. These cyclists ride not to make a political statement, but because the bicycle is the most cost efficient mode of transport. It gets them from point A to point B under most weather conditions. The roadster bicycle also does not attract the attention of bicycle thieves who value the higher-end bicycles that have more expensive components. A roadster does not have parts like a carbon fork or a mountain bike suspension to absorb the road vibrations, which makes the ride rough on pot holed roads. Further, these riders don't wear cycling gear such as helmets, cycle shorts, sunglasses, etc.

New Delhi

The usage of bicycles in New Delhi does not represent the usage across the country. In general, smaller cities with populations of less than two million have a higher percentage of bicyclists in comparison to the number of cyclists in larger cities. Even though the percentage of cyclists is lower in large cities, the absolute number of cyclists is higher since large cities tend to have populations that are five or ten times larger than the population of smaller cities. Unfortunately, the average trip length is higher in a large city, since a metropolitan area is correspondingly larger. Most large cities in India were not planned with bicycles in mind and therefore the workplace and home maybe far apart.

One suggested fix for this problem is to use the bicycle as an intermediate mode of transportation for the "last mile". So, a commuter would ride a bicycle a short distance to a train station, instead of riding the entire journey on a bicycle. This would make the commute less strenous and more feasible for a larger fraction of the population (see Section 1.5).

The other type of riders in Indian cities are the ones who are riding bicycles out of choice and not affordability. Although, such riders are still rare, their numbers are increasing steadily. The bicycles of these riders can cost several times the cost of the roadster in Figure 1.6. Carrying a load is not a primary consideration and the objective is to ride as comfortably as possible on pot holed roads and at a reasonable pace. Many of these bicycles come with shock absorbers, disc brakes, and wide tyres along with a sleek frame (see Chapter 2).

1.3.1 The Alternate Transport

Following the oil crisis in the 70s and the increasing traffic congestion in the West, there was a need for an alternative mode of non-polluting, cheap, and efficient transporation. The bicycle fit this need perfectly. Transportation planners had to add the infrastructure necessary to support a large number of riders. Considerations included segregating motorized and non-motorized vehicles and convincing citizens that riding a bicycle was safe and cost efficient. One of the big fears of a cyclist is being hit by a speeding vehicle from the rear. However, most accidents occur at junctions where a cyclist has to make a turn (see Figure 1.7).

Figure 1.7: Hazards of Turning for a Cyclist

While the Figure 1.7 shows cyclists in left hand traffic, the same issues arise in right hand traffic as well. The cyclists on the left extreme of the road cannot easily make a right turn without riding into the path of faster oncoming motorized traffic. On the opposite side of the road, a cyclist can be trapped in the middle of the road and has to be careful to move across the road to the slower lane away from the center.

Prior to the oil crisis of the 70s, transport policies focused on building infrastructure for motorized transport with wider roads, highways, and overpasses or flyovers. The idea was to reduce commute times and increase the speed of travel. However with large masses of individuals adopting motorized transport as a means of commuting, these policies led to traffic congestion, increased pollution, and effectively longer commute times. The alternative, bicycling was seen as energy efficient, non-polluting and quiet.

Is Bicycling risky?

In most countries, you do not need a license to ride a bicycle and the skill levels of cyclists can vary substantially. The skills to balance and ride a bicycle are initially sufficient to ride in a quiet neighbourhood. But, on a busy road a cyclist needs to be aware of surrounding traffic, be prepared to brake when needed, change lanes, be alert to parked vehicles, watch for a motorist suddenly opening a door, and so on. Since there is no test or training for these necessary skills, they are learnt while riding, at a slight risk to the cyclist. Additionally, cycles are not inspected periodically by any government authority and the degree of maintenance of a bicycle depends on the interest and skill level of the cyclist. Frequently, poor brakes are a cause of accidents and being able to stop within a short distance is important to avoid a collision. The brakes of most motorized vehicles are more effective than the typical rim brakes of a bicycle.

Further, cyclists sharing a common road are usually the slowest mode of transport and viewed as limiting the flow of traffic by blocking other faster vehicles. Drivers of motorized vehicles will always try to overtake a cyclist, if there is sufficient space. A cyclist who occupies the center of a lane is viewed with hostility and therefore most cyclists will ride on the edge of the road close to the pavement to prevent conflict. Even though a cyclist does not occupy a lot of space on the road and can easily ride on a relatively narrow lane, this space is preferred for on street parking by merchants and residents. Widening roads to allow for bicycles is not an option in an urban area that is fully built-up. The chances of a bicycle with a narrow tire getting stuck are high on a road with large pot holes, wide drain grills, and expansion joints.

Will Bicycle lanes reduce accidents?

If the perceived risks of riding a bicycle are high, then no amount of coaxing will encourage new riders to view the bicycle as a viable alternate mode of transport. Bicycle advocates have recommended separate lanes or paths for bicycles with at least a 2.5 meter wide lane to reduce the number of cycle accidents. Certainly, segregating traffic by vehicle speed is going to reduce the chance of collisions or accidents.

Even with wide enough lanes, motorized vehicles and cyclists have to share the same space on the road at a junction. Accidents are more common at junctions, since a cyclist maybe in the "blind spot" of a driver making a turn. Further, the driver of a car may incorrectly judge the speed of a cyclist and assume that the cyclist will not reach a junction at the same time as the driver.

Bicycle lanes will help the most in congested roads where a cyclist is more likely to have a collision. Unfortunately, those are the same roads where it is impractical to have a bicycle lane that is sufficiently wide. On other roads that are wide enough, it is not too difficult to simply mark off a lane 2-3 meters wide with little additional cost. However, the benefits of a bicycle lane on a wide road are minimal and may not substantially increase the safety of the rider. Besides, bicycle lanes are also tempting parking spots. Parked vehicles on a bicycle lane defeat the whole purpose of segregating traffic and may even make it more dangerous for bicyclists who are forced to weave in and out of lanes for motorized vehicle traffic.

Further, the use of a single bicycle lane for riders in both directions to save space is more dangerous than no bicycle lane at all. Riding against traffic and at the same time sharing a relatively narrow lane with cyclists riding in the other direction is bound to increase the number of accidents. Therefore, bicycle lanes are best located on the left and right extreme of the road (see Chapter 8).

Cyclists are not disciplined

There is a general perception that cyclists are somewhat reckless and willing to bend rules as needed in traffic. For example, a cyclist can get past a red light by simply walking the bicycle across the road and continuing on ahead. Similarly, a pavement has sufficient room to wheel a bike passed stopped traffic and it is easy to lift a bicycle across obstacles. These "benefits" are not always looked on favourably by other users of the road. Cyclists come in the middle ground between pedestrians and motorized traffic and also in most cases don't pay any road tax. Traffic police also do not take violations on bicycles as seriously as other types of traffic violations and a cyclist is rarely given a ticket. Since a cyclist is rarely penalized, it encourages more risky behaviour and a willingness to break rules.

1.4   Promoting Bicycling

Despite all the problems associated with bicycling, governments in developed and developing countries have recognized that there are significant benefits if cycling is promoted. The programs and policies promoted include new bicycle infrastructure, bike-sharing programs, pro-bicycle marketing and education, and restrictions on private automobile usage on certain days. These policies are justified by the expected benefits of reduced pollution, lower fuel consumption, improved public health through physical activity, and increased community connectivity. Still, bicycling is viewed as an optional activity by most people and if it is seen as too cumbersome or not worthwhile the effort, then few people will adopt it.

Typically a road is declared free of motorized traffic on a holiday and the community is encouraged to ride a bicycle or walk. Other public service activities are included to encourage larger participation. While these events do highlight the benefits of riding a bicycle, a minority adopt the bicycle as the primary mode of transport. Bicycling is still an optional activity and the convenience and ease of riding motorized vehicles is hard to overcome. Cyclists can be broadly categorized into three groups [27].

- **Group A:** Cyclists in this group are experienced cyclists and are comfortable sharing the road with other vehicles. They are best served if the roads are made "bicycle friendly" to accommodate shared use of the common road space. Ideally, if lane changes can be minimized and road position be maintained, the number of accidents should reduce.

- **Group B:** This group includes cyclists who know how to ride a bicycle, but are not comfortable riding on busy roads. They tend to stay close to a neighbourhood and ride on streets with relatively low volume.

- **Group C:** Children make up this group and their rides are mostly round trips to school or local shops. They ride in the daytime and with a load of books or goods.

All three groups of cyclists and pedestrians are legitimate users of the transport system and the government has a responsibility to provide for their needs. In general, **Group B/C** cyclists are more numerous and tend to use less congested roads or ride at off peak hours. Road signs like the ones shown in Figure 1.8 do help.

Figure 1.8: Common Bicycle Road Signs

Unfortunately, some roads cannot be widened without demolishing buildings that are situated on the edge of the road. Given that a bicycle lane is impractical on such narrow roads, both a bicyclist and motorist can end up being frustrated. The motorist doesn't like being slowed down by a bicyclist and has to overtake in a tight space. On the other hand, the bicyclist cannot ride fast on a narrow congested road and cannot go off the road to give room for the motorist. This problem is more severe on an uphill road where the speed of a bicyclist can drop rapidly. A motorist cannot appreciate the effort required to ride uphill and may misjudge the speed of the bicyclist, leading to a mishap.

One of the most common reasons cited for not riding a bicycle in India [28] is that it is not a safe activity on a busy road. Yet statistics have shown consistently year after year, that pedestrians and riders on motorized two wheelers are the two categories with most number of annual accidents. For a bicyclist, this simply means that the biggest threat is not from a car or bus, but from a motor cycle or scooter. These motorized two wheelers can weave in and out of traffic and accelerate faster than a bicyclist.

Keeping this in mind, the best a bicyclist can do is to leave space for the "other" two wheelers and not impede the flow of traffic. While safety will continue to be an issue and the construction of an extensive bicycle infrastructure in India is unlikely, **Group B/C** riders can move to **Group A** by acquiring skills to navigate through narrow and wide roads with mixed vehicular traffic. Due to the high density of traffic on main roads, most vehicles cannot travel much faster than a bicycle and a reasonably fit bicyclist can ride along with the other vehicles. While the number of **Group A** bicyclists may not grow significantly, many others are willing to use a bicycle for the "last mile" connectivity.

1.5   Ride Share

While the challenges and cost of building bicycle infrastructure for all roads in a city are high and often not feasible, one alternative is to encourage the use of a bicycle for distances that are slightly greater than walking distance (a kilometer or more). In this scenario, a bicycle is used for just one segment of a journey. The rest of the journey is typically via public transport such as a metro or bus. The benefits are the saved time and convenience. Compared to walking, it is estimated that cycling is about 3.5 times faster and uses less energy. It is easier to promote bicycling for the "last mile" connectivity for the following reasons -

- **Pollution:** Since the roads for "last mile" connectivity are more likely to be off the main road, the exposure to the high levels of pollution on a busy road can be avoided.

- **Safety:** Similarly, the likelihood of an accident on a side road is lower than on a busy main road.

- **Fitness:** While riding 10-15 kilometers a day may not appeal to a large majority of the population, most can easily ride a kilometer or two.

The ride share concept has been adopted worldwide from Asia to Europe to North America. The idea is fairly simple - bike stations located at strategic locations in a city have a number of bicycles for rent for a short duration (often less than an hour) at a very reasonable price. A bicycle is unlocked with a electronic key from a dock (bike rack) and returned either at the same or a different dock. The charge for the rental is based on the duration of the ride.

The types of bicycles used in bike share systems are not endurance or road bicycles with expensive components. On the other hand, they are typically a one size model and mass produced for a city in a standard colour. This helps not only keep costs of the bicycles low, but also discourages potential bicycle thieves who would find it hard to sell such bicycles. Long distance cyclists tend to buy bicycles which are of the right frame size and with other components suited for a rider willing to spend many hours in the saddle.

1.5.1 Divvy Bike Share

The bike share system of Chicago, called Divvy (divvybikes.com) was launched in 2013 and covers an area of about 260 square kilometers. Usage data for 2017 that includes the start / end stations, duration, start time, and other information per ride was made available on the Web. The four plots in Figure 1.9 show the number of rides based on the time of day, ride duration, day of year, and distance covered.

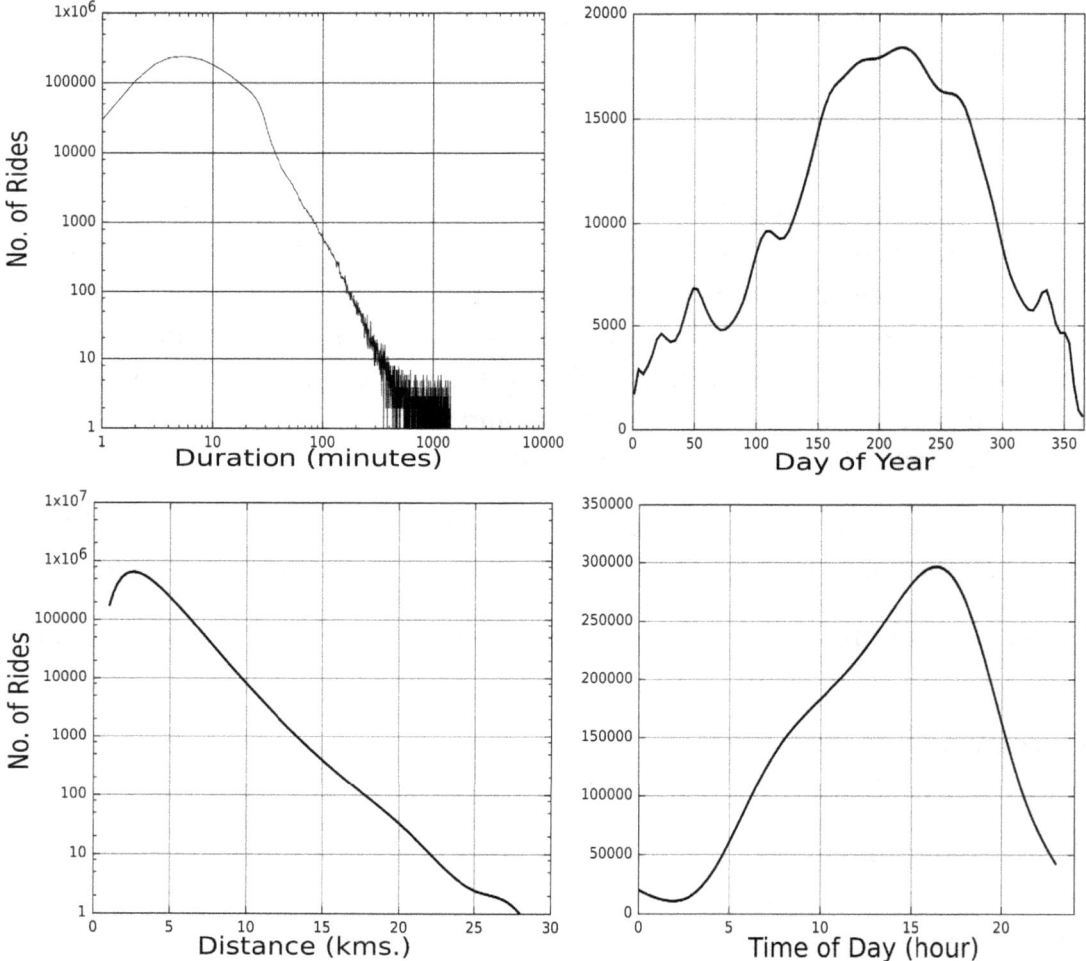

Figure 1.9: Rider Statistics from Chicago's Divvy Bike Share

Usage Statistics

Not surprisingly, most rides are of a short duration (less than half an hour). The average duration of about 4 million rides in 2017 was about 15 minutes, while the median was about 11 minutes. While most rides were for short durations (238,000 rode for just 5 minutes,), the "long tail" for rides of half an hour or more result in a higher average time than the median time (both the x and y axes for the duration plot are in a log scale).

Since most rides were of a short duration, the most frequent distance was 2 kilometers or less. The total distance for about 4 million rides was 11.3 million kms. or roughly an average of 2.8 kms. per ride. Barely 0.4% of all the rides were 10 kilometers or longer. The distance was measured using the great circle distance between the start and end station geographic locations. This of course is not accurate since a ride will follow a street route whose distance is most likely longer than the great circle distance. Still, the total distance using the great circle calculation (11.3 million kms.) is close to published distance (11.68 million kms.) on the Web [29]. The distance of 121,000 round trip rides is excluded from the total distance.

The number of riders on a given day of the year depends on the weather. The temperature of Chicago varies during the year from an average of 28°C in July to -1°C in January. The number of riders per day roughly correlates with the temperature of the day. The highest number of riders in 2017 was 22,106 on July 8th and the least number of riders was 240 on December 25th. About 55% of all the rides in a year occur in the four months from June to September. The most popular and least popular weekday for rides was Monday with 578,996 rides and Sunday with 492,564 rides respectively.

Station Network and Routes

With 585 bike stations, the number of all possible routes is 585^2 or 342,225. Of all possible routes (see Figure 1.10), about 28% (96,000) were used in a ride and a small fraction (3%) of all 4 million routes were round trip, i.e. the from and to stations were identical. A few stations tend to be far more popular than others. The busiest bike station of all was located at Streeter Drive and Grand Avenue (Navy Pier), a popular tourist attraction in Chicago, and annually accounted for over 51,000 outgoing rides and 42,000 incoming rides. A significant portion (10,042) of these rides were round trip rides.

Each edge of the network connecting two nodes represents a bicycle route between two stations. For simplicity, the edges do not have arrows indicating the direction of the route and the thickness of edges is uniform. In a high resolution plot, the popular routes can be shown with thicker edges. The dense network of routes closer to the lake is noticeable while routes further away from the lake shore are sparse. The Lake Michigan in this case is a boundary for the route network.

The network in Figure 1.10 is too dense to identify any structure, since the number of edges are close to 100 thousand. The routes that are less frequent can be removed to show the popular

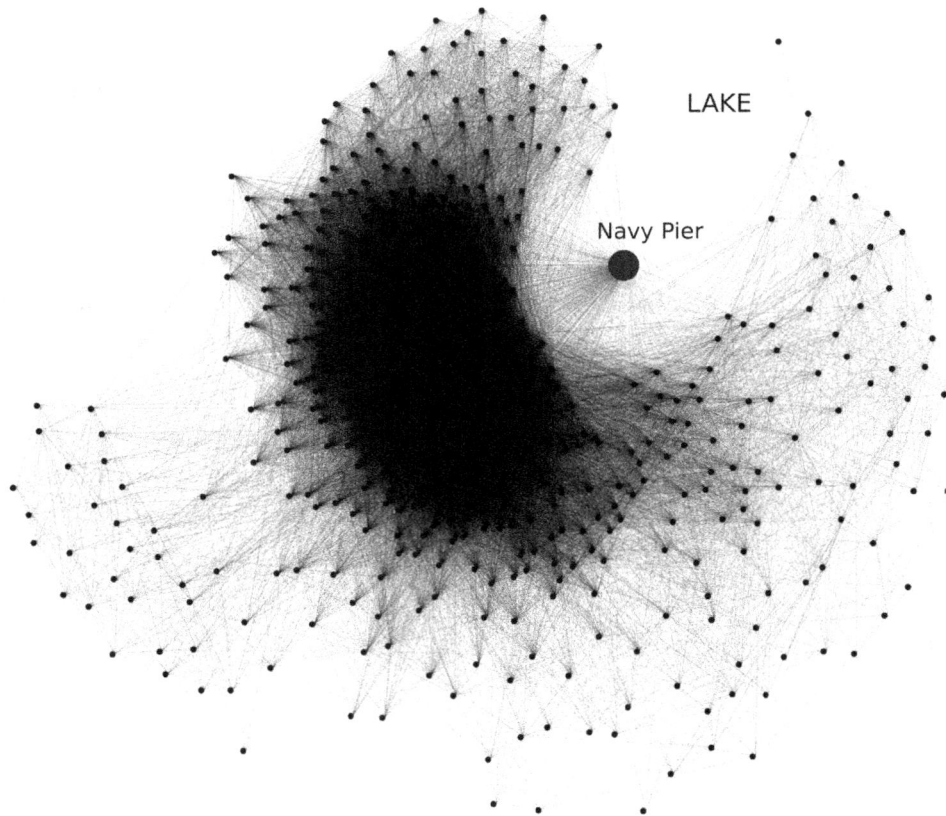

Figure 1.10: 96 Thousand Routes from / to 585 Stations in Divvy Bike Share

routes (see Figure 1.11). Instead of showing the names of bike stations, the station numbers are shown. The station number 35 is the Navy Pier station shown in Figure 1.10. The stations have been re-arranged and the edges do not represent distances. The second most popular station is 192 and is at the center of a dense sub-network. The other stations in descending order of popularity are 76, 268, and 177. The stations and their associated locations can be downloaded from the divvy Website (https://www.divvybikes.com/system-data)

It is hard to predict how a bike share network will be utilized, but clearly there are some central bike stations in the network that are busier than others. The popularity of a bike station may also change with time as new bike stations are added.

The number of incoming and outgoing rides at a bike station are roughly identical (see Figure 1.12). The number of outgoing links represents the number of rides *from* a bike station and the number of incoming links represents the number of rides *to* a bike station. A few bike stations are very busy with a high number of incoming and outgoing rides. The station number is an artificial number from a list of bike stations sorted in ascending order by the number of outgoing links.

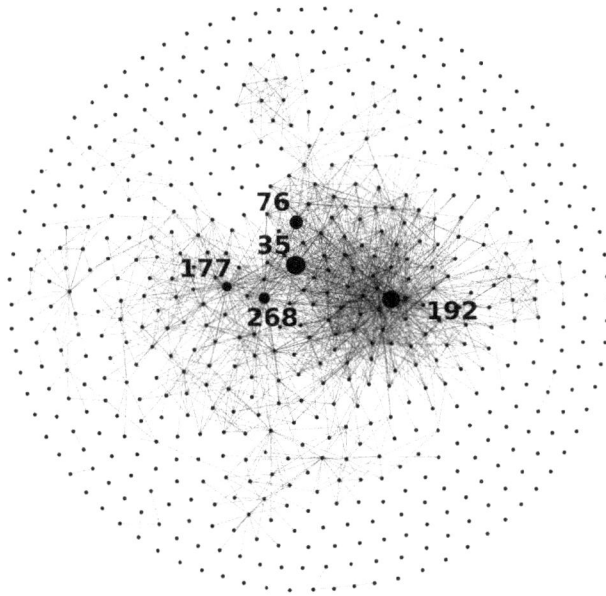

Figure 1.11: A Simplified Route Network of Divvy Bike Share

1.5.2 Metro Bike Share

Metro Bike Share (bikeshare.metro.net), a bicycle share program for Los Angeles, was launched in July 2016. The program began with 1400 bicycles spread over 140 bike stations. The city of Los Angeles is spread over about 1300 sq. kilometers while the area of Chicago is about half as much as at 600 sq. kilometers. The distances between pairs of bike stations spread out evenly across over 1000 sq. kilometers would have been excessive and therefore Metro Bike Share instead divided the 140 bike stations into three zones. All the bike stations within a single zone were located within a reasonable distance from each other (< 10 kms.) with the assumption that most rides would be relatively short within a particular zone.

Like Divvy, Metro Bike share usage data is public and can be downloaded (`https://bikeshare.metro.net/about/data/`). The following analysis is from the data downloaded for the years 2017-2018. The data contains information for each ride such as the start time, end time, start location, and end location. In a round trip, the start and end locations would be identical and therefore the distance for such rides are not possible to compute and excluded in the Figure 1.13. From the given data, we can calculate statistics such as the most frequent time to start a ride, the duration of most rides, the distribution of rides over a year, and the distance of rides.

Since Metro Bike share is relatively recent, the number of rides and distances is lower than the number of Divvy Bike share rides. In 2017, the total distance of 225,000 non-round trip rides covered about 259,000 kms. (excluding 52,000 round trips). In the following year (2018), the bike share program grew by about 36% to 306,000 non-round trip rides covering about 353,000 kms. (excluding 29,000 round trips). About 99% of all rides are less than 5 kms. long.

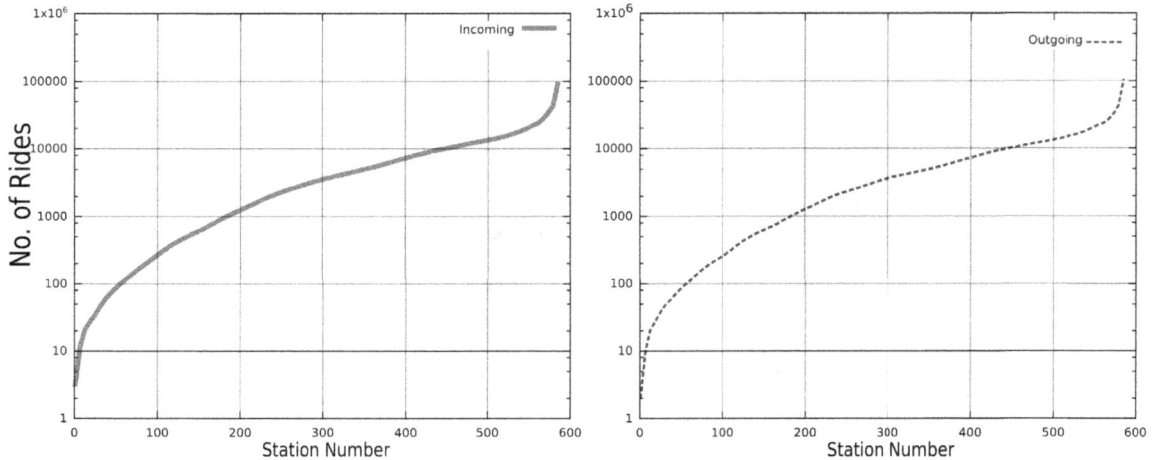

Figure 1.12: Incoming and Outgoing Rides at Bike Stations

The actual distance covered over two years is higher than 612,000 kms. since the distances are computed using the great circle distance between two locations. In reality, a rider would take a longer route and without a GPS tracker, it is hard to estimate actual distances covered. The distances of round trip rides are also not counted since the distance between identical start and end bike stations is zero.

The durations (in minutes) of rides is also correspondingly short with roughly 80% of all rides lasting for less than half an hour. However, the average duration of a ride was much higher since a significant number of rides lasted for 5 or more hours. This pattern is very similar to the pattern of the duration of rides in the Divvy Bike share program.

Since the climate of Los Angeles is milder compared to Chicago and the winters are not as harsh, the distribution of rides is wider across the year. The most popular days in 2017 and 2018 were Sunday, October 8th and Sunday, September 30th with 2219 rides and 2494 rides respectively. By contrast, the least popular days for the same two years were Sunday, January 22nd and Thursday, March 22nd with 70 and 227 rides respectively. The total number of rides were somewhat evenly distributed across all seven weekdays with the least popular on Monday (74,630 rides) and Friday (80,763 rides) the most popular. The most popular time to start a ride is around 3:00 pm and the least popular is logically late at night around 2:00 am.

In Figure 1.14 the top 1000 popular routes out of over 6500 routes from the 141 bike stations over a period of two years shows the four distinct zones - Port of Los Angeles, Downtown, Venice, and Pasadena. Since a large number (≈ 80) of bike stations are located in downtown, most of the routes are located in that zone. The remaining zones - Pasadena, Port of Los Angeles, and Venice have about 33, 13, and 15 bike stations respectively. Each of the four zones and relatively far apart (> 10 kms.) and therefore very few rides are between two different zones.

Each bike station is assigned an unique number. The numbers of a few bike stations are shown in Figure 1.14. Each edge between a pair of bike stations is a route that may contain one or more rides. The total number of rides (540,000) is not distributed evenly across routes. Some

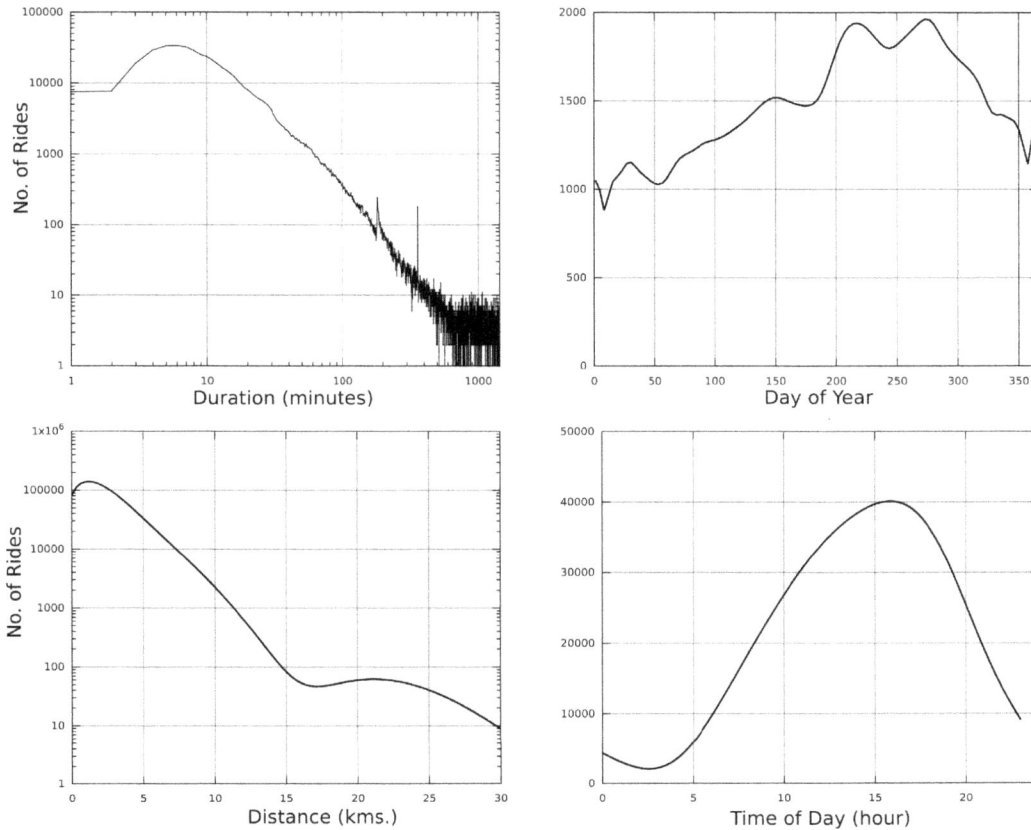

Figure 1.13: Rider Statistics from Los Angeles Metro Bike Share

routes and bike stations are far more popular than others. The most popular was bike station 3005, that is geographically centrally located within the downtown zone, with 21,367 and 23,788 outgoing and incoming rides respectively.

Within a zone, there are similar patterns for popular and seldom used bike stations. The traffic between a pair of bike stations is also not uniform. In Figure 1.14, there are far more incoming rides (5860 + 3830) to bike station 4215 from two neighbouring bike stations 4214 and 4210 than the number (2000 + 1300) of outgoing rides. A large number of rides also start and end at the same bike station. The number of round trip rides in the three bike stations, 4210, 4214, and 4215 were 6548, 9355, and 3455 respectively.

One of the goals in a bike share system is to optimize the usage of the bicycles, which does mean estimating the number of bicycles at each bike station. It would be disappointing for a potential rider to visit a bike station and discover that there are no bicycles available. On the other hand, it would be wasteful to have a large number of bicycles lying idle at a bike station. Finding the right number of bicycles for a bike station is hard to accomplish without some traffic estimates.

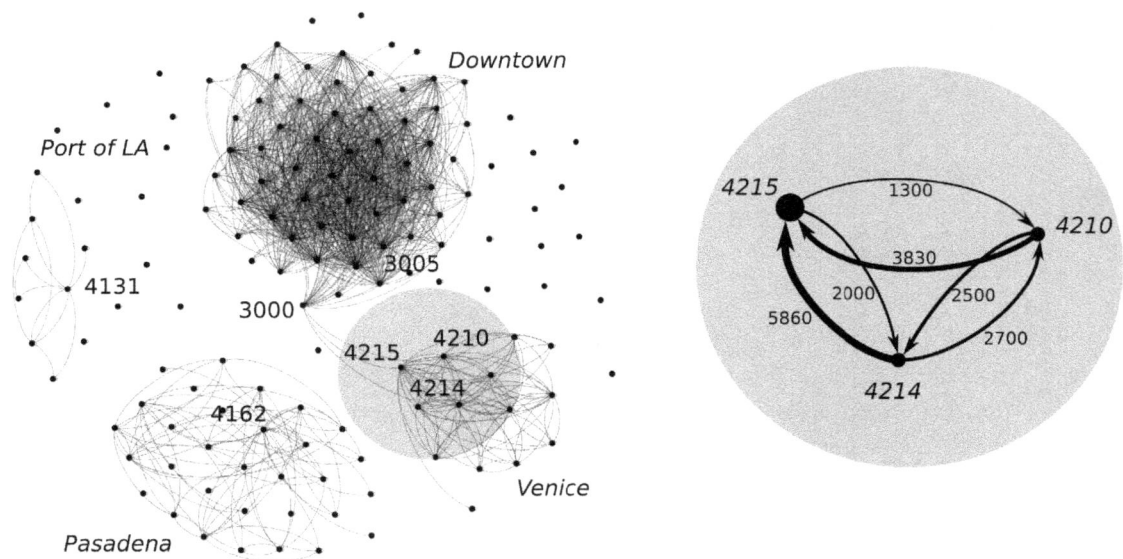

Figure 1.14: 1,000 popular routes to / from 141 stations in Metro Bike Share

1.5.3 Bike Share Issues

As bike share became more popular globally, problems that were either too minor or insignificant became noticeable when the number of riders increased. Bike share grew rapidly in popularity across Chinese cities with tens of millions of daily rides. With large investors funding bike share companies, the number of shared bicycles grew so rapidly that they led to unanticipated problems. As of 2017, two companies Ofo and Mobike, had placed 3.2 million bicycles in various Chinese cities.

One of the big concerns in a bike share system is the cost of repairs. If shared bicycles are not maintained in a good condition, then it is very likely that the number of riders will reduce and the program will not succeed. Since the number of rides is in the millions, it is not feasible to monitor the condition of bicycles before and after a rental. In the Divvy bike share, about 14,000 bicycles were repaired in 2017. The number of repairs per bicycle depends on the usage of the bicycle which roughly follows a normal distribution with a mean of 613 and standard deviation of 215. So, the average bicycle would be used for 613 rides in a year.

In Figure 1.15, the number of times a bicycle was used over a period of two years in the Metro Bike share program shows a fairly "normal" distribution. A few bicycles were used very often (800 times) and a few bicycles were seldom used (< 50 times). The average bicycle was used 362 times in two years. Although the frequency of repair would depend on how a bicycle was used, in a bike share program a repair for every 100 rides would not be unreasonable. Since the bicycle was not owned by the rider, usage maybe more harsh and repairs more frequent.

With the little regulation and infrastructure built for motorized vehicles, bicycles were parked wherever it was convenient and actually led to congestion on the pavements. Further, vandalism and misuse of bicycles grew by some who viewed a shared bicycle as a "use and throw

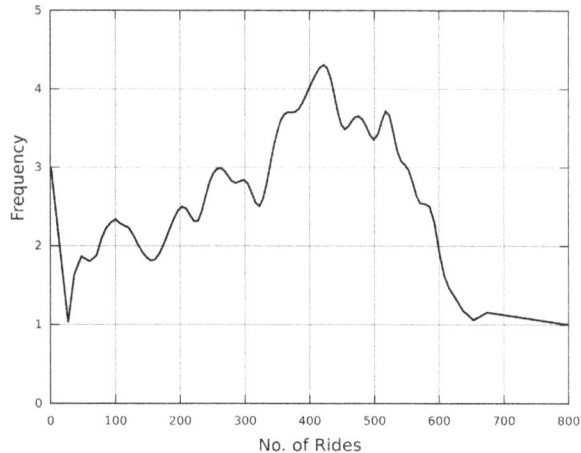

Figure 1.15: Frequency vs. No. of Rides per Bicycle in Metro Bike Share

away" product. Penalizing and enforcement by authorities may help to reduce such vandalism. Nonetheless, this issue will not be resolved easily.

At the same time, Chinese officials claim to have saved billions of dollars by reducing traffic congestion with bike share programs. While there will be problems with the implementation of bike share, the idea is here to stay. People like the convenience and the ability to get around a city without dealing with a lot of traffic and associated parking problems. Besides, it is affordable and some bike share programs even offer discounts to frequent riders.

Bike share bicycles are intentionally built to be plain and unattractive but efficient enough to ride for short distances. However, a rider of a light road bike would find these bicycles heavy and slow. But, they are suitable for tourists or a commuter who has to ride a kilometer or two. On flat terrain, the weight of the bicycle may not be a major issue. But in a city with rolling terrain, riding a heavy bicycle can be discouraging.

1.5.4 Carbon Footprint

One of the major challenges facing future generations is the rapid change in climate due to global warming. The volume of CO_2 released in the atmosphere as a result of human activities has increased steadily. The level of CO_2 in the atmosphere increased from 330 parts per million (ppm) in 2006 to over 410 ppm in 2019. This may appear to be a small quantity, i.e. only roughly 0.04% of all air molecules are CO_2 molecules. However, CO_2 is a greenhouse gas that traps heat and does not allow heat to escape the atmosphere. While there are other types of greenhouse gases in the atmosphere, CO_2 is the dominant greenhouse gas contributing to a majority of the warming of the planet. To reduce the warming rate, scientists have recommended lowering the level of CO_2 from 410 ppm to 350 ppm or lower.

A carbon footprint is the amount of CO_2 released by an individual, community, or even country. A country with a large population would logically have a larger carbon footprint than the footprint of a smaller country. In 2015 China emitted over nine billion metric tons of CO_2 compared to the next largest emitter, US which generated about five billion metric tons. While the US has less than one fourth the population of China, the CO_2 emissions per capita is much higher. This observation is fairly common in most developed countries.

Transportation is the primary human activity that accounts for over $\frac{1}{4}$ of all CO_2 released into the atmosphere [30]. This activity is a necessity in all nations. One of the ways to reduce the amount of CO_2 emitted is to consider alternate modes of transportation such as cycling.

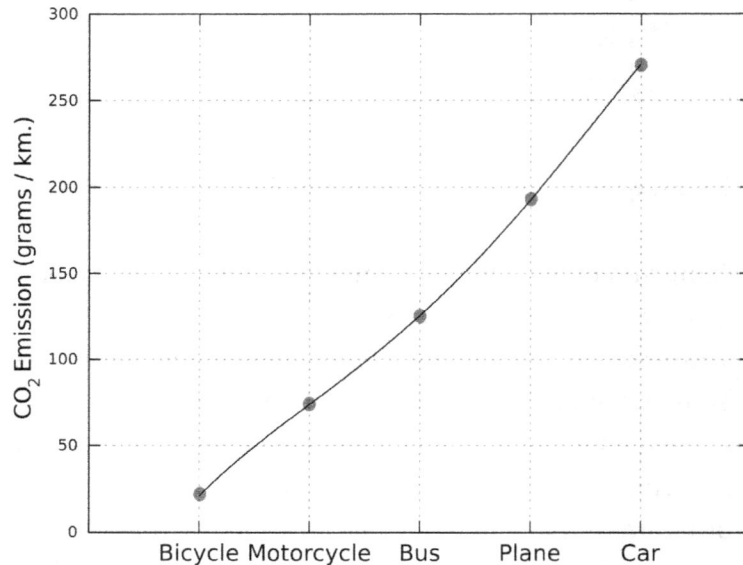

Figure 1.16: CO_2 Emissions in grams per kilometer for five transportation modes

How does cycling help reduce global warming   Accurately calculating the volume of CO_2 of a passenger vehicle is complicated due to the different vehicle efficiencies, driving styles, and fuel used. Electric, hybrid, and conventional vehicles have zero, moderate, and high emissions of CO_2 respectively. An aggressive driver who brakes and accelerates often will use more fuel and correspondingly generate more CO_2. A rough guide to calculating emitted CO_2 is based on the fuel consumed. The US EPA estimates the average passenger vehicle will generate 2.32 kgs. and 2.69 kgs. of CO_2 per liter of petrol and diesel consumed respectively [31]. This translates to about 0.116 kgs. / km. for a petrol vehicle with an efficiency of 20 kms. / liter. Other estimates [32] are higher at 0.271 kgs. / km. By contrast the estimate of CO_2 emitted by a cyclist at 0.021 kg. / km. is about $\frac{1}{10}th$ that of a motorized vehicle.

In Figure 1.16 the emissions for five transportation modes shows the bicycle has the least CO_2 emission per kilometer travelled. The values in the figure are estimates [32] and actual values can vary substantially. Although the emissions per person on a plane are less than emissions per person in a car, the distances travelled by plane are much higher. A round trip trans-atlantic trip would produce over 2 metric tons of CO_2 per passenger. In the near future, the bicycle will continue to have the smallest carbon footprint compared to the carbon footprints of other modes of transportation.

References

[1] BodyandSoul.com. Cycling advantages: 10 reasons to start now, December 2015. URL http://bit.ly/2v7t5m4.

[2] BikeRadar.com. Five reasons to cycle to work, April 2014. URL http://bit.ly/2HmccX3.

[3] Greatist.com. How to keep burning calories when your workout is over, January 2016. URL http://bit.ly/2HoafcN.

[4] Jaehyun Joo, Sinead A. Williamson, Ana I. Vazquez, Jose R. Fernandez, and Molly S. Bray. The influence of 15-week exercise training on dietary patterns among young adults. *International Journal of Obesity*, 2019.

[5] Harvard. Increase in resting heart rate is a signal worth watching, December 2011. URL http://bit.ly/2qmNj5F.

[6] NaturalNews.com. Cycling twenty miles a week cuts heart disease in half, May 2011. URL http://bit.ly/2GPt6wr.

[7] Serge C Harb, Paul C Cremer, Yuping Wu, Bo Xu, Leslie Cho, Venu Menon, and Wael A Jaber. Estimated age based on exercise stress testing performance outperforms chronological age in predicting mortality. *European Journal of Preventive Cardiology*, 2019.

[8] CNN. How to get blood pressure down to 120, September 2015. URL https://cnn.it/2qg3HWn.

[9] Bicycling.com. 40 great reasons to ride more in 2017, December 2017. URL http://bit.ly/2v7t5m4.

[10] MedicalExpress.com. Cycling does not damage men's sexual or urinary functions, January 2018. URL https://medicalxpress.com/news/2018-01-men-sexual-urinary-functions.html.

[11] De Hartog Jeroen, Johan, Boogaard Hanna, Nijland Hans, and Hoek Gerard. Do the health benefits of cycling outweigh the risks? *Environmental Health Perspectives*, 118(8):1109–1116, August 2010.

[12] Oke Olufolajimi, Kavi Bhalla, David C. Love, and Sauleh Siddiqui. Tracking global bicycle ownership patterns. *Journal of Transport & Health*, (2):490–501, August 2015.

[13] International Monetary Fund. Gdp per capita, current prices, October 2017. URL `http://www.imf.org/external/datamapper/NGDPDPC@WEO/OEMDC/ADVEC/WEOWORLD`.

[14] Union of Concerned Scientists. Cars, trucks, and air pollution, December 2014. URL `http://bit.ly/2EzZRvt`.

[15] Rahul Goel, Shahzad Gani, Sarath K. Guttikund, Daniel Wilson, and Geetam Tiwari. On-road PM2.5 pollution exposure in multiple transport microenvironments in Delhi. *Atmospheric Environment*, 123(A):129–138, December . 2015.

[16] World Bank. PM2.5 air pollution, mean annual exposure (micrograms per cubic meter), December 2016. URL `https://data.worldbank.org/indicator/EN.ATM.PM25.MC.M3`.

[17] Wikipedia. List of most polluted cities by particulate matter concentration, February 2018. URL `http://bit.ly/2qqkDZF`.

[18] Geetam Tiwari, Deepty Jain, and Kalaga Ramachandra Rao. Impact of public transport and non-motorized transport infrastructure on travel mode shares, energy, emissions and safety: Case of Indian cities. *Transport. Research*, (Part D), 2015.

[19] Planetforward.org. The most efficient transportation ever invented - The Bicycle, March 2010. URL `https://bit.ly/2LxMorn`.

[20] Bijli Nanda, Jagruti Balde, and S. Manjunatha. The acute effects of a single bout of moderate-intensity aerobic exercise on cognitive functions in healthy adult males. *Journal of Clinical and Diagnostic Research*, 7(9):1883–1885, September 2013.

[21] Stanley J. Colcombe, Kirk I. Erickson, Paige E. Scalf, Jenny S. Kim, Ruchika Prakash, Edward McAuley, Steriani Elavsky, David X. Marquez, Liang Hu, and Arthur F. Kramer. Aerobic exercise training increases brain volume in aging humans. *Journal of Gerontology*, 61A(11): 1166–1170, September 2006.

[22] savelifefoundation.org. Road statistics involving bicycle users in India, November 2017. URL `http://bit.ly/2HcMzuj`.

[23] Jerrold A. Kaplan. Characteristics of the Regular Adult Bicycle User, Master Thesis, University of Maryland, 1975.

[24] Akshat Rathi. How long can you cycle before the harm from pollution exceeds the benefits of exercise?, 2016. URL `http://bit.ly/33Rz4bZ`.

[25] Anvita Anand, Geetam Tiwari, and Rajendra Ravi. The bicycle in the lives of the urban poor, 2006. URL `http://bit.ly/2Jwurd8`.

[26] Premjeet Das Gupta and Kshama Puntambekar. Bicycle use in Indian cities: Understanding the Opportunities and Threats, December 2016. URL `http://bit.ly/2IFRNMe`.

[27] Wilkinson W. C., A. Clarke, B. Epperson, and R. Knoblauch. *Effects of bicycle accommodations on bicycle/motor vehicle safety and traffic operations.* US Federal Highway Administration, July 1994.

[28] Ministry of Road Transport and Highways. Road Accidents in India - 2017, Government of India Report.

[29] John Greenfield. 2017 was another record year for Divvy ridership, January 2018. URL `https://bit.ly/2zbu5nK`.

[30] Environmental Protection Agency. Sources of greenhouse gas emissions, 2019. URL `http://bit.ly/2ZxgKC7`.

[31] US Environmental Protection Agency. Greenhouse gases equivalencies calculator - calculations and references, 2019. URL `http://bit.ly/2ZlnfaQ`.

[32] European Cyclists' Federation. How much CO2 does cycling really save?, 2019. URL `http://bit.ly/2IeI1mJ`.

2 Choosing a Bicycle

Which bicycle is most appropriate for you? There is no single answer to this question and the best answer is that it depends. Some of the factors to consider when choosing a bicycle are - your budget, the purpose (commuting, touring, or racing), and the manufacturer.

Bicycles come in all budgets from the low priced one made up of commodity parts to the high end bicycle with light weight components and a customized frame. The more expensive bicycle can cost as much one hundred times the cost of the cheaper bicycle. A beginner bicyclist could start with a cheaper bicycle before buying a more expensive model.

2.1 Buying your Bicycle

Before you buy your bicycle, it is well worth the time to do a little research. Since bicycles come in a broad range of prices from the low cost sturdy roadster bicycle in Chapter 1 to the expensive light road bicycle, you would need to set a budget. This budget may depend on how you plan to use your bicycle. If you are going to ride 10 or more kilometers per day, then you would need a dependable bicycle. A bicycle with good components can last 10 years or more with minimal maintenance such as replacing tubes, tyres, or the chain.

A cheaper bicycle is more likely to have components that will fail sooner and you may need to spend more on maintenance. This a trade off that you need to make when evaluating bicycles. One case for a cheap bicycle is if you have to ride nearby where the risk of bicycle theft is high. Losing an expensive bicycle into which you have invested time and energy selecting good components is heart breaking. The loss of a cheap bicycle is less painful. Securely locking an expensive bicycle or using some kind of monitoring device is another option if you want to protect your investment.

How much should you spend on a bicycle? The idiom "you get what you pay for" is true for bicycles as it is for other goods. A bicycle is more complicated than it appears with more than a hundred components, small and large, that must be assembled correctly. Some of these components include the stem, handlebars, wheels, tyres, inner tubes, rim tapes, spokes, hubs, wheel rims, cables, shifters, brake levers, handle bar grips, headset, bottom bracket, pedals, cranks, chainset, sprockets, derailleurs, chain, brake arms, brake springs, brake pads, saddle, suspension, and seat post.

A bicycle does need most of the components mentioned in the list with the exception of gears, derailleurs, and suspension. The manufacturing cost of a bicycle is then the sum of the cost of these components, assembly, packaging, and shipping. A manufacturer can cut the cost of a bicycle by compromising on the cost of some components such as the saddle, pedals, grips, hubs, or quality of the frame.

2.1.1 Is a cheap bicycle worth it?

Everyone loves a bargain. When a bicycle is sold at a price that seems to be really low, you maybe getting more than you bargained for. Bicycles sold in a large store are typically inexpensive to attract a larger audience. This is coincident with the department store strategy of selling a large number of low priced products. However a bicycle is a slightly different product from clothing or electronic goods.

When electronic goods are mass produced, the cost of these products can be lowered substantially, while still maintaining reliability. The same cannot be said about a bicycle that is still a mechanical device. The parts of a bicycle are made up of rubber, steel, or some other alloy. A cheap bicycle with knobby (mountain bicycle) tyres may appear rugged and reliable, but tyres are just one of the components of a bicycle. Some cheap bicycles even have suspensions. However, a good suspension is not cheap and will add to the cost of the bicycle. Therefore, most of the suspensions on cheap bicycles are not meant for riding on rough trails.

The cheap bicycle is also known as a "bicycle shaped object" (BSO) to imply that it only looks like a more expensive bicycle with components like gears and a suspension. However, BSOs are deceptive and give the impression of being similar to a bicycle that is used to ride down a rough trail. Using a BSO as it is implied is unsafe.

The main reason to be wary of a cheap bicycle is the risk of injury. If you ride an improperly assembled bicycle, you are more likely to fall or have an accident. Poor brakes or brakes that do not work mean that you will not be able to stop and risk a collision. While cyclists do get some sympathy in an accident, riding an unsafe bicycle will not shield you from blame.

Other reasons to avoid a cheap bicycle include frequent trips to the bike shop to replace parts that break down often. To keep costs low, the quality of parts will be compromised and therefore more likely to break down sooner than later. Besides, riding on a cheap bicycle may not be fun and you maybe tempted to give up riding a bicycle, since it may appear to be a clunky and tedious affair.

Although BSOs are derided in the bicycling community, you can still ride a BSO provided it has been fully assembled and you ride on asphalt or concrete. If you are willing to make minor repairs from time to time and do not expect a smooth ride, a BSO can last for longer than you would expect.

Why are cheap bicycles being made?   Given all the problems with a cheap bicycle, you would assume that the manufacturers of such bicycles would go out of business soon enough. Consumers would stop buying such products and the loss of business would convince bicycle makers to stick with reliable components to produce a bicycle at a reasonable, but not cheap cost.

Cheap bicycles still sell well because it is the most affordable bicycle for a sizeable segment of the population. Not all cheap bicycles are bad. You cannot comfortably ride these bicycles at any speed, but slow and steady. For buyers of these bicycles, the price is the main priority, while speed and comfort are not. Cyclists on cheap bicycles are not necessarily riding to reduce pollution, but rather to travel from one place to another at the lowest possible cost.

Unlike BSOs, these cheap bicycles do not have any derailleurs, gears, or suspensions. Although they are clunky and not a lot of fun to ride, they work efficiently to transport goods and people. The roadster in Chapter 1 is one such bicycle and costs much less than a typical mountain bicycle. In 2018, an Indian roadster bicycle was priced at about 5000 INR (≈72 USD). Most of the 10 million bicycles sold in India are priced in this range [1]. A bicycle with gears may cost upwards of 8000 INR (≈115 USD) and the number of such bicycles sold is relatively small compared to the size of the market. Some fraction of the 10 million bicycles are give-aways by the government to students.

In some ways buying a roadster is simple. The bicycle comes with a standard set of accessories that may include mud guards, chain guard, stand, a bell, and a rear carrier. Riders of such bicycles would like to use the bicycle without spending anything more than buying a lock. The large number of such bicycles on the road does indicate that they are reliable and will last, provided they are used with care.

Figure 2.1: B'twin My Bike Hybrid Bicycle (courtesy Decathlon)

If the roadster looks too old fashioned, then the B'twin My Bike hybrid bicycle has a more contemporary look. For 4500 INR (≈64 USD) you get a single speed bicycle with a steel frame that appears to be reliable. The frame seems well built and should last for many years with the steel tubes that can handle all weather conditions. The 26 inch tyres are the next best components of this bicycle and suited for urban roads with some pot holes.

The rest of the bicycle is kept as simple as possible with standard components - V-brakes, grips, pedals, chain, and saddle. Since, keeping cost low is the primary concern, the ride on this bicycle is not the same as a ride on mountain bike with a suspension. However, these bicycles do not come with the accessories that are standard on a roadster such as a stand or mud guards. The addition of these accessories does not significantly change the affordability of the bicycle.

2.1.2 Beyond the Roadster

If you are prepared to buy a bicycle that costs less than 15,000 INR (≈215 USD) but which costs more than a roadster, you have more options and choices to make. The main reason to buy such bicycles is that you can choose the frame size, type of tyre, and use gears. A roadster typically comes in a one-size-fits-all frame and you also may not be able to adjust the height of the saddle on a roadster. The cheap bicycle costs less because makers invest in a single frame size and color and do not need to estimate the size of the market for large, medium, and small frames. The frame size is an average size that will fit the largest segment of the bicycle population. Short and tall bicyclists may not find such bicycles very comfortable.

2.1.3 Frame Sizes

Since adult bicyclists may have a range of heights from 4 feet to over 6 feet and also weigh from 40 kgs. to over 100 kgs., a single frame size is not appropriate for all type of bicyclists. Apart from the bicycle size, there are many other parameters to consider when buying a bicycle that will suit your body frame. But the size of the bicycle frame is the primary component that must fit your height and inseam (see Figure 2.2).

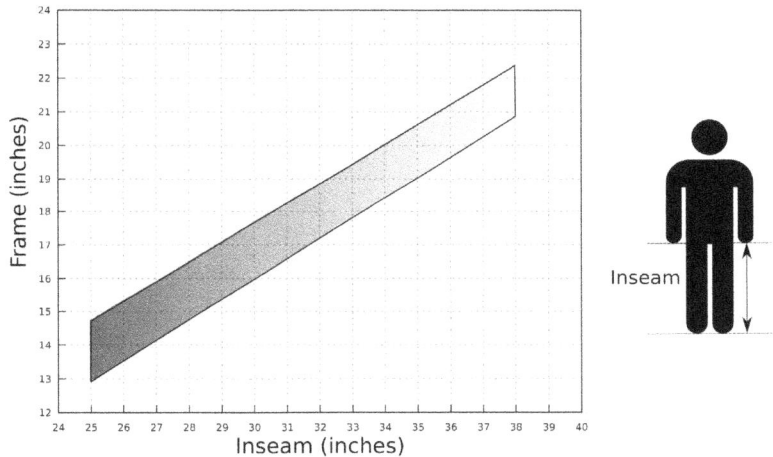

Figure 2.2: Mountain Bicycle Frame Sizes

If the frame of the bicycle fits your body structure, then you can always exchange other parts such as tyres, pedals, and saddles to build a more comfortable and appropriate bicycle. The traditional frame dimensions for a mountain bicycle and road bicycle have been in inches and centimeters respectively. So, if you know your inseam in inches, then you can find a small range of frame sizes that are most suitable for a mountain bicycle. The frame size also depends on your height. Some cyclists may have a large inseam, but small torso and therefore a frame in the smaller range maybe more appropriate (see Appendix).

The frame size is the length of the seat tube measured from the center of the bottom bracket to the top of the seat tube (see Figure 2.2). The seat post is of a smaller diameter and fits within the wider seat tube and can be raised or lowered as needed (see Chapter 3 for saddle height). The formula to calculate a mountain bicycle frame size is in the range $0.55 \times inseam$ to $0.59 \times inseam$, while the range for a road bicycle frame size is between $0.64 \times inseam$ and $0.69 \times inseam$.

Road bicycle frames are described in centimeters (see Figure 2.3) and therefore you may get a frame that is a better fit than a mountain bicycle frame which is specified in inches. For the same inseam, the seat tube for a road bicycle is longer by about 16%, since the bottom bracket is lower and closer to the ground. The bottom bracket of a mountain bicycle is located higher to provide reasonable clearance on a rough trail.

The seat tube is the best tube to use for a frame size, since the top of the seat tube is roughly where your hips would be located and the center of the bottom bracket is close to where your foot would rest on the pedal. A larger inseam would mean a longer seat tube to allow a longer leg to reach full length while pedalling.

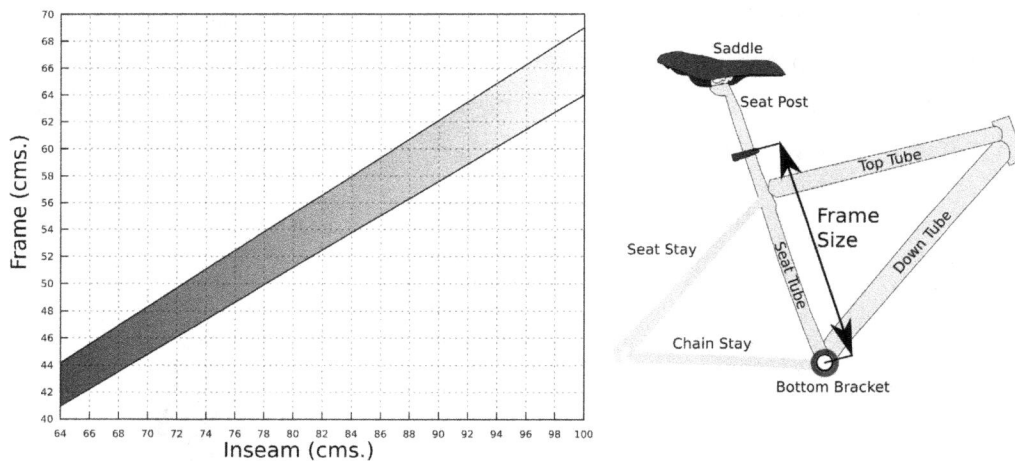

Figure 2.3: Road Bicycle Frame Sizes

The frame sizes in Figures 2.2 and 2.3 are reference sizes and may not be correct for everyone. You can refer to bicycle frame charts [2] that include height as well as inseam to get a more accurate frame size. In general, a shorter cyclist would find a smaller frame size in the range more appropriate.

As frame sizes increase, the length of the top tube along with other tubes increases as well (there maybe two top tube lengths, one for the actual length of the sloping tube and another for the smaller horizontal projection of the length). While you can compensate for a short seat tube by extending the seat post, there are limits to the maximum length of a seat post before it becomes unstable. Even though a longer seat post may be a good fit for pedalling, the top tube maybe too small for your arms. Conversely, keeping the seat post as low as possible with a long seat tube may mean that the top tube is too far for your *reach*.

Observations

- It is suprising how a few centimeters in *reach* can make a big difference. If you have to reach for the handle bar, you may find it OK for a few minutes, but after a half hour or more of stretching to hold the handle bar, you may find your back aching.

- Initially, it maybe comforting to keep a saddle low since your center of gravity will be low and therefore you are less likely to topple over. But, the ideal saddle height is when your leg has a small bend, at the bottom of a pedal stroke. One way to find your ideal saddle height is to incrementally increase the height till you feel comfortable.

- Sometimes a manufacturer may not sell models in frame sizes based on your inseam. In that case, you will have to choose between a larger or smaller frame size. In general, you can adjust to a smaller frame size more easily than a larger frame size. A longer seat post on a smaller frame can compensate for a shorter seat tube. But, most saddles cannot be moved sufficiently forward to compensate for a long top tube.

2.1.4   Frame Geometry

Unfortunately knowing your height and inseam alone may not be sufficient to find the right frame. Some makers do set the frame size based on the length of the seat tube, but others may use different lengths for the seat tube. Therefore, a 50 cms. frame from two different makers will not always be identical. There are some common specifications for a frame that you can use to compare frames of the same size.

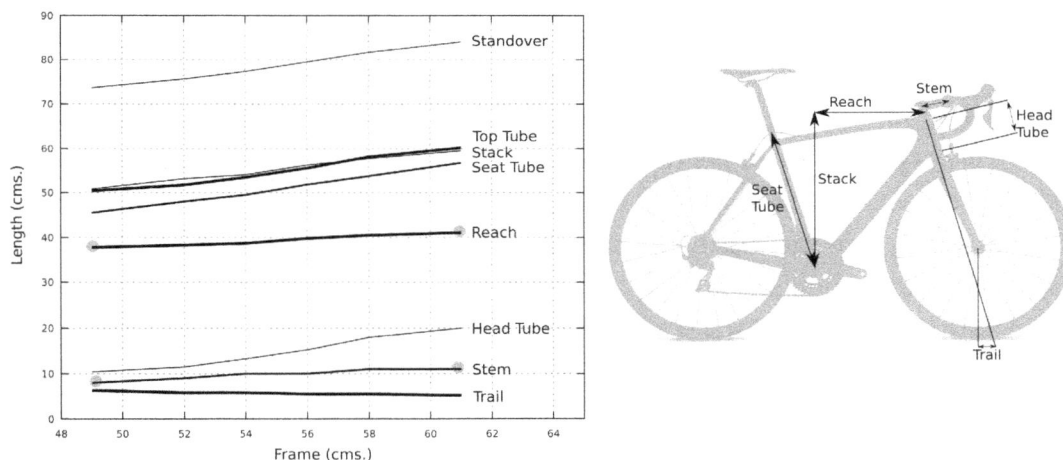

Figure 2.4: Frame Geometry (courtesy Specialized Bicycle Components)

In Figure 2.4, the parameters for bicycles from Specialized of different frame sizes show that parameter lengths do not change uniformly with frame size. For example, the trail becomes slightly smaller and the head tube becomes longer with larger frame sizes. A shorter trail improves maneuverability, but marginally reduces stability. Other parameters like the stem length and *reach* do not change significantly with frame size. The *reach* for a small frame (49 cms.) is 37.8 mm while the *reach* for a large frame (61 cms.) is 41.1 mm., a difference of 3.3 cms. However, the stem length also increases from 8 cms. for the small frame to 11 cms. for the large frame.

For a cyclist reaching for the handle bars this adds up to a sum of 6.3 cms. between the small and large frames. When buying a bicycle, you would not consider a difference of 6.3 cms. or 2.5

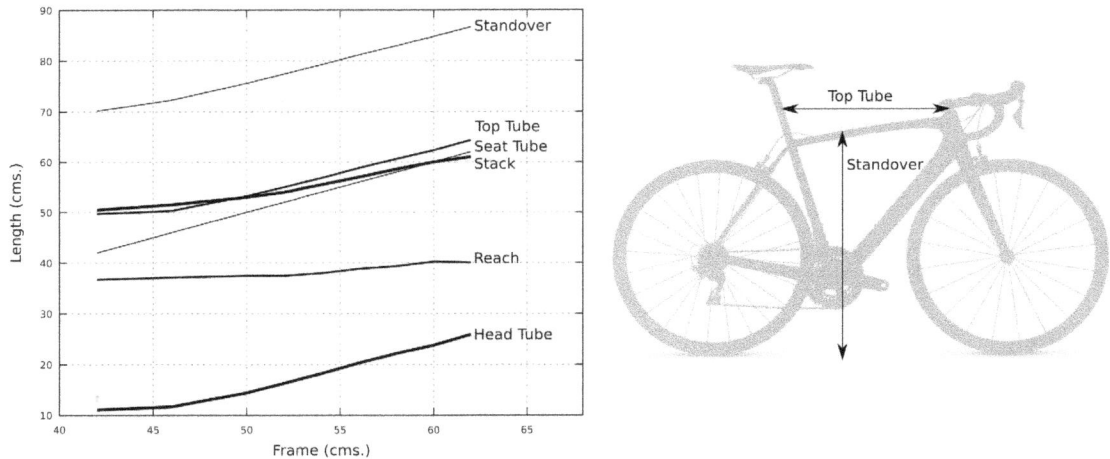

Figure 2.5: Frame Geometry (courtesy Surly)

inches as very significant. However, the small frame would suit a person of height 5 feet and the large frame a person of height 6 feet 3 inches or more. Having the correct *reach* to the handle bars is probably one of the most important measurements to make when choosing a bicycle.

The next important parameter to consider is the *stack* or height from the center of the bottom bracket to a horizontal top tube. The *stack* increases by three times the increase in the *reach* between the small and large frames. The small frame has a *stack* of 50.4 cms. compared to a *stack* of 60.2 cms. for the large frame. The difference in the increase in *stack* compared to the increase in *reach* is due to the larger difference in leg lengths compared to arm lengths.

Figure 2.5 shows the same frame geometry values from a different maker (Surly). Unlike Specialized, the length of the seat tubes matches the values for frame sizes. So, a 50 cms. Surly frame has a seat tube of length 50 cms. This implies that the length of the seat tube increases linearly in a Surly frame, while for a Specialized frame the length of the seat tube is roughly $0.93\times$ frame size. The length of the head tube increases more sharply with frame size for a Surly frame than a Specialized frame. These small differences in tube lengths and other parameters between frames of the same size from different makers mean that you need to spend more time beyond finding a frame size on a chart for a given inseam and height.

For serious cyclists, there are bike fit professionals who spend a lot of time studying many parameters to come up with a custom fit for an individual. Two reasons to consider a bike fit are comfort and performance. You are more likely to generate more power when you are riding your bicycle at some ideal position for a given bicycle.

However for the do-it-yourself cyclist, there are online bike fit calculators [3] that you can use. Another option is to rent a bicycle and test ride it for over an hour. It usually doesn't take very long to find out that your *reach* is too far or *stack* is too low.

Frame sizes also come with non-numeric titles such as X-Small, Medium, and Large. These titles usually correspond with a numeric frame size or height. Just as all manufacturers do not make a standard 50 cms. frame with the same tube dimensions, the titles small and large can mean different frame sizes depending on the manufacturer.

The Wrong Bicycle   If for some unfortunate reason you are stuck with a bicycle that does not fit you, you have a few options to get the best out of a poorly fitting bicycle. Sometimes bicycles cannot be returned after some initial period or it maybe too much effort to pack and return a bicycle that you bought online. The corrections depend on the margin of error for the bicycle - if you are 5 feet tall and the frame size of your bicycle is 58 cms., there is not much you can do to ride comfortably on a frame that is meant for a person over a foot taller.

For a frame that is too small, you can get a longer seat post. Most seat posts have an extension limit that is marked on the tube and if you extend beyond that limit you risk bending the seat post and possibly falling over. A long seat post acts like a lever and puts stress on the seat clamp when you ride. Despite these issues, a long seat post within limits can accommodate a tall person on a small frame.

If the saddle is on rails, you can move the saddle back to extend your reach. However, the changes in reach are typically limited to a few centimeters, which can make a difference. To move the handle bars still further ahead of the saddle, you can replace the stem with a longer version which will alter the steering of the bicycle, since the handle bars are away from the pivot on the head tube.

When the frame is too large, there are fewer options to make a bicycle fit a shorter cyclist. The first fix is to lower the seat post as much as possible – the seat post may not go lower than the first bottle cage bolt. You can either remove the bolt or cut the seat post to a shorter length. Similar to the change for a small frame, a saddle on rails can be moved forward instead, closer to the handle bars. A shorter stem will also lower the reach making a better fit for a short rider.

2.2 Choosing a Tyre

You may not consider tyres as important as other bicycle components such as the frame, gears, and brakes, but they are a good starting point to pick a bicycle. Besides choosing the right frame, finding the right type of tyre is also just as important. The best tyre for you depends on where you plan to ride your bicycle. If your rides will be mostly on city roads, then you do not need a mountain bike knobby tyre. A slick tyre that is reasonably wide will be sufficient.

You can ride on a wide tyre at a lower air pressure compared to the pressure on a narrow tyre. The pressure in a tyre should be sufficient to support part of the bicycle's and your weight (see Chapter 4 for the recommended pressure based on weight and width). A wider tyre at a lower pressure is more comfortable and rides just as fast as a narrow tyre at higher pressure. The range of sizes for the width of your tyre will depend on the width of the fork. Road bicycles often have narrow forks and that limit the width of the tyre that you can use.

Unfortunately, the dimensions to describe tyres, wheels, and rims come in at least three different formats. Makers used the dimensions that were common in their respective regions – so some tyres are described in inches and others in millimeters.

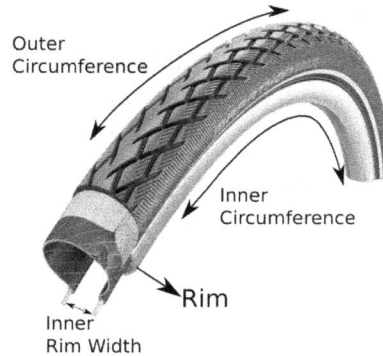

Figure 2.6: Wheel and Tyre Dimensions (courtesy Schwalbe)

The most common type of tyres called clincher tyres use a wire or beading to keep the tyre in place on the wheel (see Figure 2.6). The distance between the two wires is the inner rim width and must fit the rim of the wheel perfectly. The inner circumference is the circumference of the wheel rim and the wire diameter of the tyre must match the diameter of the wheel measured along the inner rim (the bead set diameter).

2.2.1 Tyre Sizes

The range of tyre widths that you can use on your wheel is between 1.45 to 2.0 times the inner rim width. So, if the width of the rim is 15 mm., then you can use tyres of widths 22mm. to 30 mm. The width of the tyres is the outer circumference in Figure 2.6. However, the outer circumference may change depending on the pressure within the tyre. A higher pressure may increase the outer circumference of the tyre.

To maintain consistency between tyres of the same widths from different manufacturers, the European Tyre and Rim Technical Organisation (ETRTO) has defined the pressure at which the tyre width should be measured. A tyre of width 28 mm. is inflated to 87 psi and the width excluding tread or knobs is measured after 24 hours. ETRTO uses the ISO 5775 standard to label tyres.

ETRTO also defined a standard to describe a wheel. A $622 \times 19C$ wheel means that the inner diameter or BSD of the wheel is 622 mm. and the width of the rim is 19 mm. These two values, 19 mm. and 622 mm., are sufficient to find a matching tyre. The range of the widths of tyre that will fit a $622 \times 19C$ wheel based on the formula mentioned earlier (1.45 to $2.0 \times width$) are 28 mm to 38 mm. Although you can ride a tyre wider than the range specified based on the inner rim width, the surface area of the rim to support the pressure and weight may not be sufficient. A narrow tyre on a wide rim will not be as rounded and therefore rolling resistance may increase (see Chapter 4).

On the sidewall of the tyre, you may find the same markings in inches that are not as precise as the ETRTO values (see Table 2.1). Two tyres for wheels of diameter 559 mm. and 590 mm. maybe classified as 26 inch tyres [4]. The use of inch dimensions is more common in mountain bicycles and was computed based on the position of the brakes.

ETRTO	Inches	French
$25 - 559$	26×1.0	
$28 - 584$	$26 \times 1\frac{1}{8} \times 1\frac{1}{2}$	$650 \times 28B$
$28 - 622$	28×1.1	$700 \times 28C$

Table 2.1: Sidewall markings for tyre dimensions in three formats

The French format originally used the letters "A", "B", and "C" to represent tyre widths. Now, these letters are used to maintain consistency with the original naming convention, but do not represent tyre widths. Schwalbe [4] has published a web page with over 150 tyres sizes in these three formats. The vast majority are not very common and choosing a common size like $28 - 622$ (for a road bicycle) or $50 - 559$ (for a mountain bicycle) will make it easy to find matching inner tubes. With many choices, picking the right tyre may appear complicated. To add to the complication, the sizes of tyres are described in inches and millimeters. However, if you select the type of bicycle you will be riding, the number of choices can be limited to a few.

The different systems of describing the dimensions of a tyre may appear confusing, but the standard ETRTO description of a tyre is sufficient to decide if a tyre can be used on your wheel. When you buy a bicycle, the ETRTO dimensions of the tyre $(28 - 622)$ are always given, from which you can get the inner diameter (622 mm.) and width (28 mm.) of the tyre. The inner rim width of the wheel may not be specified, but you should be able to replace a given tyre with a slightly wider tyre (32 mm.) or a narrower tyre (25 mm).

Finding the right tyre If you will be riding a road bicycle, then the $x - 622$ (or $700 \times x$) tyre where x mm. is the width and 622 mm. is the wheel (inner) diameter, is fairly standard. A wider tyre will have a larger x and there are tradeoffs between narrow tyres and wide tyres (see Chapter 4).

On a mountain bicycle, the common diameters are 26 inch, 27.5 inch, and 29 inch. The 26 inch diameter MTB tyre has been a standard size for many years and corresponds to a $x - 559$ ETRTO tyre. However, there are other 26 inch tyres which will fit a wheel with diameters greater than 559 mm. A 29 inch MTB tyre has the same inner diameter (622 mm.) as a standard road bicycle tyre. Finally, a 27.5 inch MTB tyre has a inner diameter of 584 mm. (see Figure 2.7)

The inner diameter is the bead set diameter (BSD) and the outer diameter includes the tyre width. The BSD is given for a wheel and the outer diameter is calculated as $(\pi \times BSD) + 2 \times width$. So, the outer diameter for a $26" \times 2.3"$ tyre is $(\pi \times 559) + 2 \times 2.3 \times 25.4 + 1$ or 677 mm. with an

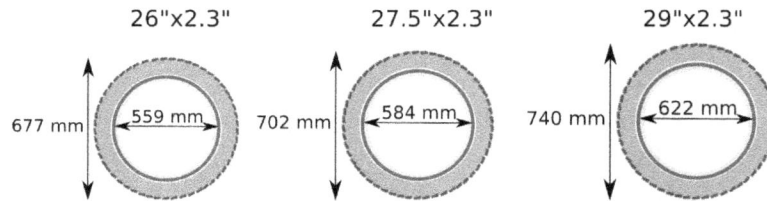

26"x2.3"   27.5"x2.3"   29"x2.3"

677 mm 559 mm   702 mm 584 mm   740 mm 622 mm

Figure 2.7: Wheel and Tyre MTB Diameters

extra mm. for the rim. The pressure of the tyre may change the outer diameter marginally. You can calibrate the outer diameter accurately by riding over some known distance with a cyclo-computer (see Chapter 7). The 2.3" width of the tyre in Figure 2.7 is just one of the tyre widths that you can use and there are a few other MTB tyre sizes as well [5].

The main advantage of tyres with a large diameter is that you can roll over obstacles more easily than with a tyre of smaller diameter. With a large tyre, you will approach an obstacle at a lower angle compared to a smaller tyre (see Chapter 4). A larger tyre also has a larger area of contact with the ground and therefore higher traction. This would imply that larger tyres are better than smaller tyres. However, smaller tyres are lighter than bigger tyres and easier to handle. They are also easier to accelerate when needed. A taller rider may find a 29 inch tyre more comfortable than a 26 inch tyre.

The tyre for a folding bicycle has the smallest diameter 20 inches or $x - 406$ ETRTO. The 406 mm. diameter of a folding bicycle tyre is about $\frac{2}{3}$ the diameter of a road bicycle tyre. You would imagine that you cannot ride fast on such a tyre, but to the contrary with a large crank, you can achieve speeds comparable to a road bicycle. The other advantage of small wheels is the quick acceleration which is useful in urban traffic where you have start and stop often. Finally a folding bicycle is easy to transport. However the pressure on these tyres is often higher than the pressure on a MTB tyre to handle the weight of an adult cyclist and bicycle.

Friction

The tyres of a bicycle are the *only* contact points with the road or trail and the type of tyres can make a significant difference in the feel of a ride. Consider the tyre rolling in Figure 2.8 on a pavement. When a tyre is rolling, each segment of the tyre makes contact with a different part of the road.

Therefore we consider *static friction*, since no part of the tyre is moving relative to the ground. The static friction between the tyre and the ground keeps the tyre stationary relative to the ground. Since you are applying a torque to the tyre while pedalling, the tyre moves forward. If this torque exceeds the static friction, then the tyre will begin to rotate over the same part of the road. The term *traction* is used instead of static friction for tyres and represents the force that resists slipping. If you want to move forward, you will need sufficient traction to prevent slipping. If you lose traction when riding at a moderate speed, you may end up skidding with the tyre moving sideways instead of forwards and the bicycle will be uncontrollable.

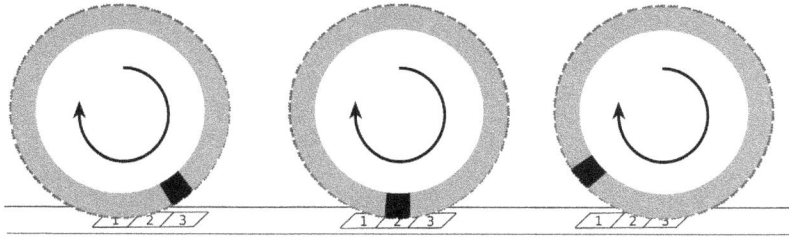

Figure 2.8: Tire Rolling on a Surface

At first this appears counter intuitive since the assumption is that friction should be reduced to gain higher speed for a given torque. But reducing the static friction between the tyre and the road surface leads to slipping and actually reduces speed. Therefore, you need adequate friction between the tyre and the ground to avoid slipping. Instead of reducing static friction, reducing the "rolling resistance" of a tyre will make your ride faster (see Chapter 4).

The type of tyres on a bicycle depends on where it will be used. If a bicycle will be used mostly in the city, then the tyres are more narrow, made of harder rubber, and will have fewer treads. A tyre for off-road use will have knobs or protrusions on the surface of the tyre. These knobs come in a variety of shapes and sizes and the main purpose of the knobs is to provide enough traction off-road. The knobs effectively "dig" into the ground to increase static friction.

The surface of a tyre for city roads will be more smooth and may have barely visible treads – also known as slick tyres. In between the knobby tyres and slicks are semi-slicks. They are mainly for roads and some off-road use. A few treads are located on the side of the tyre and the rest of the surface of the tyre is smooth. Another variation is inverted treads or knobs that are depressed in the tyre. This increases traction compared to slick tyres, but at the same time reduces rolling resistance compared to knobby tyres (see Chapter 4).

2.3 Geared or not?

A single speed bicycle has a single chain wheel and cog in the rear, while a geared bicycle may have 2-3 chain wheels and 9 or more rear cogs. So, a 27 speed bicycle may have 3 chain wheels and 9 rear cogs giving you 27 different gear ratios (speeds). Having more gear ratios will not make you ride any faster - the power that you can generate ultimately determines your speed. However, using gear ratios appropriately can make it easier to ride uphill at a slower speed and downhill at a higher speed. With a single speed, you have only one gear ratio and therefore changing your speed depends on the force that you can apply to the pedal and your cadence (rpm).

How do gears work? The early bicycles without gears such as the Penny Farthing used wheels with large diameters to ride faster. The maximum rate (cadence) that an average cyclist can pedal is about 100-120 rpm and it is difficult to sustain a high cadence over a long distance. If the diameter of a wheel is 0.7 meters, then with a maximum cadence of 120 rpm, speed is limited to 15.8 kmph. which is relatively slow. Without gears, the only way to increase speed was to increase the diameter of the wheel.

Gear	High	Medium	Low
Cog	Small	Medium	Large

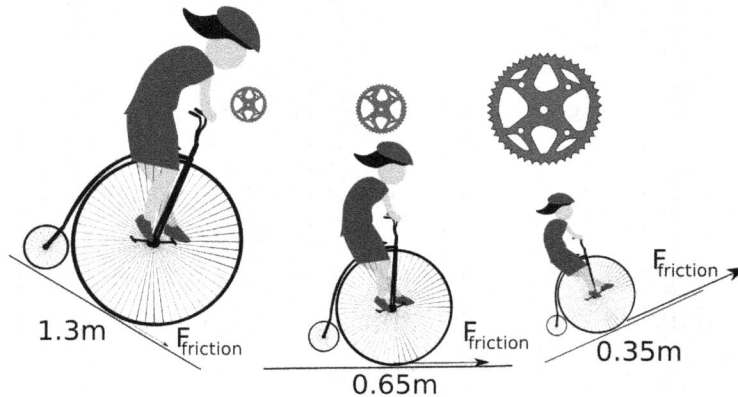

Figure 2.9: Using gears to change wheel diameters and force

In Figure 2.9, three bicycles of wheel diameters 1.3 m., 0.65 m., and 0.35 m. will have maximum speeds of 29.4 kmph., 14.7 kmph., and 7.9 kmph. respectively (for a cadence of 120 rpm). While a large wheel was good for riding fast and to ride over bumps, the brave cyclist had to sit high up and risk a fall from some height.

The use of gears solved this problem and allowed you to ride on a wheel with a small diameter, yet effectively change the diameter as and when needed. If you were riding downwhill and wanted to change to a large diameter wheel (a higher gear), you would move the chain to a smaller rear cog. Similarly if you were climbing uphill, you could change to a smaller diameter wheel by moving the chain to a large rear cog.

Although it appears counter intuitive, the frictional force of the ground on the wheel $F_{friction}$ increases as you shift to a lower gear. In a low gear the torque, $F_{chain} \times R_{cog}$, where F_{chain} is the force on the chain and R_{cog} is the radius of the cog is higher for the same F_{chain} compared to the torque generated in a high gear. The radius of the cog in a high gear is smaller and therefore, the torque is correspondingly smaller. This is an useful feature, since you need more force on the wheel to ride uphill than downhill. In general, there is also a tendency to apply a greater force (F_{chain}) when riding uphill which further increases $F_{friction}$.

Gear Ratios In Figure 2.10 on a bicycle with gears, you could use three different gear ratios to ride downwhill, on a flat road, and uphill. The higher the gear ratio, the faster you will ride. With a gear ratio of 4.54, a single spin of the front crank will rotate the rear cog 4.54 times. In other words, if you could pedal the front crank once a second (cadence) and the circumference of your rear wheel was 2 meters, then you would be riding at 2×4.54 or about 9 meters per second (32.4 kmph.).

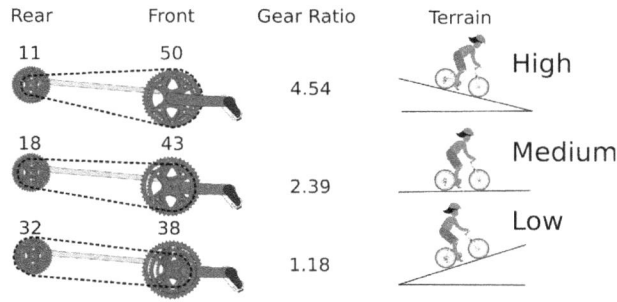

Figure 2.10: Gear Ratios for Riding Downhill, Flat, and Uphill

This is not very strenuous on a downwhill since you have gravity assisting you and the force that you need to apply on the pedal is not excessive. On a flat with a lower gear ratio of 2.39 and the same cadence of 1 per second, your speed would be 2×2.39 or 4.78 meters per second (17.3 kmph.) On a steep uphill, the gear ratio is even lower at 1.18 corresponding to a speed of 2.36 meter per second (8.5 kmph.).

If you buy a 3×9 geared bicycle, then you have 27 gear ratios to choose from. Consider a bicycle with 3 front chain rings of 48-38-28 teeth and a rear cassette with 9 rings of 12-13-15-17-19-21-24-28-32 teeth (see Figure 2.11).

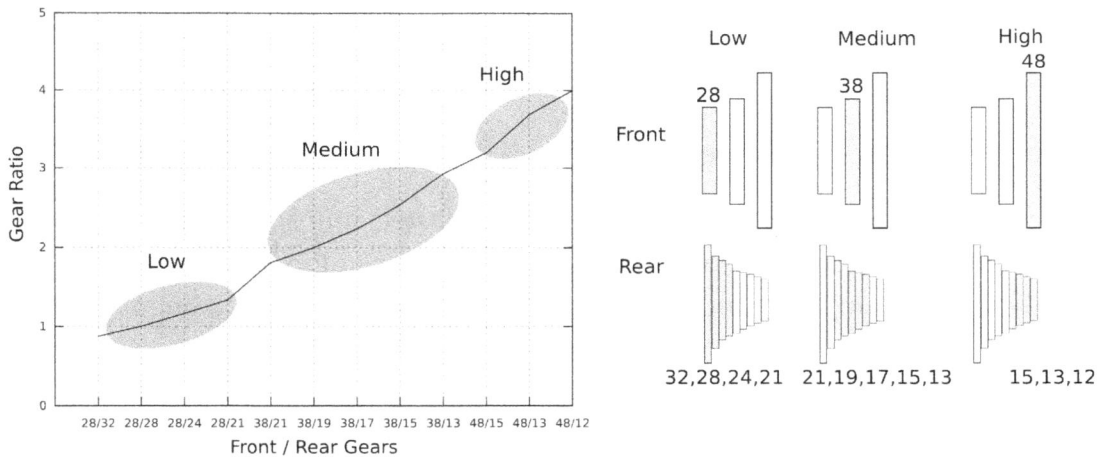

Figure 2.11: Recommended rear sprocket combinations for front chain rings

Although you have 27 possible gear ratios, several of them are not recommended since they run cross chain. A cross chain gear ratio is when the chain is on a small front chain ring and a small rear spocket or on a large chain ring and a large rear sprocket. In both cases, the chain runs across at an angle and besides being inefficient in transmitting chain tension to the hub, it can also decrease the life of the chain and cassette.

A few of the gear ratios are almost identical. A 48-24 ratio is the same as a 38-19 ratio and similarly the ratios 38-21 and 28-15 are almost identical. This means that in both gear ratios, one revolution of the crank will give you the same speed. Shifting front chain rings is slower and also may make a sharp change in the gear ratio.

The 12 recommended front gear combinations in Figure 2.11 increase the gear ratio linearly without a sharp increase between any two consecutive combinations. The combinations also ensure that the chain is relatively parallel to the plane of the cassette and the front chain ring.

The gain ratio (see Chapter 3) more accurately reflects the forces necessary for different gears. It takes into consideration the radius of the crank and the wheel. Although all 26 inch wheels will have the same circumference, the thickness of the tyre will change the "real" circumference. A 26 inch wheel with a $1\frac{1}{2}$ inch width tyre will have a circumference of 2.01 meters, while the same wheel with a 3 inch width tyre will have a circumference of 2.17 meters. These slight differences in the circumference will add up in the calculation of speed over an hour or more.

Observations

- Finding the right gear depending on the terrain is a matter of trial and error. If you find your cadence dropping below 60 rpm, then a lower gear will be more efficient. Conversely if you don't feel any resistance when pedalling at a high rpm, you will ride faster in a higher gear.

- If you are approaching an uphill, it is better to change to a lower gear before you begin the climb. Although it maybe tempting to accelerate into the climb at a high gear and then switch to a lower gear when your cadence slows, the tension in the chain will be high leading to a noisy and slow shift.

- Reducing the force on your pedals before you shift gears will make the change smoother.

- A bicycle with more gears will not make you ride faster. Your speed depends on the power that you can generate and not on the number of gears in the bicycle. Riding in the right gear based on your power output will make your ride more efficient, but not faster.

How fast will I go on a bicycle? This is a fairly common question for a beginner and most answers vary from 15 kmph. to 25 kmph. However, speed depends on many factors including the condition of the road, the grade of the road, wind speed and direction, condition / type of bicycle, and your weight. A more appropriate question would be how much power can you generate on a bicycle? Speed is directly proportional to the power generated. However for the same power, speed will differ depending on the total weight and the aerodynamics of the bicycle and rider.

2.4 Bicycle Types

The ideal bicycle should be both comfortable and fast, but in most cases there are tradeoffs to be made based on whether you prefer comfort or speed. The following is a broad categorization of most bicycles and there are some other types that are not included here.

- City Bicycle for riding on streets

- Mountain Bicycle for riding off road and on streets

- Hybrid Bicycle for riding on city streets with features to ride faster than a city bicycle

- Road Bicycle for riding on highways and city streets at the highest speed possible

- Folding Bicycle for riding on streets, relatively lightweight and foldable to carry on public transport

- Electric Bicycle or E-Bike for riding on streets with a battery powered motor

While choosing a bike, you could decide where you want to ride and pick an appropriate bike category. However, a different approach is to look at the type of tyres on the bike before picking a category. There are a broad range of tyres and sizes that you can pick from when selecting a bicycle. The reason for choosing a tyre before the bicycle is that once you have selected a bicycle, the frame of the bicycle will limit the size and width of the tyre that you can fit on the bicycle.

Typically wider tyres will be more comfortable to ride than narrow tyres (see Chapter 4). Not only does a wide tyre have more air (cushion), but may even roll faster than a narrower tyre at a lower pressure. Road bicycles designed for speed and less air drag have narrow forks that do not allow wide tyres. Mountain, hybrid, and city bicycles have wider tyres than road bicycles, since they are designed more for comfort than speed and allow you to ride on a rough trail.

A folding bicycle has smaller tyres than all of the other types of bicycles. At first, you may assume that these bicycles look like bicycles for children and will be slow. Surprisingly, a folding bicycle with a large crank and small wheel can be as fast as a road bicycle. Besides, you can accelerate faster on a small wheel bicycle, which is a very useful feature, if you ride in the city where you need to stop and start often.

Finally, electric bicycles are becoming popular for commuting, since a ride to the office will not become a workout and you will not sweat as much. Electric bicycles have options to use motor assisted pedalling or to ride manual. If you need to ride uphill, assisted pedalling is very useful because the weight of an electric bicycle may range from 20-40 kgs. An electric motor also makes it possible to ride at a more uniform speed and therefore maybe safer than a manual bicycle on which maintaining a constant speed is hard if the terrain changes.

2.4.1 The City Bicycle

The City bicycle as the name suggests is for riding around town with some specific purpose (shopping or commuting). It is does not appear as attractive as a road bike (see Section 2.4.3)

Figure 2.12: A City Bicycle (courtesy Pure Cycles)

and typically costs much less as well. As a functional bicycle, a standard model comes with accessories such as bumpers (mud guards or fenders), a stand, and maybe even a bell. This type of bicycle is also called an urban or commuter bicycle.

Many of the components of this bicycle are mass produced and therefore not available in a wide range of sizes. Since the components are mass produced, the cost of the components and subsequently the bicycle is lower than other types of bicycles below. However, it is not necessarily true that *all* such bicycles will have frequent breakdowns or are low cost.

A bicycle from Pure Cycles in Figure 2.12 illustrates a typical such bicycle. It includes many of the accessories such as rear and forward fenders, a kickstand, chain guard, a rear carrier, and a bell. The tires are 35 mm. ($\approx$ 1.4 inches) wide and are sufficient to ride on a road with minor pot holes. The frame is made of steel and the complete bicycle weighs 13.6 kg. ($\approx$ 30 pounds). The rear and forward brakes are standard and the wheel rim uses 36 spokes to handle moderately heavy loads.

Although, this particular bicycle does not have visible gears, it uses a 3 speed hub in the rear which is assumed to be sufficient for riding in the city. This intentional design makes the bicycle more functional and simpler since there are no rear and front derailleurs to change gears.

2.4.2 The Mountain Bicycle

The mountain bicycle or MTB is distinct from the city bicycle with wider knobby tyres. A MTB is designed to handle rough terrain with a suspension to smooth out the bumps. The tyres besides being wide and at a lower pressure (30 - 50 psi) compared to the pressure on a road bike tyre also have a knobby appearance. The protrusions or knobs on the tire increase the friction on a dirt road and therefore a bicycle with a MTB tyre tends to have a better grip than a bicycle with a smooth tyre. The knobs are triangular, round, or in some other shape that the tire manufacturer prefers.

Although mountain biking has existed for a long time, it was not termed as such, till more recently. Most early cyclists rode on dirt roads or trails. The distinguishing feature of a MTB is the wide knobby tyre. With a wider tyre, not only can you ride on a larger air cushion, you can also ride at a lower pressure to reduce vibrations from an irregular surface, and you also have more traction with a wider contact area between the tyre and a surface. Riding on a tyre with pressure that is high enough to avoid punctures, but low enough compared to a road bicycle tyre has several advantages -

- The loss of energy due to vibrations to the body is minimized as the tyre rolls over most bumps and does not vibrate.

- Riding at a low psi is more comfortable than riding at a high psi.

- Since pressure is inversely proportional to area for a given force, a lower pressure means that the contact area increases, leading to more traction.

Figure 2.13: A Mountain Bicycle (courtesy Specialized Bicycle Components)

A typical mountain bicycle like the one in Figure 2.13 has fat tyres and a front suspension in the fork. The fat tyres connote rides away from an asphalt road and closer to nature on rough trails and soil. A mountain bicycle with a single front suspension is called a hard tail and a bicycle with a front and rear suspension is called a dual (or full) suspension MTB.

Typically, a good MTB also comes with efficient disc brakes. A disc brake is more effective than a rim brake and works well in all weather conditions (see Chapter 6). You can also get a wide range of gears on a MTB to handle all kinds of terrain.

Observations

- If a majority of your rides will be on asphalt roads, then a hybrid bicycle will work as well as a MTB.

- The suspensions of a MTB do help make a ride smoother, but they are really needed in a rough trail. You can manage to ride over bumps and pot holes on a road with wide and large tyres alone.

2.4.3 The Road Bicycle

If a MTB is like an SUV for bicycles, then a road bike is like the "sports car" of bicycling. It is designed and built for speed and made to look as attractive as possible. Of course, not everyone needs it. But just like sports cars, road bicycles are appealing. The most expensive road bicycles are made of the lightest components and ride fast (see Figure 2.14).

At speeds over 20 kmph. air drag becomes significant (see Chapter 3). Road bicycles are designed to minimize air drag with an aerodynamic frame design and drop down handle bars to keep the rider low. The "front area" of a rider is one of the biggest components of air drag. Reducing front area is key to maintaining speed for some given power that a rider can generate.

Figure 2.14: A Road Bicycle (courtesy Trek)

However there are limits to how much you can minimize your front area. Since a road bicycle is built for speed, aerodynamics and reducing drag is key. Next to air drag, the rolling resistance and slope resistance are two other drag factors. Both resistances are based on the combined weight of the bicycle and rider.

If you can generate 100 watts of power and the total weight including the weight of the bicycle is 80 kgs., then without a head wind you should be able to ride at about 24 kmph. on a flat road (see bikecalculator.com). If the total weight is reduced by 10 kgs., then speed increases to about 24.5 kmph. You will notice slightly larger changes in speed in a light bicycle on a slope or with a head wind.

With the same scenario at 150 watts of power and on a slope of 5%, the speed at 80 kgs. is 11.3 kmph. compared to 12.6 kmph. at 70 kgs. The other ways to increase speed include lowering rolling resistance with better tyres (see Chapter 4).

Since air drag depends on the square of the velocity, reducing air drag will have a large impact on your speed. At 100 watts on a flat road with a weight of 80 kgs. and 78% of the earlier aerodynamic factor, speed increases from 24 kmph. to about 26 kmph. At higher speeds, the air drag is a major portion of the total drag. If you can lower the upper part of your body, you can reduce your front area (air drag) substantially.

Drop Handle Bars   The drop handle bars in Figure 2.15 are almost always found on road bicycles. One of the big advantages of these handle bars is that you have multiple positions for your hand. In an aggressive position (less front area) for high speed, you could place your hands on the drop bar and bend low. In this position, your front area is minimized and you can ride as fast as you can based on your power.

Figure 2.15: Positions to hold a drop handle bar

Being in an aggressive position is not very comfortable for long and you will want to relax a bit at some point. The other two positions on the hoods and the handle bar are more relaxed and suitable for long rides.

Observations

- Adding drop down handle bars alone will not convert a "regular" bicycle into a road bicycle.

- Just as you wouldn't drive an expensive sports car on a mud road, most road bikes are built exclusively for a hard surface like a road. Some manufacturers like Trek and Specialized have introduced proprietary devices (Isospeed coupler and Future Shock) to ride a road bicycle on rough surfaces.

- There is a certain indescribable thrill in riding at 40-50 kmph. on a smooth road on your own power.

2.4.4   The Hybrid Bicycle

If a road bike is too aggressive for your taste and you are not into racing, then a hybrid bicycle combines features from a MTB and road bike. The frame geometry is not designed exclusively for speed and will accommodate tyres wider than road bicycle tyres. The handle bars are often straight and slightly raised for comfort (see Figure 2.16). The wheel size is usually the same as the wheel size of a road bicycle (622 mm) but with a wider fork for wider tyres.

A hybrid bicycle is a general purpose bicycle that can be used for commuting as well as unpaved trails. However a hybrid bicycle cannot replace a MTB that you can ride on rough off-road trails. Some other features of a hybrid bicycle include a suspension, disk brakes, and mounts for a rack and fenders.

Observations

- While weight is one of the primary factors in a road bicycle, a hybrid bicycle may have a heavier frame. The reason weight is not as important in a hybrid bicycle is that speed is not the only consideration.

Figure 2.16: A Hybrid Bicycle (courtesy Specialized Bicycle Components)

- Other features such as a suspension, disk brakes, and options to carry some luggage are found in hybrid bicycles.

- A hybrid bicycle is a good option for someone who is not interested in investing in multiple bicycles - separate bicycles for the road and off-road.

2.4.5 The Folding Bicycle

Bicycles are shipped from most manufacturers to retailers in boxes that are in roughly 45 inches ×30 inches ×11 inches. This is a large bulky box (see Figure 2.21) mainly because the frame and wheels are significantly larger compared to the other components of a bicycle. Besides the issues with shipping full size bicycles, they also take up a lot of space in a train or a bus. A rack in the front of a bus can possibly hold only 2-3 full size bicycles.

To make it easier to carry a bicycle in public transportation without annoying other commuters, manufacturers have designed the folding bicycle. The two large components, the frame and the wheel, were reduced in size. The frame was made foldable and the wheel size was reduced from 26 inch diameter to 20 inch diameter or less. The dimensions of the folding bicycle shown in Figure 2.17 is about 23 inches ×22.2 inches ×10.6 inches.

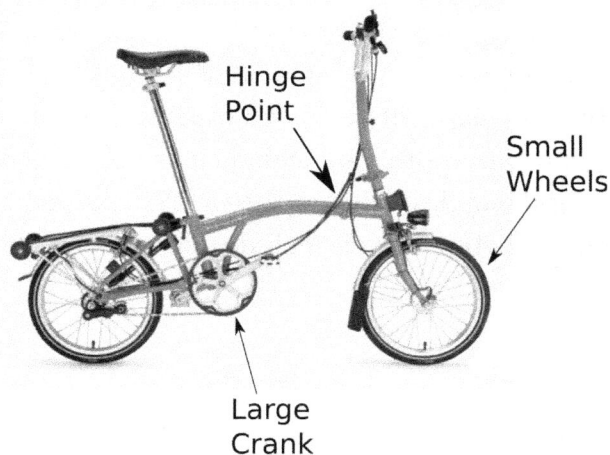

Figure 2.17: Folding Bicycle (courtesy Brompton Inc.)

Small Wheels   A common reaction to a bicycle with a 20 inch wheel is that it is a bicycle for children and not adults. In other words, you cannot ride enough fast enough to keep up with riders on full size bicycles. Contrary to that perception, a bicycle with small wheels can be really fast [2] . A folding bicycle will in most cases have a large crank with about 45-55 teeth compared to the crank of a normal bicycle. The rear sprocket may have 13-16 teeth. If the front crank has 52 teeth and the rear sprocket has 13 teeth, then one crank revolution will correspond to 4 revolutions of the rear wheel. For a wheel with a 20 inch diameter, this corresponds to a reasonably high speed of 38 kmph. at 75 rpm.

While small wheels do not limit your speed, the tyres of small wheels must be inflated to a higher pressure than the tyres of a full size bicycle. The volume of air in a small tyre will be less than the volume of air in a full size bicycle tyre and therefore the pressure must be higher to handle the weight of an adult and the bicycle. The ride on a tyre with high pressure is usual bumpier, but maybe acceptable on asphalt roads.

Foldable Frame   The frame of a foldable bicycle has a "hinge point" where the frame can be folded to reduce the length of the bicycle. This may not appear as critical as you would imagine, but designing a frame that can be folded is non-trivial. A frame must be rigid enough to sustain loads without bending or cracking and adding a hinge point introduces a weak point in the frame where the stresses due a load will be high.

The designs for making a folding frame are typically proprietary and one of the popular technologies is the Ritchie break away frame with couplers to fold a full size frame that will fit in a suitcase. A foldable frame is also desirable for touring bicycles that have to be shipped by air. Each airline has its own regulations for large cargo and being able to fit a bicycle into a standard size suitcase is a big plus.

2.4.6   The E-Bike

If you have been riding a bicycle for even a short period of time, you probably know that riding uphill is the toughest part of a ride and takes the most energy. Not only does your speed drop, you start sweating and generating the power to sustain a reasonable speed becomes harder. The steeper the slope, the harder it is to ride.

For a 70 kg. rider on a 10 kg. bicycle with no wind on a flat road, it would take about 110 watts to ride at 25 kmph. On a 1% grade uphill and 3% grade uphill, it would take 166 watts and 281 watts respectively to sustain the same speed. The power to ride at the same speed increases rapidly with an increase in the grade since the slope resistance, the gravitational force due to the total weight of the bicycle and rider opposing your motion, becomes much larger. It takes about 50% more power to ride at the same speed on a 1% grade uphill compared to a flat road.

This problem has deterred many a potential rider from using the bicycle as a regular means of transportation if the local roads are uphill. The same problem occurs in a strong head wind,

---

[2]The speed record of 295 kmph. was set on a bicycle with 17 inch wheels.

although this problem maybe intermittent. A lighter bicycle does not solve the problem. A 70 kg. rider on a 5 kg. bicycle still has to generate 162 watts and 270 watts for the same 1% and 3% grades respectively.

Figure 2.18: An Electric Motor Assisted Bicycle (courtesy Specialized Bicycle Components)

The e-bike or electric motor assisted bicycle has largely solved this problem. An e-bike is a regular bicycle with an electric motor to assist you when you need more power (see Figure 2.18). The battery is typically integrated with the frame. In Figure 2.18, the battery is located on the down tube, which is the longest tube in the bicycle and a convenient location as well. The battery in the Specialized Turbo Vado is a 604 Watt hour battery. Theoretically, if all the 600 watts of the battery was used to power the bicycle, then you could ride at about 47 kmph. on a flat road with no head wind or 38 kmph. on an uphill road with a 3% grade.

In practice, an e-bike has modes to limit the power drawn from the battery. The "economy" mode will use less power than the "turbo" mode. You also have losses in the conversion of electric energy from the battery to mechanical energy supplied to the hub. The losses depend on the efficiency of the electric motor.

Ideally, you would like a battery that would last for several hours, so that you can ride without the fear of losing battery power. But a large battery will add substantially to the weight of the bicycle. A heavy battery must also be matched with a heavier electic motor as well. So, most manufacturers seek a compromise between the size of the battery and its duration. The battery on an e-bike is not extremely heavy, but heavy enough to power a ride for about an hour or so. The assumption is that most e-bikes will be used for commuting and not for long endurance rides of a 100 kms. or more.

The average weight of an e-bike may be roughly 22 kgs., of which 10 kgs. is the combined weight of the battery and electric motor. In future, this may change with lighter batteries and motors. Many e-bikes also come with an attached computer to monitor battery levels and the predicted range of the bicycle before the battery runs out of power.

E-Bike Motors The motor in an e-bike converts electrical energy (from a DC battery) to mechanical energy. The two locations where a motor can assist a rider are the rear wheel hub and the crank. The hub motor is placed in the rear wheel's hub and when powered will generate torque usually in proportional to the power that you generate. If you are riding leisurely, then the hub motor will run at a lower capacity compared to a more intense [3] ride.

While many hub motors are relatively inexpensive, they can weigh 10 kgs. or more. This means that the rear wheel and tyre which support roughly 60% of the combined weight of the rider and bicycle, must also be sturdy enough to handle the additional weight of the motor. Fixing a flat on a rear wheel will be a little more tedious.

Mid drive motors by contrast are smaller and placed on the crank or the chain wheel. The batteries associated with mid driver motors are also smaller and typically add less weight to the bicycle. They also allow you use a rear derailleur which can make climbing hills easier. Since it takes a lot more energy to climb a hill, this is a big plus. Reducing the tension in the chain when shifting gears makes the gear change smooth and lowers the wear / tear on the cassette as well as the chain. A mid drive motor may not detect when you are changing gears and maintains a steady tension on a chain, leading to a slightly noisier gear shift.

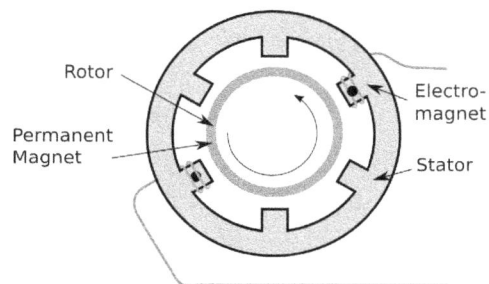

Figure 2.19: A DC Motor with an Inner Rotor and Outer Stator

The mid drive and hub motors differ in how the motor rotates the hub or crank. The two main parts of a DC motor are the stator and rotor (see Figure 2.19). One of them is stationary and the other rotates. In a hub motor, the rotor (a permanent magnet) spins about an axes while the stator is stationary. In some motors, the rotor maybe placed in an outer ring unlike the rotor in Figure 2.19.

The force driving the rotation is due the changing magnetic field of the electro-magnets which alternately attracts and repels the permanent magnet when current runs through the stator in a circular sequence. In a hub motor, the electromagnet is held stationary while the rotor connected to the hub rotates providing additional power to the rider. The roles of the stator and rotor are reversed in a mid drive motor. The rotor remains stationary in a mid drive motor while the stator rotates the crank. E-bikes are still evolving and they will likely become more popular over time for commuting (see [6] for more on the different types of motors and e-bikes).

Observations

---

[3]The intensity of your ride is detected from a cadence sensor.

- An e-bike will definitely make a ride that is an hour or less faster and much easier compared to a regular bicycle. Most e-bikes come with different modes that will generate more or less power depending on the mode. You can set the degree of assistance that you need from the electric motor.

- The range of an e-bike can be extended by being stingy with the use of battery power. Conversely, you can quickly drain the battery, if you constantly use the maximum power that the battery can provide.

- While an e-bike is faster, it is also a lot heavier. If for some unfortunate reason your battery dies, you can still ride the e-bike. However, an e-bike then becomes equivalent to a regular bicycle with an additional weight of about 10 kgs.

- Batteries are still relatively expensive and after many cycles of charge and discharge, you will need to replace the battery. The lithium in a used battery also may not be recycled, since it is more expensive to recycle than to extract lithium from raw mineral deposits.

2.4.7 Common Features of Bicycles

Figure 2.20 shows how the six bicycle types share some common features. For example, a mountain bicycle is more likely to have wide knobby tyres, be relatively slow, be suitable for rough terrain, and weigh more than most other bicycles. By contrast, a road bicycle will be lighter, have narrower tyres, be more aerodynamic, and have a frame geometry that requires the rider to be positioned more forward.

A folding bicycle will also be light, but have a more upright position, be reasonably fast, and have tyres suitable for a road surface. An electric bicycle will be significantly heavier than most other bicycles with wide tyres for city riding and fast enough for the city, but not as aerodynamic as a road bicycle. The hybrid and city bicycles have features that are in the mid-ranges and are in general for daily use or commuting.

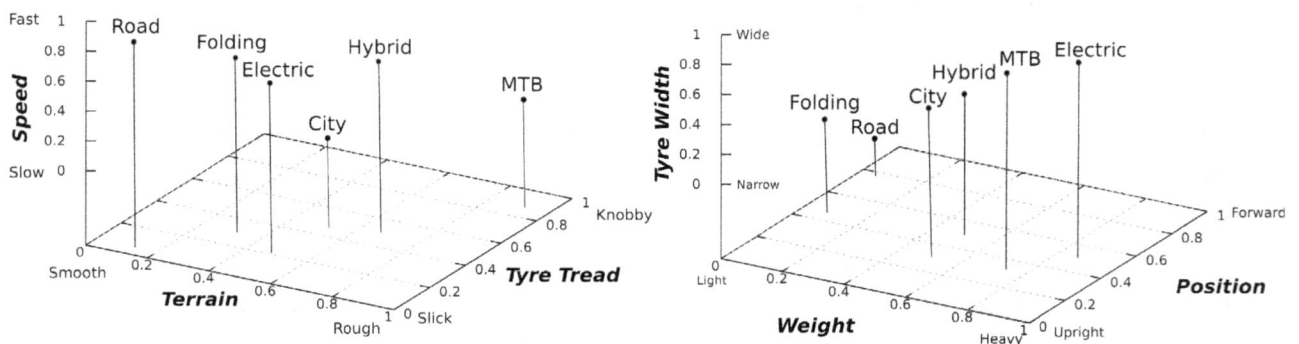

Figure 2.20: Common Features of Bicycle Types

Other Bicycles   There are a range of other bicycles that don't fit into one of the seven listed categories. The fat bicycle features tyres that are extremely wide compared to a road or mountain bicycle. These bicycles with 3.5 inch or wider tyres are used on soft terrain such as snow or sand. The triathlon bicycle is designed to be slightly more aerodynamic than a traditional road bicycle. The frame tubes and geometry of a triathlon bicycle are designed to reduce air drag. While reducing air drag is important for a road bicycle as well, a road bicycle is a more versatile bicycle for long rides and climbing hills.

The cruiser bicycle is like the roadster bicycle built for comfort and slow riding. A typical cruiser bicycle is relatively heavy with accessories such as fenders, padded saddles, and suspensions. The cyclo-cross bicycle is a type of road bicycle designed to ride on rough terrain with wider tyres than a normal road bicycle. Unlike a mountain bicycle, a cyclo-cross bicycle is lighter and may not include a suspension. The bottom bracket and chain stays of a cyclo-cross bicycle maybe higher than a road bicycle to handle muddy terrain. A hybrid bicycle maybe called a fitness bicycle to appeal to riders who plan on buying a relatively inexpensive bicycle to maintain fitness. A fitness bicycle can also be a stationary bicycle.

A touring bicycle is often a variant of a road bicycle designed for comfort, stability, and ability to handle larger than normal bicycle loads. The wheels of a touring bicycle will have 36 or more spokes to handle a heavier load. The wheelbase or distance between the front and rear wheels will be slightly larger for better stability. Also the tyres on a touring bicycle will be wider than the tyres on a normal road bicycle for more comfort. The frame of a touring bicycle will have mount points for racks, fenders, and multiple bottle cages. With the addition of these accessories a touring bicycle will be significantly heavier than a normal road bicycle. The light weight touring bicycle is a compromise between a road bicycle and a full fledged touring bicycle. It is designed for long rides that may not span more than a day or two.

2.5   Buying a Bicycle on the Web

It maybe tempting to buy a bicycle on the Web and then assemble it at home. If the components of the bicycle are well made and the instructions to assemble the bicycle are clear, it is possible to build your own bicycle at a reasonable cost. You would also need a basic set of tools to assemble your bicycle.

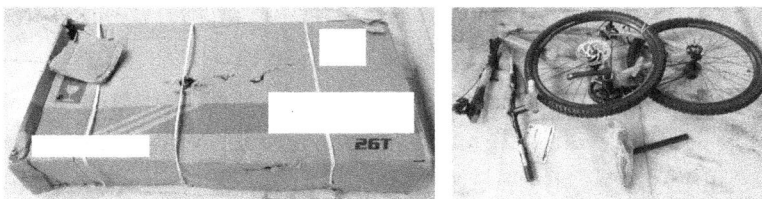

Figure 2.21: Bicycle ordered from the Web

In Figure 2.21, a bicycle ordered from the Web and delivered home is shown. Bicycles are usually packed well in a box with bubble wrap to protect components from damage during transportation. If you are not as fortunate, you may receive a box like the one shown in the Figure 2.21. Parts like the spokes or derailleur maybe bent when not packaged properly. Some of the spokes on the wheels may not have enough tension and therefore rattle when you ride the assembled bicycle.

The web page describing the bicycle may mention that the bicycle comes 85% assembled and that you would have to spend 15-30 minutes to complete the assembly. In some cases where the instructions are clear, it is possible to complete the assembly within 30 minutes. But, often it will take much longer to make sure that all the parts are in the right position and tight enough. In practice, the idea that you can unpack the bicycle out of the box, make a few adjustments, inflate the tyres, and be ready to ride in a short time is often misleading.

Step by step instructions to install parts like the seat post, wheels, and handle bars do help. The clarity of the instructions depend on the manufacturer's efforts to make the assembly as easy as possible. If you have not assembled a bicycle before, then you can expect to spend a lot more than 30 minutes to assembly the bicycle. Among the problems that you could face include

- Loose cables: The length of the cables to change gears or adjust brakes may have to be corrected.

- Brakes: The brake pads may not be centered, too far apart, or too close to the rim.

- Seat: The seat post and saddle should sit firmly in the frame.

- Wheels: The wheels and tyres should rotate smoothly without any obvious noise.

- Pedals: The pedals should not be lose and must move smoothly.

- Chain: A slack chain may not create enough tension to rotate the hub of the rear wheel.

Even though there are a number of problems you may encounter when assembling a bicycle, the benefits of buying a bicycle from a reputed maker without travelling a long distance to collect the bicycle outweigh the disadvantages. Bicycles are shipped worldwide to stores and a professional can quickly assembly a bicycle out of a box in a short time. If you have the patience, the tools, and order a bicycle from a maker with a good reputation, you can build your own bicycle.

It is still very difficult to ship a fully assembled bicycle in a box that is ready to ride. Either the box would be too large or the extra packaging would make it too expensive. The other alternative is to take the box to a local bicycle shop and have it assembled for some cost. The additional assembly cost may still make buying on the Web affordable. The risk of buying on the Web is that you cannot judge the quality of the bicycle or the ride until you actually receive the bicycle. If you are unhappy with the bicycle after assembling it, then it is inconvenient to disassemble, pack, and ship the bicycle back to the vendor (if you are lucky enough to have a return refund option). For these reasons, a new bicyclist would be better served buying a bicycle from a local bicycle shop than from the Web.

Also using your local bicycle shop (LBS) to find the right bicycle and then trying to find the same bicycle online at a lower price is legal but unethical. Even if you can do most of your repairs on your own, you will need your LBS at some point. An LBS can provide a part when you need it urgently.

2.6   Buying an Used Bicycle

An used bicycle can be a bargain if you are familiar with the components and brand names of common bicycles. A logical method to estimate the value of a bicycle is to simply sum the values of the individual components taking into account brand name, condition, and age. However, this is hard to do and a simpler method is to just evaluate the overall condition of the bicycle, age, and use the original price to compute the estimate current value.

The simplest method to calculate the value of an used bicycle is the linear depreciation model. If a bicycle has an estimated life of 10 years, then annually a bicycle will loose roughly a tenth of its value. At the end of 10 years, the bicycle will have some salvage value. For a bicycle with an initial cost of 500 USD , salvage value of 50 USD, and a life of 10 years, the annual depreciation is $\frac{500-50}{10} = 45$. So after 7 years, the bicycle is worth $500 - (45 \times 7) = 185$ USD.

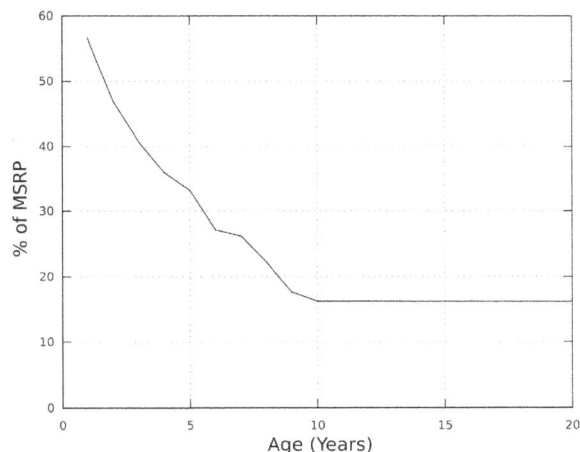

Figure 2.22: Percentage of Original MSRP vs. Age in year

A more sophisticated model changes the rate of depreciation depending on the age. The linear depreciation model uses an uniform depreciation regardless of the age. A more accurate rate of annual depreciation for a bicycle would be larger for bicycles that are a few years old. A bicycle that is a year old has the maximum rate of depreciation.

If you know the brand, model and age of the used bicycle, the website bicyclebluebook.com is a good place to find an initial value. Once you have found the bicycle, you will find the original retail price. Interestingly, the retail price does not steadily increase with year. For example, the

2016 Trek 520 sold for 1260 USD while the 2013 model of the same bicycle sold for 1490 USD. You will also have to judge the condition of the bicycle.

In Figure 2.22, the change in the price of an used bicycle as a percentage of the original manufacturer suggested retail price (MSRP) vs. age in years, shows that a bicycle looses a large fraction of its value in the initial years. Following 10 years, the value of the bicycle stays roughly the same. If the MSRP of a 2013 Trek 520 was 1490 USD, then five years later in 2018, its value would be a third (497 USD) of its original value. This value is appropriate if the condition of the bicycle is excellent.

Evaluating the Condition of a Bicycle   The purpose of evaluating the condition of the bicycle is to adjust the value based on its condition. The bicycle bluebook website uses four categories for the condition of a bicycle - *excellent, very good, good,* and *fair*. Deciding which of the four categories to assign is a subjective decision.

One approach is to start with an evaluation of the most expensive component. The frame is often the most expensive part of a bicycle and a damaged frame can be expensive to fix. A few scratches and dents may not matter much on a steel or aluminium frame, but a crack in a carbon fiber frame is hard to fix.

Bicycles do not come with odometers and so when estimating mileage, you may have to take the word of the prior owner. A high mileage bicycle may have a loose chain or worn teeth in cogs besides brake pads that are not effective. The rim of the wheel may have many scratches and the tyre may look worn out as well. Many of these parts can be easily replaced, but will add to the cost of the bicycle. Some parts such as a tyre or pedal will not cost as much to replace as a broken wheel, crank or fork.

The bicycle bluebook website assigns a bicycle in *very good* condition about 95% of the price of a similar bicycle in *excellent* condition. Similarly, *good* and *fair* bicycles are assigned roughly 87% and 67% the price of an *excellent* bicycle respectively.

Observations

- Buying a used bicycle will always have some risks. One, you want to be certain that the seller is legitimate and the cycle has not been stolen. Annually an estimated 1.5 million bicycles are stolen in the US and often these bicycles are sold for bargain prices.

- The price a seller sets may be higher or lower than the actual value of the bicycle. A desperate seller may lower the price and conversely an eager but somewhat ignorant buyer will pay a higher price.

2.7 Groupsets and Wheelsets

There are a large number of manufacturers for bicycle parts such as the frame, fork, stem, handle bars, and saddle. In general, you will not go wrong with a frame from a known firm. However, known firms may source their components from a third party.

The number of manufacturers for other components such as derailleurs, shifters, and cranks is fewer. The term *groupset* is used to refer the collection of gear levers or shifters, brake levers, front and rear brakes, front and rear derailleurs, bottom bracket, crankset, chain, and freewheel or cassette (see Figure 2.23).

Figure 2.23: Groupset components (courtesy Shimano)

Knowing the groupset is important when you want to replace a particular component such as a shifter which may not be compatible with a shifter from a different groupset. The top three companies that make groupsets for road bicycles are Shimano, SRAM, and Campagnolo. Of the three, Shimano is probably the largest, with a wide range of groupsets for budget to high end bicycles.

Broadly, groupsets can be divided into three categories - high performance, mid-range, and budget. As you would expect, the more expensive groupsets are found on the high performance bicycles and budget groupsets on low cost bicycles. Groupsets on high performance bicycles are lighter, have smoother gear shifting, better brake performance, and a larger range of gears (see Table 2.2). Within each category, a manufacturer may have several models - the Shimano Dura-Ace groupset has four or more models.

You can expect to find the groupsets at the bottom of the table in budget road bicycles. The lower end groupsets perform the same function as the high end groupsets. The differences lie mainly in the weight, durability, and feel. The more expensive groupsets are more likely to be lighter than the less expensive groupsets. However, this is not significant when compared to the weight of the rider and the bicycle. Shifting gears maybe more smoother with the higher end groupsets and braking may also be stronger. To reap the benefits of a groupset, you also need to ensure that cabling is precise and not too loose or tight. You will notice a difference in performance if you replace a Tourney groupset with a Tiagra groupset.

Cost	Shimano	SRAM	Campagnolo	Shimano (MTB)
$$$	Dura-Ace	Red	Super Record	XTR
	Ultegra		Record	Deore XT
	105	Force	Chorus	Deore LX
	Tiagra		Athena	SLX
$$	Sora	Rival	Centaur	Deore
	Claris		Veloce	Alvio
$	Tourney	Apex	Mirage	Altus

Table 2.2: Groupsets for Road and Mountain Bicycles

Groupsets may also have electronic and mechanical versions of the same model. Manufacturers claim the electronic version makes shifting gears more precise and easier than on the mechanical version. Changing the gear in the front derailleur can make large changes in the gear ratio (see Section 2.3) and you may suddenly find yourself spinning the crank with little to no effort on a much lower gear or it may become much harder to pedal at a high gear.

The electronic shifter is a single shifter and will change gears with a somewhat uniform change in the power needed at a higher or lower gear. A shift in the front derailleur to a larger chain wheel maybe smoother if the shift is simultaneously accompanied by a shift to larger rear cog. This is somewhat harder to do manually and simply pressing a button to move to a higher gear is easier and requires little thinking.

Bicycle builders may also mix and match groupset components from different categories and manufacturers. A bicycle may have a SRAM crankset with Shimano derailleurs or the front derailleur and the rear derailleur maybe from the Shimano Sora groupset and Deore XT groupset respectively. The choice of components maybe based on cost and compatibility. Components from a high end groupset will be used for a more frequently used part like the rear derailleur.

2.8   Spending Wisely

A common question that arises is whether you will be able to ride faster on a more expensive bicycle than a cheaper one. More expensive bicycles tend to have lighter components and use composite materials that are light but strong enough to handle the load of a rider. It maybe even reassuring as you ride along on your less expensive bicycle to assume that a cyclist overtaking you is able to ride faster because of the superior quality and the lower weight of the bicycle.

In general, this is true. A lighter bicycle does require less power to ride than a heavier bicycle at the same speed. Manufacturers of expensive bicycles go to great lengths to reduce weight by using lighter frames and associated components. Bicycle makers vie for the title of selling the lightest production cycle with lightest materials for the essential components of a road bicycle. As of 2018 [7], the lightest bicycle weighed less than 5 kgs. and doubtless the competition for the lightest functional bicycle will continue.

Since the weight of a bicycle is one of the first questions from a customer, most manufacturers do try to keep the weight of a bicycle as low as possible within a given budget or do not mention the weight in the list of specifications. The listed weight typically does not include the weight of optional components such as a stand, bell, reflectors, etc. In Figure 2.24, the rough cost and corresponding weights of bicycle types shows that lighter bicycles can be substantially more expensive than bicycles that weigh more than 10 kgs.

2.8.1 Bicycle Weight

While it is obvious that a lighter bicycle will require less power to ride at the same speed as a heavier bicycle, how much do you actually gain by reducing the weight of a 12 kg. bicycle to say 7 kgs. Often, the cost of a 7kg. bicycle is substantially higher than the cost of a 12 kg. bicycle. Manufacturers of light bicycles do make other tweaks to the frame, wheels, and other components to reduce drag, that adds to the cost. However, a light carbon fiber frame usually costs significantly more than a similar steel or aluminium frame of the same dimensions.

If you assume that an average cyclist can generate about 200 watts for a short duration, then a rider weighing 70 kgs. on a 10 kg. bicycle with clincher tyres on a flat road with no wind riding in an upright position would be able to ride at about 31.5 kmph. Reducing the weight of the bicycle by 5kgs. will increase your speed from 31.5 kmph. to about 32 kmph. under the same conditions and power (see `http://bikecalculator.com` and Chapter 3).

While it may seem not worthwhile to invest in an ultra-light bicycle if the gains in speed are minimal, there are other benefits. If you are considering a light bicycle, then speed is probably a high priority compared to comfort. Most light bicycles are sleek and aerodynamic. The benefits of better aerodynamics will probably make a larger difference than weight alone. When your hands are on the drops (of the handlebars), your speed increases to 34.5 kmph. with the same 200 watts.

Weight is just one factor to consider when buying a bicycle. The other factors include power that you can generate and the purpose of the bicycle (to race, commute, ride long distances, or ride on rough terrrain). The average height and weight of the top 10 riders in the 2013 Tour De France was 177 cms. and 63 kgs. respectively or a body mass index (BMI) of 20.1. The BMI range for a healthy person is between 18.5 to 25.0 and therefore professional riders have BMIs in the lower side of the range. Besides lower BMIs, professional riders can also generate substantially higher power relative to their weight.

The Figure 2.24 is just a rough model of the cost by bicycle type. You can find a light mountain bicycle that costs many thousands of dollars and similarly there are low cost road bicycles that cost less than a thousand dollars.

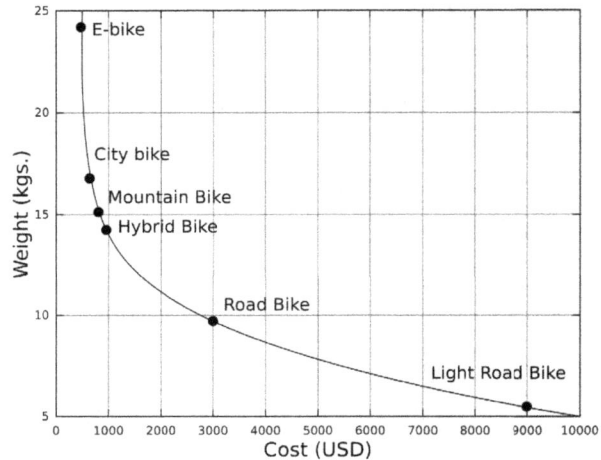

Figure 2.24: Weight vs. Cost of a Bicycle

2.8.2 Are Road Bicycles Expensive?

High performance bicycles use light weight components and frames that are more expensive, since they may not be mass produced. The vast majority of bicycles cost less than a few thousand dollars – a few professionals and enthusiasts are willing to pay more than 5000 USD for a bicycle.

Manufacturers often make two versions of the same model, one in aluminium and one in carbon fiber. The aluminum model may weigh two kilos more, but will cost 30% or less than the carbon fiber version. Although a lighter bicycle will make your ride faster, a two kilo reduction in weight compared to an estimated combined weight of 80 kgs. for a bicycle and rider is not significant (see Chapter 4).

Road bicycles like sports cars are designed to appeal to the desire to travel fast. The term road bicycle is reserved for a bicycle that prioritizes performance over utility or even comfort. Since at any speed over 20 kmph. air drag becomes dominant, aerodynamics is a priority in the design of a road bicycle. All the components from the wheels, tyres, handle bars to the saddle are designed to reduce the "front area" of the cyclist.

Although you may not be interested in the kind of expensive road bicycle used in the Tour De France, some of the features of a road bicycle such as the drop down handle bars are found in non-road bicycles. These bicycles are not made for racing and cost less than a racing bicycle. They will not be designed exclusively for better aerodynamics. The "fast leisure" rider who is not racing but wants to ride on a bicycle that is faster than a roadster would use such bicycles.

The gains in upgrading to a moderately expensive bicycle from a cheap bicycle will exceed the benefits of upgrading to a really expensive bicycle. There is a substantial difference between a ride on a cheap (< 200 USD) bicycle and a moderately expensive (500-3000 USD) bicycle. The expensive (> 3000 USD) bicycles have light weight components and frames that make a difference in a race.

A large part of the cost of a high performance road bicycle goes into the composite materials used in the frame and other parts as well as other light components. Some of the high end bicycles also have luxuries such as electronic gear shifting that are not essential for the average rider. Mechanical gear shifting has worked for decades and is possibly less expensive to repair.

Probably, the best way to compare bicycles to is to ride both an expensive and moderately expensive bicycle. Your local bicycle shop may hesitate to let you ride a new expensive road bicycle and an alternative is to rent a road bicycle and try riding it for an hour or so.

Spending more on tyres will pay off in the long run since you are more likely to have better puncture protection as well as less rolling resistance. The rolling resistance of a tyre is not proportional to its weight (see Chapter 4). If you plan on riding more than an hour or more, then investing in a comfortable saddle will also pay off. Spending on other components such as the pedals, helmet, jersey, and gloves has marginal benefit.

Observations

- The feel and experience of riding a road bicycle of reasonable quality is much superior to the ride on a budget bicycle that has a clunky feel and is slower.

- You don't need to buy an expensive road bicycle to enjoy the ride. A more enjoyable ride means that you will ride the bicycle more often and you should recoup your investment before long.

- Acceleration from a stop is often easier on a road bicycle than other types of bicycles. In an urban environment, this is a big plus since you will have to start and stop often.

- You will not ride significantly faster with a light road bicycle and an expensive groupset. The decision to spend a lot more on a bicycle depends on affordability and whether you consider the "better" ride worthwhile.

- A well maintained bicycle can last for many years. The frames of most bicycles made of durable material such as steel or aluminium and can sustain minor mishaps on the road. A dent or rust on a frame can be repaired and the bicycle will be usable.

References

[1] Akash Krishna Srivastava, Shashank Mishra, and Debalina Chakravarty. Analysis of bicycle usage in india: An environmental perspective. *International Journal of Innovations & Advancement in Computer Science*, 6(8), August 2017.

[2] ebicycles.com. Bicycle frame size charts, January 2018. URL `https://bit.ly/2wgx0u1`.

[3] competitivecyclist.com. Bicycle fit calculator, January 2018. URL `https://bit.ly/2wG3gWE`.

[4] Schwalbe Tyres. Tyre size markings, January 2018. URL https://bit.ly/2BZdKHM.

[5] Planet X Bikes. MTB Wheel Sizes Guide: 650+ and 29+ Explained, January 2018. URL https://bit.ly/2oHpkOH.

[6] Jan Roe. E-bike motors, explained, 2019. URL https://bit.ly/2RIJQyS.

[7] Mat Brett. Six of the lightest road bikes, bike makers challenge the scales with exotic materials, April 2018. URL https://bit.ly/2Iljme1.

3 Forces

One of the purposes of understanding bicycle physics is to evaluate if you are riding efficiently, i.e. given the power you can generate, are you riding at an optimal speed. The power that you can generate when riding a bicycle is dependent on age, physique, and weather among other factors and you may have natural limits on your maximum power. An elite rider produces about 5 watts per kilogram, so a 70 kg. elite rider would be able to produce 350 watts for an hour.[4]. A sprinter maybe able to generate 1500 watts when accelerating for less than a minute. Even if you are unable to generate the power of an elite rider, you can still ride at reasonable speeds for a long duration and cover distances ranging from 10 to hundreds of kilometers. The two main forces that you have to overcome are aerodynamic drag and rolling resistance. First, the force that you need to generate to ride at a given speed[5] is calculated and then the natural forces (such as gravity and wind) that you need to overcome is estimated.

3.1 Pedal Forces

An approximation of the forces acting on a bicycle on a level terrain can be divided into horizontal and vertical forces. The horizontal forces are rotational forces or torques are shown in Figure 3.1 assuming a constant speed. The downward force on the pedal is converted into a forward frictional force moving from left to right which propels the bicycle forward.

The vertical forces are the weight of the rider and the bicycle acting downward that is balanced by the ground reactionary force acting upwards. If the bicycle is not flying over a bump, then both these upward and downward forces are equal and the rider is not moving vertically. A larger fraction of the weight is supported by the rear tyre since a rider is seated close to the rear tyre. Therefore, a logical solution is to increase the pressure of the rear tyre by about 10% higher than the pressure of the front tyre (see Section 4.5).

The process of moving the bicycle forward begins by applying a downward force F_{pedal} on the pedal. This downward force gets converted into a leftward tension force F_{chain} that opposes the force on the pedal. This is very similar to a pulley where the direction of the applied force is translated into a tension force in a different direction. Since the chain is connected to a cog in the rear, the same tension force F_{chain} is applied in the rear in the opposite direction.

The rotational force of the cog in a clockwise direction is transferred to the wheel through the spokes and attempts to push the wheel from right to left. This force is opposed by the static friction force $F_{friction\_r}$ generated by the ground and acts in the direction of movement. Oddly enough, the bicycle moves forward even though the part of the tyre in contact at any given time with road, is stationary. A stationary contact area of the road with the tyre provides the friction force $F_{friction\_r}$.

---

[4]350 watts is like lifting a 35 kg. mass up at the rate of 1 meter per second.

[5]The vector velocity is used to calculate drag forces, however in many contexts the equivalent scalar speed is adequate since the magnitude of velocity alone is required.

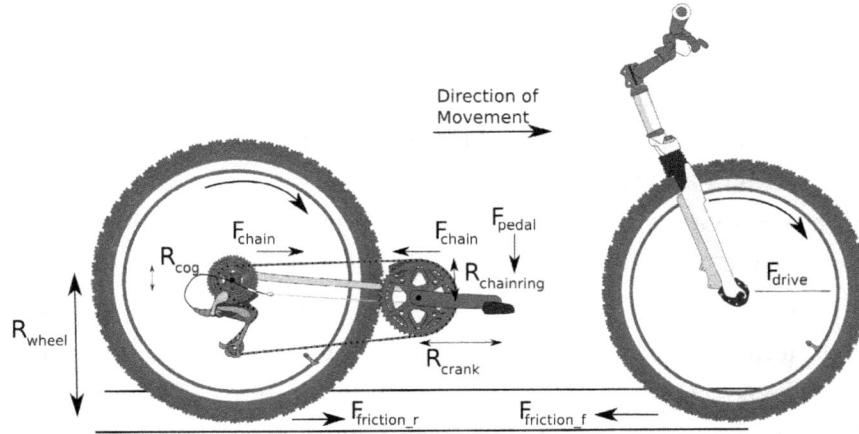

Figure 3.1: Rotational Forces on the Rear Wheel

At first it may appear odd that friction is being used to move forward, since friction is considered a force that resists an applied force. In this case, friction is a reactive force and is somewhat analogous to paddling an oar backwards to move forwards in water. The applied force or torque is countered by the static friction of the tyre.

While friction $F_{friction\_r}$ acts from left to right on the rear wheel in the direction of movement, $F_{friction\_f}$ acts in the reverse (right to left) direction on the front wheel. The force F_{drive} is transmitted through the frame to the fork pushing the front wheel forward in the direction of movement. Therefore, on the front wheel, friction acts from right to left. The effective propulsive force is $F_{friction\_r} - F_{friction\_f}$.

The rotational forces are defined as torques, i.e. $force \times radius$ (the perpendicular distance to the center of rotation) and are assumed to be equal for simplicity (a well maintained drive train usually has minimal losses). So, $F_{pedal} \times R_{crank} = F_{chain} \times R_{chainring}$ and similarly $F_{chain} \times R_{cog} = F_{friction\_r} \times R_{wheel}$. The ratio of F_{pedal} to $F_{friction\_r}$ is also called the *gain ratio* [1]. If $F_{friction\_r}$ is defined as $\frac{F_{chain} \times R_{cog}}{R_{wheel}}$ and F_{pedal} as $\frac{F_{chain} \times R_{chainring}}{R_{crank}}$, then the gain ratio is defined as

$$\frac{F_{pedal}}{F_{friction\_r}} = \frac{R_{chainring} \times R_{wheel}}{R_{crank} \times R_{cog}}$$

3.1.1 Gain Ratio

The gain ratio expresses the conversion of the crank force from the pedal to the friction force on the wheel. The values for $R_{chainring}$ and R_{cog} can be stated in terms of the number of teeth on the chainring and cog. If the the number of teeth in the chainring is 53, the radius of the wheel including the tyre is 340 mm., the radius of the crank is 170 mm. and the number of teeth in the cog is 19, then the gain ratio is $\frac{53 \times 340}{170 \times 19} = 5.58$. This means that when the pedal travels 1 meter, the wheel will travel 5.58 meters.

The typical gain ratio ranges from 1.5 to 7.5. Since the values of R_{wheel} and R_{crank} are fixed during a ride, the values of R_{cog} and $R_{chainring}$ can be changed in a geared bicycle to alter the gain ratio. In a lower gear where R_{cog} and $R_{chainring}$ are 20 and 38 respectively with the same wheel and crank as before, the gain ratio is 3.8. If the values of R_{cog} and $R_{chainring}$ are changed to 14 and 50 respectively, then the gain ratio becomes 7.14.

The ground velocity V_{ground} can be calculated as $2\pi R_{wheel} \times n_{wheel}$ where n_{wheel} is the number of wheel revolutions per second and the pedal velocity V_{pedal} is $2\pi R_{crank} \times n_{crank}$ and n_{crank} is the number of crank revolutions per second. The ratio of the ground velocity V_{ground} to pedal velocity V_{pedal} is the same as the gain ratio

$$\frac{V_{ground}}{V_{pedal}} = \frac{R_{wheel} \times n_{wheel}}{R_{crank} \times n_{crank}}$$

Since the chain has to traverse the same number of teeth on the cog and the crank, the product $n_{wheel} \times R_{cog}$ must be the same as $n_{crank} \times R_{chainring}$ where the radii of the cog and chainring are proportional to the number of teeth. The ratio $\frac{n_{wheel}}{n_{crank}}$ becomes $\frac{R_{chainring}}{R_{cog}}$ and therefore

$$\frac{F_{pedal}}{F_{friction\_r}} = \frac{V_{ground}}{V_{pedal}} \equiv F_{pedal} \times V_{pedal} = F_{friction\_r} \times V_{ground}$$

In theory a rider could apply a high F_{pedal} to ride at high speed. But it is the generated power that finally determines speed and a high F_{pedal} must be accompanied by a high V_{pedal} which is hard for an average cyclist. If there are no losses, then the power P_{pedal} generated at the pedal is the same as the power P_{wheel} generated on the ground. The real value of P_{wheel} is between 85% - 97% of the generated power P_{pedal}[2].

3.1.2   How do Gears help?

In a single speed bicycle, the front chain ring and rear cog have a fixed number of teeth. The front chain ring may have 42 teeth, while the rear cog will have 18 teeth. Since the number of teeth is fixed, each rotation of the crank will move the wheel forward by the same distance regardless of the gradient. On a downhill, this is not an issue. However, on a climb it can take more effort to ride with the same ratio of teeth for the front chain and rear cog.

Early bicycles like the "Penny Farthing" used pedals and a crank attached to the hub of a large wheel of diameter of about 1.3 meters. This meant the rider was quite high off the ground. Getting on and off the bicycle needed some skill that took more effort to acquire than the skill needed to ride current bicycles with smaller wheels. The main purpose of the large wheel was to increase speed, since a single revolution of the pedal would move the bicycle forward by about 4 meters. However, the aerodynamic drag (see Section 3.2.1) would be higher and an accident with a rider higher on the saddle maybe more serious. The introduction of geared bicycles made it possible to effectively ride with large wheels, but with much less risk on smaller wheels that are less than half the diameter.

A geared bicycle makes it possible to ride faster downhill and climb uphill with less power and at a slower pace. Each rider has the capability to generate forces and torque based on age, muscle development, and build. Based on the force that you can generate, you can appropriately set a gear ratio. In Figure 3.2, the difference between a high and low gear is illustrated.

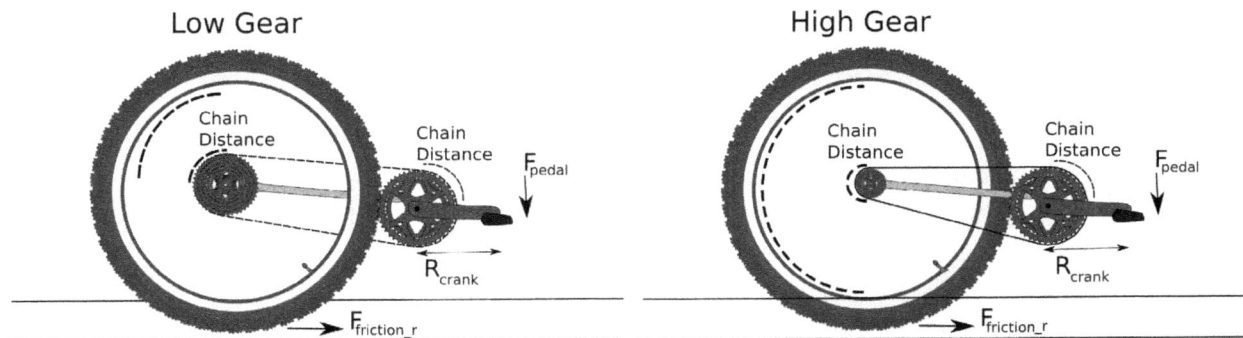

Figure 3.2: Using Low and High Gears

In a low gear, if F_{pedal} is 100 N and the crank radius is 170 mm., then the work done by a cyclist to move the crank a quarter revolution (0.267 m.) is 26.7 N-m. The chain moves x links clockwise and the rear cog similarly moves x links in the same direction. In a low gear, the rear cog is large and corresponds to a quarter revolution of the wheel. Assuming no losses, the work done on the crank is completely transferred to the wheel, which also rotates a quarter revolution. If the diameter of the wheel 670 mm., then the generated force $F_{friction\_r}$ is 50.7 N.

In a high gear, the work done by the cyclist is the same, i.e. 26.7 N-m. However, the rear cog is smaller and the rotation of x links corresponds to a half rotation of the wheel. Now, the generated force $F_{friction\_r}$ is 25.35 N or half of the generated force in a low gear. In a high gear on a downhill gradient, you do not need to generate a high $F_{friction\_r}$, since the force of gravity will add to your applied force, F_{pedal}.

In a low gear, you will not travel as far as you would have in a high gear for the same applied force. This is however useful when climbing uphill, since you can generate more force on the wheel, though you are applying the same force to the pedal in both the low and high gears. Even though your speed is reduced in a low gear, climbing uphill becomes easier since you do not have to dramatically increase applied force. Conversely when riding downhill, you can change to a higher gear and travel faster with the same applied force.

When you ride uphill, you can increase $F_{friction\_r}$ by changing to lower gear. Assuming the power P_{pedal} generated from pedalling is completely transferred to the rear wheel as P_{wheel_r}, then $P_{pedal} = P_{wheel\_r}$ or $F_{pedal} \times V_{pedal} = F_{friction\_r} \times V_{wheel\_r}$. If you generate the same power on a flat or uphill, then in a low gear V_{pedal} will be higher and F_{pedal} will be lower. On the right hand side of the equation, V_{wheel_r} will be lower since you are riding at a lower speed and $F_{friction\_r}$ will be higher, to make the climb less strenuous with the same power.

Another way of looking at gears is via the effects of chain tension on $F_{friction\_r}$. In a low gear (large rear cog), the ratio of the radii of the cog and wheel is about 1:5, meaning that $\frac{1}{5}th$ of the chain tension is transferred to $F_{friction\_r}$. With a high gear (small rear cog), the ratio becomes 1:12 or $\frac{1}{12}th$ of chain tension and therefore $F_{friction\_r}$ is less than half the force in a low gear.

To summarize, using a geared bicycle does not reduce the work load, but instead distributes the work load over time such that you can ride faster downwhill and slower uphill for the same applied force. If you want to ride faster uphill, then you will have to generate more power, which implies that even if you apply the same force, you will have to increase your cadence. This is much easier to accomplish when you are riding downwhill since you have gravity assisting you to generate the required $F_{friction\_r}$.

Assuming you can generate 100 watts of steady power, then you could vary your cadence (rpm) and correspondingly change gears based on the gradient (see Table 3.1). Assuming a crank with radius of 0.17m., when cadence is high (90 rpm), V_{pedal} is high as well and therefore for a given power (100 watts), F_{pedal} is lower. This is appropriate when you are climbing, since you are *not* applying a high pedal force that would lead to fatigue. When going downhill, you have gravity assisting you and can switch to a higher gear and apply a higher F_{pedal}.

Cadence (rpm)	F_{pedal} (N)	V_{pedal} (m/s)	Gradient	Gear
90	62.4	1.60	Uphill	Low
70	80.2	1.25	Flat	Medium
60	93.6	1.06	Downhill	High

Table 3.1: Applied Force for Low to High Gears for 100 Watts

To translate the V_{pedal} to ground speed, we need the gain ratio. If the gain ratio for a low gear is 2.5, then the effective ground speed is 14.4 kmph. and similarly for a gain ratio of 5.58 with a high gear, the ground speed is 21.2 kmph. This implies that when you set an upper limit on the steady power that you can generate, your highest speed is also limited.

If you could generate enough power, then you could ride at a high cadence *and* with a high gear. While riding in the highest gear possible, you are actually straining your muscles and more likely to becoming tired soon. Even though you can travel further on a high gear with fewer pedal strokes, it becomes unsustainable and leads to exhaustion. While there is no perfect number for cadence, most recommendations are to maintain 60 rpm and higher. If your cadence starts dropping, then you can switch to a lower gear to maintain a high enough rpm.

Another common recommendation is to change to a lower gear earlier than later. If you are approaching an uphill you can change gears as soon as you detect some resistance instead of waiting till your cadence drops significantly. Changing gears when you have started riding uphill will be more noisy since the chain tension is higher. This is less of an issue when riding downhill since your cadence will increase with the same effort and you can switch to a higher gear without a large change in cadence.

Cross Chaining   If you have a front and rear derailleur, then ideally you want to keep the chain aligned parallel to the chain stay and not skewed. If the chain is on the largest chainring in the front and the largest rear cog, then the chain is skewed away from the chainring towards the rear cog. (see Figure 3.3).

The chain meets the cog and the chainring at an angle increasing friction and the wear on the chain. Besides the deterioration of the chain links, the teeth of the cog and chainring are also likely to erode faster than if the chain was aligned. A simple solution is to switch to a smaller chainring and a smaller cog as well, which typically should give the same gear ratio as when the chain was mis-aligned.

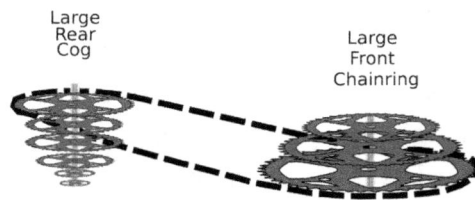

Large
Rear
Cog

Large
Front
Chainring

Figure 3.3: Cross Chaining with a Large Chainring and Large Rear Cog

You can also have cross chaining with a small front chainring and a small rear cog. Changing to a smaller or larger chainring is slower and usually done less often than changing to another cog. If you prefer to stay on a large front chainring for as long as possible and shift between rear cogs alone, then a strong and flexible chain may reduce the inefficiencies due to cross chaining.

Finding the Right Gear   While there is no perfect gear for every situation, a rule of thumb is to maintain a cadence of close to 70 rpm. If your cadence falls below 60 rpm, then you may find it easier to ride in a lower gear. This typically happens when you are riding uphill. If you know that you will soon begin a climb, you can change to a lower gear before the climb instead of after.

However, if your gear is too low, then you will be pedalling ferociously, but crawling uphill. On the other hand if your gear is too high, then you will be exerting more force on the pedal than you can sustain and eventually ride uphill slowly as well, since your cadence will be low. Finding the right gear to ride uphill with a sustainable power output is a matter of trial and error. In addition gradients during a climb may change meaning that you will have to change gears during a climb.

Riding uphill is when you can actually appreciate the benefits of a geared bicycle and having a large number of gears makes it possible for riders of different weight to power ratios (see Section 3.2.10) to ride comfortably. At some point if your speed (< 5 kmph.) slows down to a crawl, besides looking out of place, it maybe better to walk the bicycle with less effort (see Section 3.2.11 for limits).

Do you Really Need Gears?   The "keep it simple" principle is based on the idea that most systems work best when they have less complexity. A geared bicycle with a front and rear derailleur does add more components to a bicycle with levers to change gears, additional cabling, a chain ring, and a cassette. If you are certain that you will be riding on roads with few steep uphills, then you could ride without gears.

If you ride with a single chainwheel and cog, then you can pick the combination that best suits your needs. A 2:1 ratio implies that the number of teeth in the front chainwheel is twice the number of teeth in the rear cog. One rotation of the crank mean two rotations of the rear wheel. A lower ratio of say 1.5:1 means pedalling will be easier compared to a ratio of 2:1. With higher power, you can increase the cadence and correspondingly your speed as well. The wheel size is also a factor and using the gain ratio described earlier maybe more accurate in evaluating different gear combinations. Regardless, riding a single speed bicycle with fewer components is simpler than riding a geared bicycle – it is appealing and does achieve the goal of efficient transportation.

3.1.3 Pedal Force and Pedal Angle

Since the driving force and correspondingly speed is proportional to pedal force, it would be logical to examine how and when to apply force on the pedal. There are many theories on the best pedalling techniques [3, 4]. It is not raw power or fitness alone that determines speed but instead it is the pedalling technique that is most efficient for you.

A lot of research has been invested in understanding the best foot positions to generate power because the pedal stroke is repeated hundreds or even thousands of times. If you ride for an hour with an average cadence of 60 rpm, then you would have pedalled 3600 crank revolutions. So optimizing your pedal stroke is worthwhile, since the benefits will multiplied based on the duration of your ride.

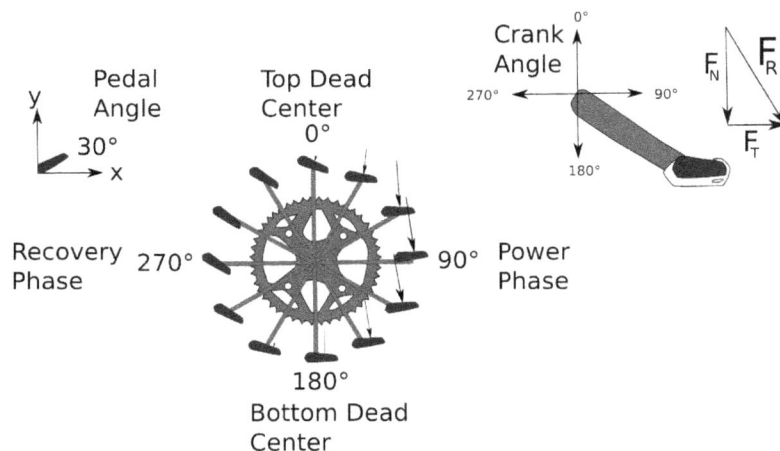

Figure 3.4: Applied Pedal Force in a Single Stroke

Earlier a fixed F_{pedal} was assumed, but in reality the value of F_{pedal} is dynamic and depends on the position of the pedal. A single stroke starts from the top dead center (see Figure 3.4) or 0° to 180° (bottom dead center) and back to 0°. The two angles in the pedal stroke are the crank angle and the pedal angle. The pedal angle can be positive or negative. In the power phase (from $0°$ to $180°$) the pedal angle is mostly positive and in the recovery phase (from $180°$ to $0°$), it is mostly negative.

The pedal angles and applied force at a crank angle were calculated in a study of 14 elite cyclists [3]. Applying force uniformly to the crank during the entire stroke from $0°$ to $360°$ is not efficient, since in the first half of the stroke (power phase) you can use gravity to your benefit and during the second half of the stroke (recovery phase) the torque from a downward force will be negative. From a study of the pedalling technique of elite cyclists, we can look at how and when they apply torque to the pedals. Although, a non-racer may not have the muscles to generate the power of an elite cyclist, the same pedalling technique that elite riders adopt can be used. It is assumed that racers have honed their pedalling skills by locating the optimal pedal positions to generate power. Since each racer may not generate power identically, the average pedal angles and applied force of 14 cyclists shows patterns in Figure 3.4 that a non-racer can adopt.

In the figure, the total applied force (F_R) applied to the pedal is the sum of the tangential force (F_T) and the normal force (F_N). The reason the applied force is highest between $60°$ to $120°$ degrees is that the maximum force a muscle can generate is when it is not contracted or stretched [4]. At $0°$ and $180°$, the muscles to apply force on the pedals are contracted and stretched respectively. Between these two angles are crank positions where you can apply your maximum downward force.

There is a minimal upward force applied to the pedal in the second half of the stroke from the bottom dead center to the top dead center. This applied force is not visible in the figure, because the total upward force is a small percentage ($< 3\%$) of the total downward force. You do need cleats or toe clips to generate this upward force (see 3.1.5). Figure 3.5 shows the change in torque vs. the crank angle.

The torque generated is calculated from the normal force (F_N) and the tangential force (F_T). The F_x and F_y force components along the x and y axes are computed for the both F_N and F_T. The resultant force on the pedal F_R is the sum of the F_x and F_y forces. F_R is applied to the pedal which is located at a distance l_c from the center of the crank. Therefore the torque is the cross product of these two vectors - $l_c \times F_R$. In Figure 3.5 the maximum torque is generated when the pedal is near $90°$. Roughly 80% of the positive torque is generated when the crank is between $40°$ and $140°$. Over 96% of the total positive torque is generated in the downward pedal stroke, between $0°$ and $180°$. To generate this torque, you also need to be seated in the right position on your saddle.

The large change in torque applied to the pedal from $0°$ to $180°$ is due to the circular shape of the crank. Some have experimented with an elliptical crank to maintain a more uniform torque on the pedal, but the circular crank is still very popular.

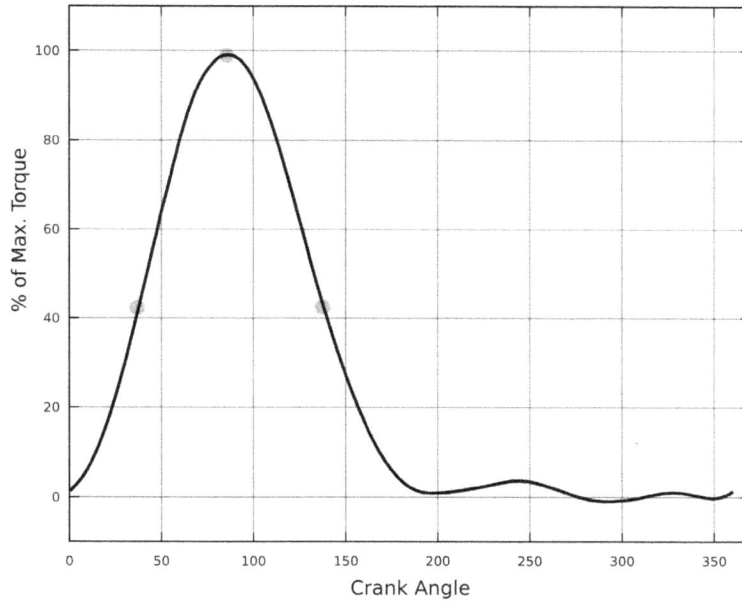

Figure 3.5: Percentage of Max. Torque vs. Crank Angle

3.1.4 Saddle Height for Optimal Pedal Force

Your position on the saddle determines whether your muscles are contracted or stretched. If you are too low on the saddle, then your muscles are likely to be contracted, similarly when you are too high on the saddle your muscles are stretched and it will be hard to generate the power that you could, if your saddle was at the "right" height.

What is the best saddle height for a given rider such that the power generated will be optimal? The saddle height is one important measurement along with others that defines the saddle position in three dimensions. The saddle nose should be pointed directly towards the front wheel. Similarly, the saddle should not be tilted upwards or downwards by more than a few degrees, which would make it hard to stay in position to pedal.

Saddles also have rails and you can position a saddle closer or further from the handle bar. Since a saddle is not very large, you cannot make large adjustments by moving the saddle along its rails. Instead if your bicycle is the right frame size for your height, then you can make finer adjustments by moving the saddle forwards or backwards.

The saddle height is measured from the center of the saddle to the center of the bottom bracket (see Figure 3.6). The optimal saddle angle is between $140°$ and $150°$, when the crank is at $180°$ [5].

There is no formula to come up with precise numbers for the saddle height and angle based on your height – instead, you can tinker moving the saddle higher or lower and backwards or forwards till you feel comfortable. If you move the saddle too high, then you may have to tilt the bicycle when you stop riding to get off safely. Similarly you will have to tilt the bicycle to get back on the saddle and the ride maybe uncomfortable.

Figure 3.6: Adjusting the Saddle Height

If the saddle is too low, then it will be easier to get on and off the bicycle. However, your knees may begin to hurt and you will not be in a position to apply maximum downward force on the pedal when the crank is around $90°$. Therefore, your speed is reduced and the likelihood of knee pain is higher.

A saddle at the "right" height allows you apply force to the pedal as in Figure 3.4. A simple solution to find the "right" saddle height is to start low and then raise the height in increments till you are comfortable.

The saddle height and saddle angle are related. If you adjust the saddle height, then the saddle angle will also change. If you move the saddle forward to change the saddle angle, then the saddle height will also change. These changes are incremental and you can keep tinkering with these adjustments till you have your optimal height and angle to generate the highest force on the pedal. Cleats are protrusions on the sole of a shoe that clip on to a pedal and make it possible to pull up the pedal when the crank is between $180°$ to $360°$.

3.1.5 Do Cleats Help?

The argument for the use of cleats is that since your foot is attached to the pedal, it does positive work on both the upward and downward stroke of the pedal. This would imply that the total power you generate will increase since it is being generated all through the pedal stroke and not just one half from $0°$ to $180°$.

However, empirical results [3] have shown that during a low workload, 99% of power was generated during the downward stroke. Even when the workload was high, less than 4% of power was generated during the upward stroke. Another argument against generating power in the upward stroke is that using a different set of muscles from $180°$ to $360°$ results in a loss of power output from the most efficient muscle groups used in the downward stroke [4].

Although, the power gains from using cleats may not be as much as perceived, there is another added benefit. At a high cadence, the foot is firmly attached to the pedal and not likely to slip off the pedal. At a low cadence, if your foot slips off the pedal it maybe just an inconvenience, but at a high cadence, there is a risk of the pedal hitting your ankle or foot.

Further, a rider used to cleats typically finds flat pedals not as comfortable and secure. On the other hand with a flat pedal, you can adjust your foot position as needed and are not locked into a single position. In summary, racers will use cleats to get a small percentage gain in power while non-racers may use cleats for comfort as well as generating more power.

3.2 Drag Forces

Before you can ride your bicycle, you have to overcome drag forces. These forces act against the movement of the bicycle and the pedal force F_{pedal} must be larger than the drag forces to accelerate the bicycle. Wilson [2] defines four types of drag forces - the aerodynamic drag force, rolling resistance, slope resistance, and bump resistance. The magnitude of these forces depends on velocity, terrain, and road / weather conditions. The power required to ride a bicycle at a velocity of V_{bike} is defined as $P = \frac{V_{bike}}{\eta} \times (sum\ of\ drag\ forces)$. This is called the steady state power or the simply the power required to keep riding at a steady velocity. The value of η is an efficiency factor to account for frictional losses in the drive train and is estimated at 0.95. To ride at a higher velocity, you will need additional power beyond the power required to overcome drag forces.

3.2.1 Aerodynamic Drag

When you ride your bicycle it may seem like the wind is always against you no matter what direction you ride. If you ride in the opposite direction of the wind, then the force of the wind will seem much stronger and it takes more effort to ride at the same speed that you would have achieved without any wind. A tail wind blows in the same direction that you are riding and will help you ride faster.

Since the direction of wind changes over time, you will sometimes benefit and sometimes lose from its effects on the power you need to generate. Besides wind, there are other forms of drag that can slow you down. An uphill will always require more power to ride at a given speed compared to a ride on flat terrain.

On flat terrain, the two primary drag forces are aerodynamic drag and rolling resistance. At low speeds (< 15 kmph.) rolling resistance is dominant and subsequently aerodynamic drag increases rapidly (see Figure 3.19). For simplicity, the speed of wind is assumed to be constant over time (a steady flow). In reality, wind speeds may change rapidly (a gust) and in direction as well. On a highway, wind direction is fairly steady, whereas in an urban area wind may reflect and bounce of buildings changing direction over time.

In the absence of wind and at walking pace (5-6 kmph.), air resistance is not noticeable. Air is considered a fluid and has the same properties as water, albeit with a much lower viscosity compared to water. A lot of research and work has gone into making vehicles more aerodynamic, i.e. less drag forces. The study of fluid flow (computational fluid dynamics) in wind tunnels and in mathematical models is critical for vehicles such as cars and planes. For a bicycle, aerodynamic drag becomes a significant factor at relatively high speeds (> than the average commuter speed).

Figure 3.7: An Aerodynamic Bicycle

Some of the characteristics of an aerodynamic bicycle are a light frame, tubes shaped like foils to smooth air flow, light and thin wheels, and handle bars to place your arms for the least resistance (see Figure 3.7). Even the helmet of the rider has a longer profile to reduce drag. The rear wheel is often solid to reduce air drag further (this works as long as there is no cross wind). A solid front wheel may make it harder to handle the bicycle, if wind direction changes.

These designs are not adopted in the more common road bikes or mountain bikes since being aerodynamic means sacrificing comfort. The tyres of common bicycles are wider and allow you sit more comfortably. Even if you don't plan on riding at speeds of a racer or using an aerodynamic bicycle, keeping the drag forces low is important when you ride in a head wind.

3.2.2 Calculating Drag

In a non-aerodynamic bicycle, the energy of wind observed on contact is dissipated in the form of kinetic energy in the wake or at the rear of the bicycle. This loss of kinetic energy must be compensated from opposing energy generated by the cyclist. An aerodynamic or streamlined shape has a leading round edge followed by a gradual tapering. While shapes for enclosed vehicles can be made aerodynamic, the air flow in a bicycle with riders of different shapes and sizes is more complex.

A bicycle and rider as shown in Figure 3.8 is not streamlined to allow the smooth flow of air. A bicyclist collides with a certain volume of air over a given period, accelerates the volume of air to the speed of the bicycle, and then pushes it aside. The volume of air encountered depends on the speed of the bicycle and the frontal area. The volume of air encountered is measured in $m^3/second$ and is the product of the density of air ρ, the frontal area A, and the velocity v. If we can reduce the volume of air encountered, then drag will also reduce.

The density of air ρ depends on the altitude and is relatively constant on most rides. A higher velocity v implies a proportionately higher volume of air. The frontal area A can be limited by crouching and reducing the width of the bicycle.

The mass of air encountered per second is ρAv and the product of air mass encountered per second and the velocity gives the drag force F_a which is defined as ρAv^2. Since, F_a is proportional

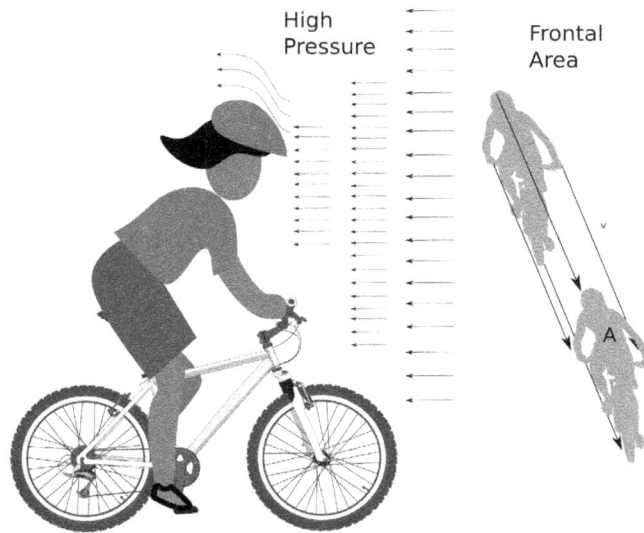

Figure 3.8: Aerodynamic Drag of a Bicycle and Rider

to the square of the velocity v, the drag force becomes larger than rolling resistance at velocities > 15 kmph. [6]. The drag force must take into account the loss of air pressure in the collision with the front area and the shape and material of the front area. The term "dynamic pressure" defined as $\frac{1}{2}\rho v^2$, represents the loss of air pressure. The product of the dynamic pressure and front area gives the drag force. Finally, a coefficient of drag force C_d is added to the expression to account for the "smoothness" and shape of the front area. The expression for drag force becomes $\frac{1}{2}C_d\rho A v^2$ and the power required to overcome air drag is the product of the velocity of the bicycle and drag force. The drag force is essentially the kinetic energy / unit distance lost in the encounter between some mass of air at a velocity v.

The description so far explains the primary drag force caused by the perpendicular collision of the front area with a mass of air. However, following the collision, air must move around the surface area of the bicycle and rider. Skin friction on the surface is the secondary drag force and is typically much smaller than the initial drag force caused by the collision.

Both the primary and secondary drag forces depend on the clothing of the rider. Racers in general wear tight fitting clothes to allow air to move around smoothly without losing a lot of pressure. In summer, clothing maybe loser to allow air to circulate to cool the body. Typically on a flat road, tight fitting clothes will help since speeds maybe high enough for air drag to be a significant part of the total drag. On a climb, speeds are lower and tight fitting clothes do not help as much.

3.2.3 Reducing Aerodynamic Drag

Since aerodynamic drag is the main drag force for speeds above 15 kmph., reducing this drag is a subject of research and bike manufacturers and riders have studied methods that work at minimizing air drag. The first attempt is to simply reduce the value of the front area A by altering the saddle position (see Figure 3.9).

A typical cyclist has a front area of $0.3 - 0.6 \ m^2$ [2]. The front area can be reduced by moving from a commuter to a racer position. On a bicycle with drop bars, you can move your hands between bar tops, hoods, and drops to change the aerodynamic factor K_a. The value of K_a is defined as $\frac{1}{2}C_d\rho A$ where C_d is the coefficient of drag, ρ is the density of air and A is the front area. If the density of air is $1.058 \ kg./m^3$, then the value of K_a can be calculated for each of the four positions (see Table 3.2).

Rider Position	C_d	$A \ (m^2)$	K_a	$Power \ (W)$
Commuter	1.15	0.50	0.30	155
Relaxed	1.00	0.40	0.21	108
Aggressive	0.95	0.30	0.15	77
Racer	0.88	0.25	0.11	59

Table 3.2: Aerodynamic Factor K_a for Different Rider Positions at a speed of 21.6 kmph. and power

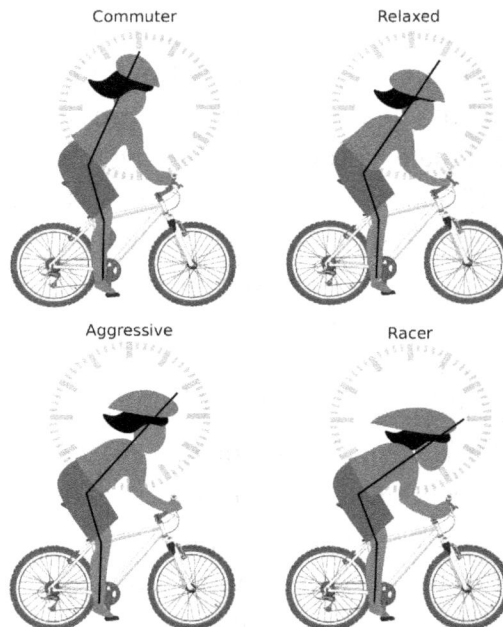

Figure 3.9: Rider positions to reduce air drag

The front of aerodynamic vehicles such as a bullet train, a plane, or a fast car typically has a shape that reduces the drag coefficient, C_d. The value of C_d depends on the shape of an object. An object shaped like a three dimensional tear drop has a much lower C_d than an object shaped like a short cylinder. Air molecules that approach a shape with high C_d will bunch up in the front of the object and increase air pressure opposing the movement of the object. An object with low

C_d allows air molecules to move smoothly around the front surface and reduces the opposing air pressure.

The power calculated in Table 3.2 assumes a bicycle is moving at 21.6 kmph. (6 m / sec.) with a head wind of 10.8 kmph (3 m / sec.) - other resistances such as the rolling resistance and slope resistance are ignored. Lowering K_a reduces the required power to ride at the same speed proportionately. With a low K_a you can increase your speed substantially on a downwhill ride (see Chapter 4 for the maximum speed you can reach on a downhill).

Observations

- Interestingly, the aerodynamic drag force is not dependent on the mass of the bicycle or the rider. The effects of mass are significant in the calculation of slope and rolling resistance.

- Consider an ideal case where you are riding on a flat (0° slope) with no head or tail wind, a tyre of rolling resistance C_r (see Section 3.2.8) of 0.003 and a K_a of 0.3. In Figure 3.17, the speed at 200 watts with the given conditions and zero slope is almost constant at 30 kmph., even though weight increases from 50 kgs. to 110 kgs. The slight decrease in one kmph. is due to the increase in rolling resistance which is dependent on weight.

3.2.4 Wind and its Effects

The Figure 3.8 shows wind blowing head on towards the cyclist. In reality, wind can blow from any direction and its effects on a cyclist depend on the "apparent" wind. Since we have two objects - the cyclist and some volume of air - moving simultaneously, we must calculate the apparent wind or the wind velocity relative to the cyclist. If $\vec{V_{bike}}$ and $\vec{V_{wind}}$ are the velocities of the bicycle and wind respectively, then the apparent wind is $\vec{V_{wind}} - \vec{V_{bike}}$ (see Figure 3.10).

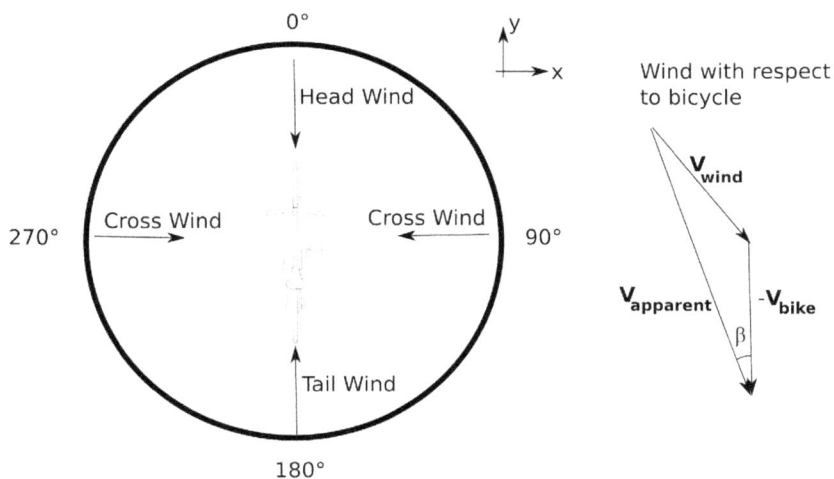

Figure 3.10: Calculating Apparent Wind with Respect to Bicycle

Since wind can flow from any direction, its effects are either helpful, a hindrance, or neutral. For a bicyclist headed directly north (0°), wind blowing south (180°) is a pure head wind. The apparent wind velocity is the difference between the wind and bicycle velocities. If a bicyclist is headed 0° at 6 meters per second ($\vec{V_{bike}}$) and wind is blowing from 0° at 3 meters per second ($\vec{V_{wind}}$), the magnitude of the apparent wind is $(-3 - 6)$ or -9 meters per second. The square of the magnitude of this apparent wind contributes to the drag force. When the wind blows from other directions, then the vector sum of the wind and negative bike velocities gives the apparent wind velocity. This is the velocity of wind that the rider feels and it is calculated relative to the velocity of the bicycle. However, only the component of the apparent wind that is in the same direction as the bicyclist contributes to the drag force.

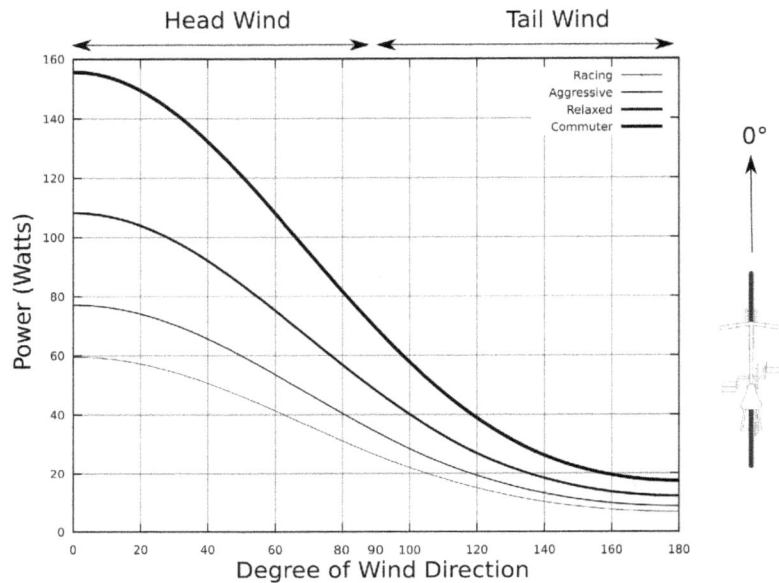

Figure 3.11: Power to Overcome Air Resistance for Different Rider Positions

In Figure 3.11, power required to overcome air resistance is maximum when wind originates directly from the North (0°). This is logical since the wind directly opposes the movement of the bicycle in the north direction. At an angle $\beta = 0°$, the head wind of 3 meters per second increases the power required to ride at 6 meters per second. For a racer, the required power is the product of the magnitude of the bicycle V_{bike} and the air drag force

$$V_{bike} \times (V_{apparent} \times V_{apparent} \times cos(\beta) \times K_a) = 6 \times (-9 \times -9 \times 0.1164) = 56.57 \; (w)$$

The magnitude of the apparent wind $V_{apparent}$ is -9 meters per second and β is the angle between the $V_{apparent}$ and V_{bike} (bike velocity). Since this a drag force against the direction of movement, it is a positive drag force. The value of K_a for a racer is 0.1164 (see Table 3.2). If an efficiency of 0.95 is included in the computation to take into account the transfer of power from the crank to the wheel, then the required power is 59.5 watts. At 90°, the cross wind of 3 meters per second does not contribute to the head wind faced by the cyclist which remains at the speed of the bicyclist (6 meters per second).

How does wind change your speed?   A head wind will slow you down and attempting to ride at the same pace or gear that you would on a flat road is going to eventually slow you down without additional power. A head wind is a drag force much like slope resistance on an uphill and one way to counter this additional drag is by switching to a lower gear. While slope resistance is not proportional to front area, the drag from a head wind depends heavily on the shape of your front area. Being aerodynamic can greatly reduce air drag in a head wind. Other attempts to reduce air drag include using tight fitting clothing as well as riding in a group.

One method to counter a cross wind blowing at 90° to your direction of movement is to lean into the wind. This method does not use any extra energy. At some point if the cross wind becomes strong, it could push the bicycle sideways or lead to drag in the direction of movement. In real life, a cross wind will slow you down since it rarely blows at the exact angle that has a zero component in the apparent wind direction.

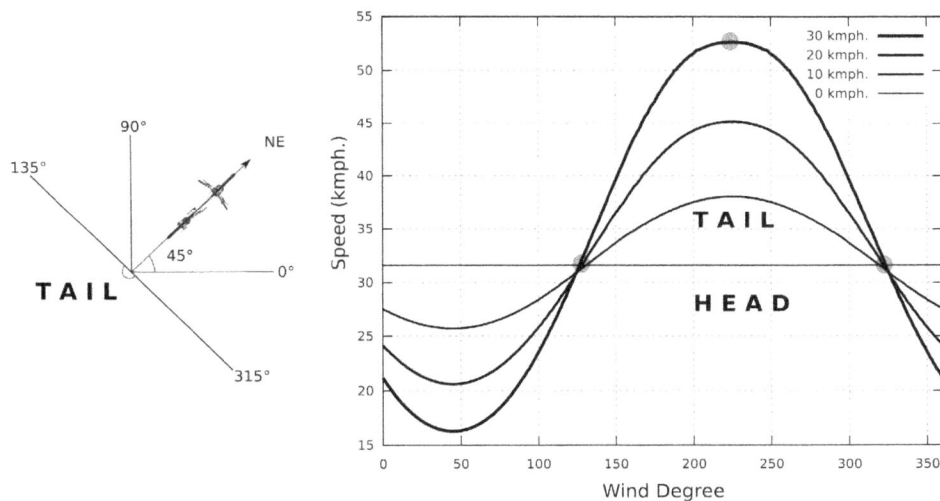

Figure 3.12: Bicycle Speed in kmph. with 200 watts vs. Wind Degree for Four Wind Magnitudes

Figure 3.12 shows the change in speed for a given bicycle configuration (Total weight of 80 kgs., K_a of 0.23 for a relaxed position, flat road, on clincher tyres headed north east at 45° and power output of 200 watts). When there is no wind, the expected speed is 31.6 kmph. Wind blowing between 135° and 315° will generate a tail wind with a peak tail wind at 225°. If the magnitude of the wind is 30 kmph., then the expected speed is 52.6 kmph., an increase of 21 kmph. over the speed with no wind. Bicycle speed cannot increase linearly with increase in wind speed, since air drag is proportional to the square of the velocity. Similarly speed is the least at 16.3 kmph., a decrease of 15.3 kmph. over the speed at no wind, when the wind blows at 45° with a magnitude of 30 kmph. directly opposing the direction of travel.

Observations

- If you face a head wind on one leg of a ride and a tail wind on the other, you will not recover the time lost due to the higher air resistance from the head wind. On a round trip ride with wind, you will spend more time riding with a head wind than with a tail wind. If you ride 10 kms. at 150 watts with a K_a of 0.3 and a head wind of 15 kmph., your speed will be (ignoring other drag forces) 19 kmph. for about 32 minutes. On your return journey, the tail wind of 15 kmph. will increase your speed to 38 kmph. and you will cover the same distance of 10 kms. in 16 minutes. Your average speed will be $\frac{20 \times 60}{(32+16)}$ or 25 kmph. For about 66% of the journey with a head wind, your speed is below the average. If you cover the same distance of 20 kms. without any head wind at the same power (150 watts), then your speed will be 28 kmph. and the ride would take 43 minutes instead of 48 minutes.

- Increasing power to counter a strong head wind will not increase your speed in proportion to the additional power. Since power is proportional to the cube of velocity, doubling your power will increase your velocity by 1.26 times. Doubling your power from 100 watts to 200 watts at 100 kgs. with zero slope and zero rolling resistance, increases speed from 24 kmph. to 30 kmph. Unfortunately speed does not double with twice as much power. Therefore generating twice as much power will only increase your speed by $2^{\frac{1}{3}}$ or 1.26 times.

- Accurately calculating the drag force based on external parameters such as the wind velocity and front area is hard to accomplish due to errors in estimating wind velocity and direction. In an urban environment with numerous buildings, wind will be reflected and change directions. Weather forecasts give a single wind direction and degree for a fairly large area.

- The front area is also not constant since a rider keeps changing positions. Roughly $\frac{3}{4}$ of the total front area is due to the rider alone and so moving from a commuter position to a racer position can change the front area substantially. Using a frontal image of a rider on the bicycle will give a more accurate estimate of the front area (see Appendix). Others [7] have also used computer aided design software to compute front area.

3.2.5   Slip Streaming

Air drag is dominant when you have to face a head wind alone, but drops by 30% or more when you are trailing a fellow cyclist. You can save your energy by riding in the "slip stream" of a rider ahead. The "slip stream" is a region a few meters behind a cyclist where the air drag is about 1/3 less than the air drag faced by the lead rider (see Figure 3.13)

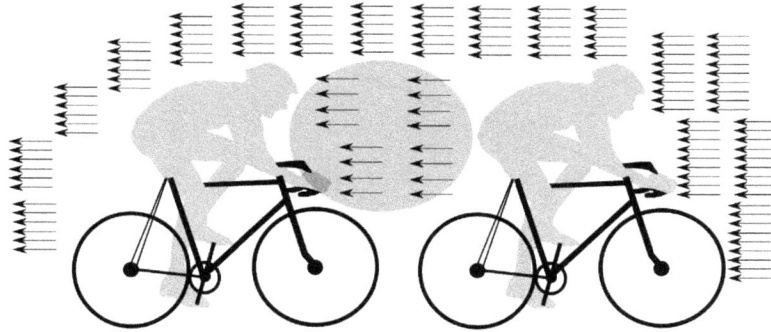

Figure 3.13: Slip Streaming to Reduce Air Drag

When the front cyclist pushes through the head wind, the air is moved aside and around the cyclist. This leaves a vacuum immediately behind the front cyclist and the rear cyclist can take advantage by riding in this slip stream region. This region will have less head wind and will be easier for the cyclist behind to ride at the same pace as the lead cyclist.

You could extend this same idea to a long line of cyclists, each one riding behind another in the slip stream of the cyclist immediately ahead. This idea works well on a flat road with a head wind, but loses value in an uphill climb. Slope resistance becomes dominant on an uphill and speeds reduce substantially, which limit the benefits of slip streaming.

Since slip streaming works for a line of riders, it also works for a large group of cyclists (peloton) with the riders on the perimeter shielding the riders in the peloton from air drag. Riders in the middle of the peloton may reduce air drag by 90% or more [8]. Slip streaming is so effective that it is used to set bicycling speed records. The record speed of 295 kmph. on a bicycle was set in 2018 using slip streaming.

How much speed do you gain by slip streaming?   Chester Kyle [9] came up with an equation to compute the savings in air drag by slip streaming. Essentially, the closer you followed the cyclist ahead, the more you could reduce your air drag. There is a risk in following a rider too closely and risking a crash. At the same time, if you are more than three meters behind the cyclist ahead, then it is the same as riding without any slip streaming, i.e. there is no reduction in air drag.

According to the formula the percentage of air drag that you would face was $(0.62 - 0.0104 \times d - 0.0452 \times d^2) * 100$ where d is the distance behind the lead cyclist. When $d = 3$ meters, the percentage of air drag is 100% and beyond three meters the formula does not apply.

In Figure 3.14 the percentage of reduction in air drag drops from about 35% at one meter behind to 20% at two meters to almost 0% at three meters. Since you have lowered air drag by about one thirds when you are close (1 meter behind), it follows that your effective speed for a given power will be the highest compared to trailing behind at two or three meters.

The benefits are also the highest when power is also high. This is logical since high power translates to high velocity and a corresponding high air drag. Therefore, there is a greater reduction in the force due to air drag when power is high leading to a higher effective speed for

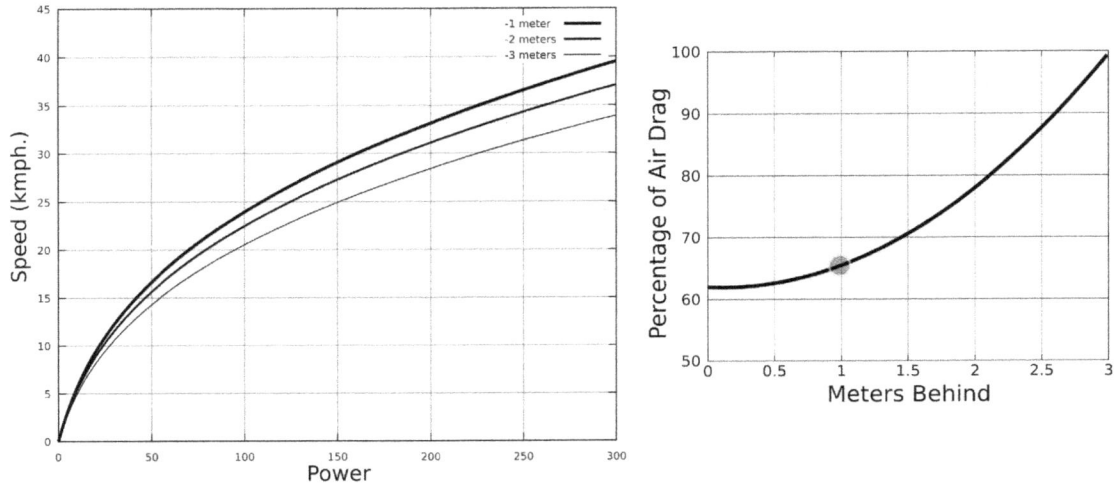

Figure 3.14: Speed gained from Slip Streaming vs. Power and % of Air Drag vs. Trailing Distance

the same power. At 300 watts, there is a gain of roughly 6 kmph. from 34 kmph. to 40 kmph (assuming a rider and bicycle weight of 80 kgs., a rolling resistance of 0.005, slope resistance of 0.0, a head wind of 10kmph. and a frontal area of $0.3m^2$). This gain in speed diminishes at low power to 1-2 kmph. since air drag is not as substantial as it is at 300 watts.

3.2.6   How much speed do you gain with less weight?

A light bicycle feels good and you would expect to ride faster than on a heavier bicycle. However, the gains may not be as much as you would expect. A 70 kg. cyclist on a 10 kg. bicycle on a flat road with no head wind generating 100 watts would ride at about 24 kmph. while the same cyclist generating 200 watts can expect to ride at about 32 kmph. (bikecalculator.com) under similar conditions (hands on the hoods and with clincher tyres at 25°C and 100 m. elevation). The left side of Figure 3.15 shows the increase and decrease in speed with lighter and heavier bicycles. At 100 watts on a flat road decreasing the weight of the bicycle from 10 kgs. to 6 kgs. will increase speed by 0.15 kmph. while at 300 watts the increase in speed for the same decrease of weight by 4 kgs. is about 0.10 kmph. However on a gradient of 10%, the benefits of a lighter bicycle are substantially higher. At 300 watts on a 10% gradient, you would be able to ride 0.55 kmph. faster on a 6 kg. bicycle than on a 10 kg. bicycle.

The loss in speed due to a heavier bicycle mirrors the gain in speed from a lighter bicycle. You will similarly lose more speed on a heavier bicycle on a 10% gradient than on a flat road for the same power. The loss in speed is about 0.53 kmph. for 300 watts on a 10% gradient for a 10 kg. vs. a 14 kg. bicycle. The loss in speed may not seem much, however over a period of an hour, you will be about $\frac{1}{2}$ kilometer or 165 seconds behind on a heavier 14 kg. bicycle.

If the gains in speed do not appear significant with lighter bicycles, you can ride several kilo-meters faster per hour simply by being more aerodynamic. The difference in speed between

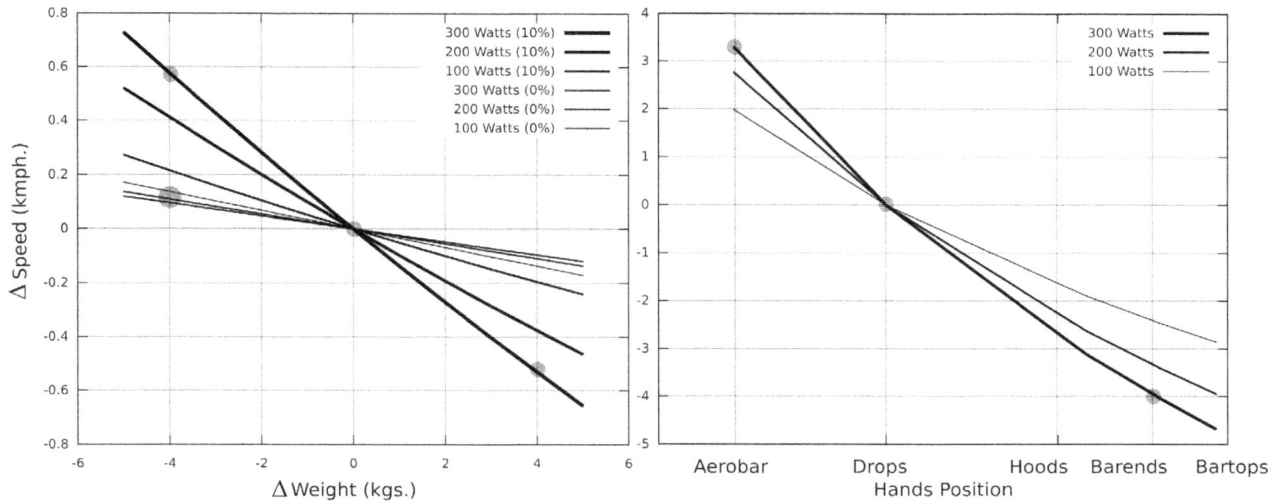

Figure 3.15: Change in Speed for 10% and 0% gradients and Riding Position for Power from 100 Watts to 300 Watts

keeping your hands on an aerobar and on the barends is over 7 kmph. at 300 watts. Even at a lower power of 100 watts, the difference in speed is over 4 kmph. Most light bicycles tend to encourage an aerodynamic position which would make you ride faster, but reducing weight alone does not make as large a difference in speed as changing your position. However, it is difficult to stay in an aerodynamic position for very long.

3.2.7 Slope Resistance

If you have been riding a bicycle, you know that a looming uphill will take more power to climb than on a flat stretch. The steeper the climb, the harder it becomes to generate the power needed to stay in motion. On an undulating terrain, you will get a break and some time to recover before another climb begins. Generating and sustaining the power needed on a long climb is as hard as riding with a strong head wind.

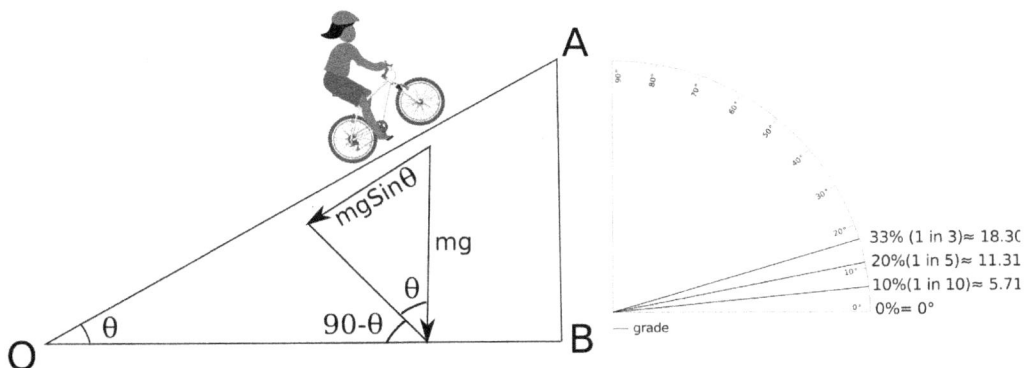

Figure 3.16: Slope Resistance when Riding Uphill

In Figure 3.16 on a ride up a hill with a slope of $tan(\theta)$ or $\frac{AB}{OB}$ you will need to overcome the slope force $mg \times Sin(\theta)$ directly opposing the direction of movement. The combined weight of the bicycle and rider (m) acts perpendicular to OB and generates a force mg.

The downward force $mgsin(\theta)$ is measured as the fraction of your weight that opposes your generated force. The slope can be defined as the tangent of the angle θ, a percentage, or in terms of the rise to run ratio (1 in 10). The percentage of slope or gradient can be positive or negative and describes the steepness and direction of the slope. A positive gradient is an uphill.

A negative 10% gradient means that for every 10 meters in the horizontal, there is a drop of 1 meter in altitude. This also means that 10% of your weight supports your generated force. The angle θ can be calculated by taking the $tan^{-1}(-0.1)$ or $-5.7°$ (-0.1 radians). The steepest uphill gradients that most bicyclists could ride are below 30% (or 16.7°).

For small angles ($< 16.7°$), the values of $tan(\theta)$, $sin(\theta)$, and θ (in radians) are all approximately the same. The calculation of slope resistance then becomes simply $mg\theta$ where θ is in radians. For a 10% gradient $tan(5.7°)$, $sin(5.7°)$ and 5.7° in radians, are all approximately 0.1. The drag force due a slope is then mgs where s is the slope fraction. On a downhill, a heavier bicycle will help with a negative slope resistance.

By how much will a slope slow you down?   An increasing slope will clearly slow you down since slope resistance (mgs) is proportional to the slope fraction (s). Assuming you can generate a steady 100 watts of power with a K_a of 0.34, a low head wind of one meter / second, gradient of 0%, and a total weight of 80 kgs., then you could ride at 19.3 kmph. The coefficient of rolling resistance (see Section 3.2.8) is 0.005.

Slope	Speed (100w)	Speed (150w)	Speed (200w)
0.00	19.3	22.9	25.7
0.01	15.8	19.7	22.8
0.02	12.8	16.8	20.0
0.04	8.7	12.2	15.3

Table 3.3: Speeds (in kmph.) for 100W, 150W, 200W and Slopes from 0.00 to 0.04

In Table 3.3, the drop in speed is higher with an increase in slope. When the slope is 0.04, the speeds at 100 watts and 200 watts is 8.7 kmph. and 15.3 kmph respectively, a drop in almost half. As slope increases, the slope resistances becomes dominant and the effect of air drag on speed is minimal.

Reducing Slope Resistance   Since changing the slope θ of the terrain on your route is beyond your control, the gradient cannot be altered. Further the acceleration due to gravity g is a constant. Therefore, the only factor that can be changed to alter slope resistance is m the sum of your and the bicycle's weight (a zero head and tail wind is assumed).

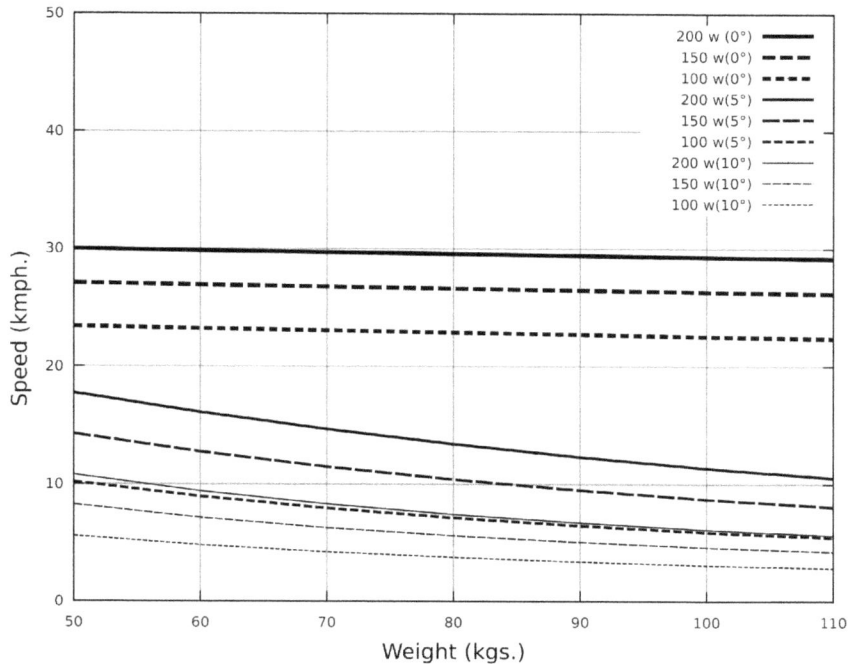

Figure 3.17: Speed vs. Weight for given Power and Slope

In Figure 3.17 as weight increases, speed does decrease. But the decrease in absolute speed is the maximum from 17.7 kmph. to 10.5 kmph. (a decrease of about 40%) for a cyclist generating 200 watts on a 5° slope. Without a strong head wind, the air drag force at low speeds below 20 kmph. is not significant compared to the slope resistance. When the slope is increased further, the percentage drop speed for a cyclist generating 200 watts on a 10° slope is about 50% (from 10.8 kmph. to 5.6 kmph.).

The decrease in speed is not linear with the increase in weight. On a 5° slope, if you increase the total weight of the bicycle and cyclist from 60 kgs. to 70 kgs. for a cyclist generating 200 watts, the slope resistance increases from 30.8 N to 35.9 N, an increase of 5.1 N. This in turn reduces the generated speed from 17.74 kmph to 16.11 kmph., a reduction of 1.63 kmph. which may not seem much, but is close to a difference of 3 minutes on a 7 km. route with the same slope of 5° and power of 200 watts.

Slope Resistance vs. Air Drag   Slope resistance and air drag are the main drag forces that constitute most of the total drag force. A steep slope will slow down even the most fit cyclist since it is equivalent to a large air drag force. In Figure 3.18, the grade percent is shown vs. the equivalent head wind in kmph. and meters per second. For example, a gradient of 10% is approximately equivalent to a head wind of 54 kmph (15 m/sec.). The drag force due to a slope of 10% is mgs or $80 \times 9.8 \times 0.1$ or 78.4 N for a 70 kg. rider and 10 kg. bicycle.

The equivalent head wind for the same slope of 10% is 15 m/sec. and the air drag is $K_a v_{head}^2$ or 0.35×15^2 (assuming a K_a of 0.35 for an average rider on a commuter bicycle) which is a drag

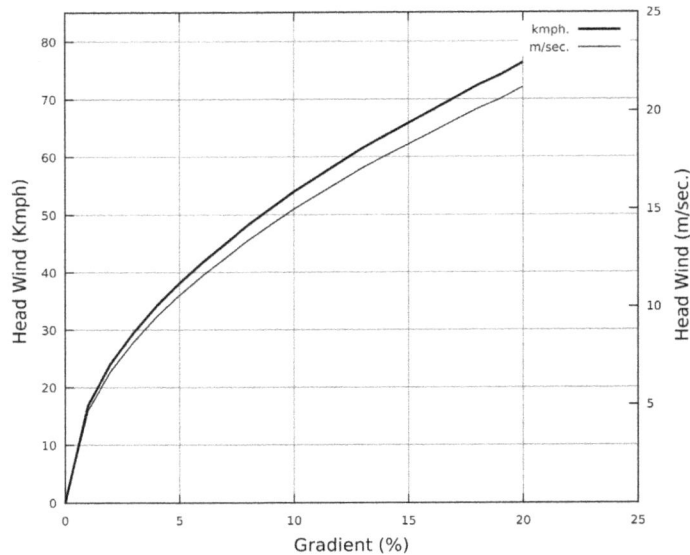

Figure 3.18: Gradient % vs. Equivalent Head Wind

force of of 78.75N. Groeskamp [10] defined an incline-equivalent wind velocity to translate a gradient to the equivalent head wind for the same power.

Observations

- Riding on an undulating terrain of some distance x kms. at a given power will take longer to cover the same x kms. on a flat terrain. The time lost in the climb due to a lower speed is not recovered when descending at a higher speed. Riding uphill and downhill is analogous to riding with a head wind and tail wind respectively.

- In most cases, slope resistance will be a more common drag force, since it is not dependent on weather conditions. However, a strong head wind is as good as a steep climb and the worst case is a climb with a strong head wind.

- A heavier bicycle will help you ride faster downhill since a fraction of your weight is a drag force acting in the same direction. However, on a uphill ride, that same fraction of your weight is now a drag force opposing your ride uphill. Since you will be spending more time riding uphill than downhill, a lighter bicycle will shorten the duration of your journey besides reducing the power needed to climb uphill.

- If you are able to generate more power, then it will have a much greater effect on your speed when you climb uphill than if you face a head wind. This is because slope resistance is not dependent on speed and therefore increasing power proportionately will increase speed if other drag forces are minimal.

- You are more likely to be slowed down by an uphill than a strong head wind. Few locations are perfectly flat (0% gradient). Gradients of 1% and 3% correspond to head winds of 17 kmph. and 30 kmph. respectively.

3.2.8 Rolling Resistance

Since tyres are your only contact points on the road, they play a major role in the speed of the bicycle. The assumption that lowering friction between the tyre and the road will make you ride faster is contrary to the necessity of maintaining traction. Traction is the frictional force necessary to maintain contact between the tyre and the road surface without slipping. If you could coat your tyre with a slippery liquid like oil, then you would reduce friction between the road and the tyre. But you will not be riding any faster, since the lower friction leads to slipping and a tyre spinning in the same location is not going anywhere. If the propulsive forward force exceeds the frictional force between the tyre and the road surface, then your tyre will slip and the bicycle maybe become uncontrollable. On a typical dry city road with asphalt and a bicycle weighing about 15 kgs., you would need substantial power to generate a propulsive force that exceeds the traction force.

The frictional force between two surfaces in contact depends on the material of the surfaces. In the case of a bicycle on a road, it depends on the type of rubber of the tyre and the road surface material. If the texture of these materials is rough, then the friction will be higher leading to a better road grip. Materials such as water, ice, or gravel on the road will alter the friction between the tyre and the road. This is a problem for vehicles that travel at high speeds and are effectively "hydroplaning". When speeds are high enough, the tyre loses contact with the road surface since the water between the tyre and road cannot be removed fast enough and therefore the vehicle is riding on a thin film of water. With lower friction between water and the tyre compared to the road surface and the tyre, the chance of slipping is much higher. However this is not an issue in bicycles, since the speeds necessary for hydroplaning are above 50 kmph.

Even though the tyre is in motion, static friction and not kinetic friction represents the traction to necessary to keep a tyre on the ground. At any given time, a unique section of the tyre is in contact with the road (see Figure 2.8). Static friction is measured as $F_s = \mu_s N$ where μ_s is the coefficient of static friction and N is the normal force or effectively the weight due to gravitation acting perpendicular to the ground in newtons. The rolling resistance is the drag force on a bicycle tyre and measured in the same units (Newtons) as static friction with a similar formula, $F_{roll} = C_r N$ where F_{roll} is the rolling resistance and C_r is the coefficient of rolling resistance. The drag force caused by rolling resistance can be reduced to increase efficiency provided there is sufficient traction (static friction) to prevent the tyre from slipping. Although the units for rolling resistance are the same as the units for force (N), it is represented in a scalar form as unit energy loss.

Rolling resistance (F_{roll}) is the loss of energy per unit distance travelled. Pedalling a bicycle consumes energy and some of it is lost (E_{roll}) when the tyre repeatedly deforms and regains its shape while rolling on a road. The energy required to deform the tyre is greater than the energy required to return to it to its original shape. This is similar to a rubber ball that bounces lower with each bounce. Energy is usually lost in the form of heat and depends on the type of rubber compound. Tyre companies mix artificial compounds with rubber to create a tyre with low rolling resistance.

Ideally, if you were riding on a hard surface and a tyre had enough pressure to prevent any deformation, then rolling resistance would be 0. The next best real scenario is the railways where a steel wheel rolls on a steel rail. Both surfaces are made of hard steel and deform far less than a rubber tyre even with heavy loads. Therefore, not surprisingly the railways can claim to be 3.5 times more efficient in transporting freight than a truck [11]. The propelling force F_{propel} available from the generated force F_{gen} (from the combined F_{pedal} of both pedals) can be calculated as

$$F_{propel} = F_{gen} - (F_{aero} + F_{roll} + F_{slope})$$

To get the most of out of your generated force F_{gen}, you would need to minimize the three quantities - F_{aero}, F_{roll}, and F_{slope} which represent aerodynamic drag force, drag force due to rolling resistance, and the slope drag force respectively. For simplicity, F_{slope} is assumed to be zero. A cyclist cannot change the slope on a ride, but has some control over the air drag and rolling resistance.

At low speeds, the rolling resistance F_{roll} is a large fraction of the total drag and at higher speeds, the air drag F_{aero} is dominant (see Figure 3.19). Other types of resistance such as the ball bearing and chain resistances (minor compared to air and rolling resistances) are not included in the total resistance.

At low speeds (< 10 kmph), the rolling resistance is more dominant than air resistance. But at speeds over 20 kmph, air resistance is substantially higher than rolling resistance. If you ride a bicycle around the city at low speeds, then reducing rolling resistance over air resistance would make you more efficient. At racing speeds (> 45 kmph), close to 90% of all drag forces will be due to air resistance.

The plot in Figure 3.19 shows the percentage of total resistance for just two of the three drag factors (a ride on a flat road). If the drag due to slope resistance is included, then percentage of total resistance due to rolling resistance may drop to 10% or less. Even a moderate climb with a slope of 0.03 will dominate the total resistance.

The value of E_{roll} depends on the tire width, tire pressure, tread design, and characteristics of the tyre rubber and the terrain. If you ride on a city road, the road surface will be harder than a bicycle tyre and therefore the bicycle tyre will deform due to the weight of the rider and bicycle. The more a bicycle tyre deforms, the greater the rolling resistance.

Reducing Rolling Resistance   Several factors affect rolling resistance including tyre width (narrow vs. wide), tyre pressure (high vs. low), tyre tread (rough vs. smooth), tyre weight (light vs.

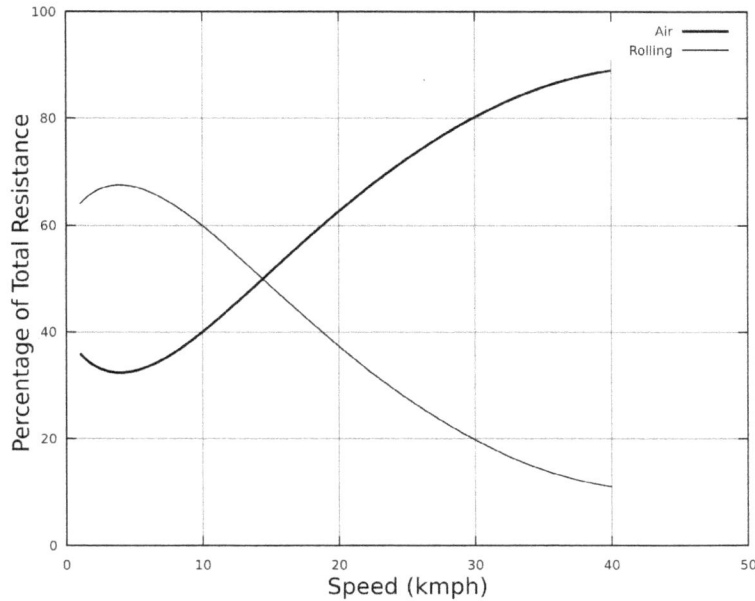

Figure 3.19: Percentage of Total Resistance for Rolling and Air Resistances

heavy), tyre diameter (large vs. small), and terrain (rough vs. smooth). These factors are related and the affects of these factors on rolling resistance can be compared if we keep the other factors constant (see Chapter 4).

3.2.9 Acceleration

Finally after overcoming all the drag forces, you can accelerate your bicycle. A cyclist needs to generate translational as well as rotational energy to move the cycle forward. The translational energy is needed to accelerate the bicycle along a linear path, while the rotational energy is required to accelerate the speed at which the wheels rotate. The power that you can generate will determine the time it takes to produce the total energy required.

The force required to generated this energy is divided into the translational and rotational forces. The translational force is the mass of the bicycle $M_{bicycle}$ times the linear acceleration a of the bicycle while the rotational force is the rotational mass I_{wheel} of the wheels times the angular acceleration α.

$$Force = M_{bicycle} \times a + I_{wheel} \times \alpha$$

The value of $M_{bicycle}$ is constant and a can be computed from two readings of the velocity at different times on a ride. The rotational mass I_{wheel} has two parts - the weight of the rim plus tyre and the weight of the spokes. The part of the rotational mass due to the rim weight is calculated as $M_{rim} \times r^2_{wheel}$ where M_{rim} and r_{wheel} are the mass of the rim and the radius of the wheel respectively. The weight of the hub is not considered since it is very close to the axis of rotation and therefore the moment of inertia of the hub will be low compared to the wheel.

For a rim of mass 1.5 kgs. and wheel radius 0.33 meters, the rotational mass due to the rim alone is 1.5×0.33^2 or $0.163 \, kgm^2$. The rotational mass due to each spoke is calculated as $M_{spoke} \times L_{spoke}^2 / 3$. If the wheel has 36 spokes and each spoke weighs 0.006 kgs. and the length of the spoke is 0.3 meters, then the rotational mass due to the 36 spokes is $36 \times \frac{0.006 \times 0.3^2}{3}$ or $0.0065 \, kgm^2$. The value of I_{wheel} becomes $0.163 + 0.0065 = 0.1698 \, kgm^2$. For two wheels, this becomes about $0.34 \, kgm^2$.

The angular acceleration α is rate of change of the angular velocity ω (in radians per second). If you know the radius r of the wheel including the tyre, then one rotation of the wheel corresponds to a linear distance of $2\pi r$. For a speed of 25 kmph. or 6.9 m/s, a wheel of radius 0.33 meters will have to make 3.3 revolutions or $3.3 \times 2\pi$ (or 20.9) radians per second. If speed increases to 30 kmph. or 8.33 m/s in 4 seconds, then the number of revolutions per second increases to about 4.0 or equivalently 25.2 radians per second.

Therefore the change in angular velocity is $25.2 - 20.9 = 4.3$ radians per second over an interval of 4 seconds. This corresponds to an angular acceleration of $\frac{4.3}{4}$ or $1.075 \, radians/sec^2$. Finally, the rotational force becomes 0.34×1.075 or $0.365 N$.

The force for linear acceleration is the product of the mass of the bike M_{bike} (80 kgs.) and the acceleration $a = \frac{8.33 - 6.90}{4} = 0.3575 \, m/s^2$ which becomes $80 \times 0.3575 = 28.6 N$. The acceleration force is then the total of the translational force ($28.6N$) and the rotational force ($0.365N$) or $28.965N$. Since the weight of the wheels and spokes (3 kgs.) are light compared to the rest of the bicycle (80 kgs.) including the rider, the rotational force is not significant.

How much power do you need to accelerate?   As seen earlier, additional power will help you ride faster uphill and raise the maximum gradient of a slope that you can climb (see Figure 3.24). If you assume a flat road and no head wind, then the additional power to ride faster depends on your initial speed.

In Figure 3.20 the additional power that you will need to accelerate by 3, 5, and 7 kmph. at different speeds is shown. While a higher power to ride 7 kmph. faster than 3 kmph. is expected, the difference in the required additional power depends on the current speed. When the current speed is high (> 40 kmph.), it takes a lot more additional power to increase speed to 47 kmph. than to increase speed from 10 kmph. to 17 kmph.

At low speeds, the air drag is significantly less and therefore the power required to ride faster is less than at higher speeds. Since it becomes harder to generate additional power beyond your natural capacity, the other option to increase speed is lower the front area. Racers are known to ride extremely low in contorted positions to reduce front area as much as possible at high speeds. This is most visible on downhill rides where speeds are high.

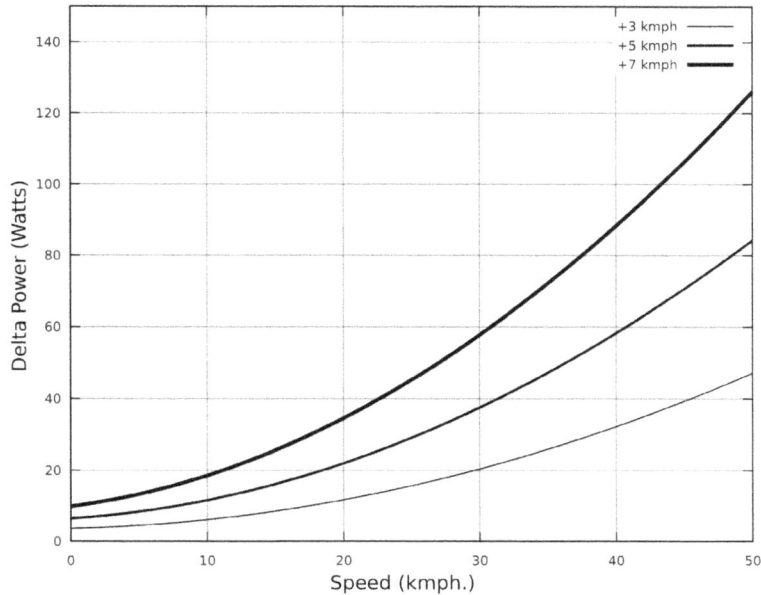

Figure 3.20: Additional Power to Increase Speed vs. Current Speed

How Fast can you Ride?   This is a question that you would expect from a young rider but is also of interest to adult riders as well. Since power is proportional to speed, the fastest speed at which you can ride will depend on the power that you can generate. A light aerodynamic bicycle will help to some extent, but at high speeds (above 25 kmph.) resistance from air drag is the dominant drag force that will slow you down.

If a rider weighs 70 kgs. and the bicycle weighs 10 kgs., with tubular tyres (low rolling resistance), and an aerodynamic drag factor (K_a) of 0.14, then the maximum speed you could theoretically reach with 200 watts would be 37.6 kmph on a flat road with zero wind conditions (see Figure 3.21). If you can generate double the power (400 watts), then your maximum speed would increase to 48.5 kmph. At 800 watts you could ride at 62 kmph. and the gain in maximum speed diminishes with still higher power.

On a downhill, the speeds are logically higher with steeper gradients. Even a small (-5%) downhill gradient at 200 watts is sufficient to increase the maximum speed from 37.6 kmph. to 66 kmph. The steeper gradients of -10% and -20% have maximum speeds of 88 kmph. and 122 kmph. at the same power output. At lower (< 300 watts) power ranges, the gains in speed from riding downhill are greater than at higher (> 500 watts) power ranges.

The dashed lines in Figure 3.21 show the same results for an average bicycle with a K_a of 0.24 with MTB tyres (high rolling resistance). Clearly, you will ride faster on a more aerodynamic bicycle with low rolling resistance. At 200 watts on a flat road, the difference is about 9 kmph. which increases to a difference of about 30 kmph. on a -20% gradient. This seems logical since at higher speeds, air drag is roughly 90% or more of the total resistance.

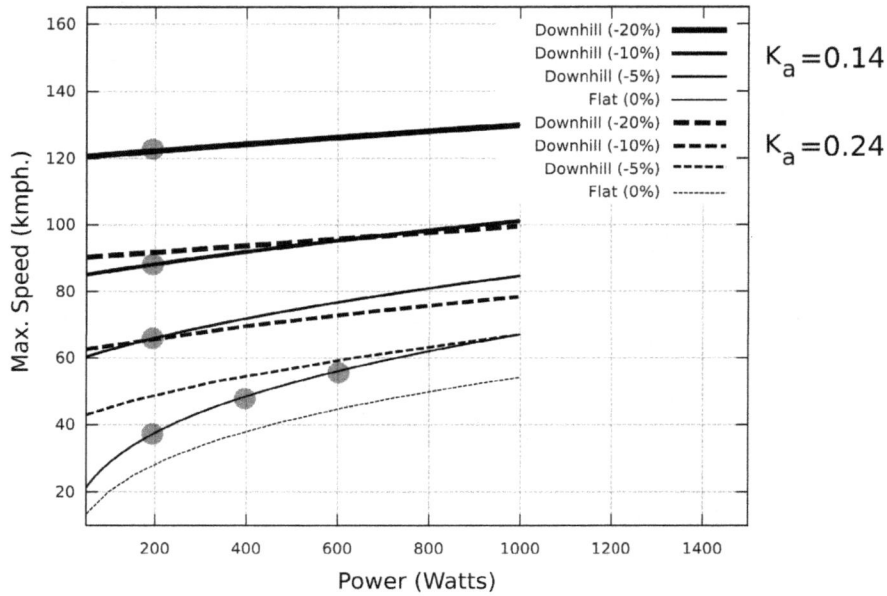

Figure 3.21: Maximum speed in kmph. vs. Power in watts for four different gradients and two drag factors (K_a)

Although bicycle speed records have far exceeded 130 kmph., most of these records above 150 kmph. are assisted with a leading motorized vehicle to reduce drag and to tow the bicycle to an initial speed. Since air drag is over 90% of the resistance at high speeds, slip streaming behind a fast moving vehicle can substantially increase your maximum speed.

Observations

- It takes more power (effort) to start riding a bicycle. When you begin riding a bicycle your initial velocity (v_i) is zero and you will need to generate power to produce the F_{accel} force. This acceleration force is needed to ride at some final velocity (v_f) and F_{accel} is inversely proportional to the duration to reach v_f. Since you need some minimum speed (5 kmph.) to stay balanced on a bicycle, you have to accelerate to at least 5 kmph. from 0 kmph. in a few seconds. The power P that you generate will be used to produce the sum of all drag forces and F_{accel}. Once you are riding at some velocity v_f, then you can continue riding by generating power less than P that is sufficient to overcome drag forces alone.

- When you are coasting (freewheeling) you are not generating any power, yet the bicycle is moving. This is due to the bicycle's existing momentum that is gradually reduced to zero by the drag forces.

3.2.10 Calculating Power and Velocity

If you know your velocity and drag forces, then you can calculate the power that you are generating using the formula - $P = V_{bike} \times (K_a \times V_{total} \times V_{total} + F_{roll} + F_{slope})$ where V_{total} is the sum of V_{bike} and V_{head} velocities. If you know the power P that you can generate, then you can also estimate V_{bike} for a given V_{head}. Since the right hand side of the power formula uses a cube of V_{bike}, its value can be estimated by taking a differential of the equation with respect to V_{bike} and then minimizing the difference for a given power (see Appendix).

The force to accelerate is not included in this power equation. Acceleration is harder to compute since it is a differential ($\frac{dv}{dt}$) and can change substantially during a ride. Whereas air drag is dependent on a few parameters like front area and the velocities of the head wind and bicycle, acceleration depends on the rider generating sufficient power to increase V_{bike}. An empirical value for acceleration is easy to calculate from GPS readings since the current and prior velocity / timestamps are known (see Chapter 7).

Measuring Power Accurately One way of measuring the power that you generate is to measure the force that you apply on the pedals. Usually, the pedals with power meters are installed just as a normal pedal would be installed, except that the pedals must be paired with each other (using Bluetooth ®). So, the right pedal maybe paired with the left pedal which is in turn paired with a device mounted on the bicycle frame. The left pedal combines the data from the right pedal and transmits it to a device. The length of crank arm may also need to be set on the receiving device (typically a smartphone).

Since the pedals are the closest to the source of power generation, a power meter on the pedal will accurately measure power without including the drive train efficiency which is estimated at 95% or more. Other locations to install a power meter include the chain ring and the hub. Currently, professionals and serious athletes interested in calculating power use these devices since they are still relatively expensive. Power meters were first used in road bicycles and now are available for mountain bicycles as well. On a MTB, there are issues with measuring power accurately, since not all of the power to ride on a rough trail is generated at the pedals and energy lost due to vibrations and steering will not be visible at the pedal.

The accuracy of the measurements from a power meter is within $\pm 1.5\%$ of the actual power generated. More importantly, the power meter's accuracy should be consistent over time. Trying to get this level of accuracy from external measurements such as the wind, bicycle weight, rider weight, gradient, and rolling resistance is very hard.

At high speeds (> 20 kmph.) air drag becomes significant and the accuracy of the value of head wind V_{head} will largely determine the accuracy of the power calculated. Head wind is hard to measure without installing a specific sensor dedicated to measuring wind alone. One alternative is to use a source (like OpenWeatherMap) which gives the magnitude and direction of the wind at a specific latitude and longitude.

However since the resolution of the lat/lon on OpenWeatherMap is currently limited to two decimal digits, this is equivalent to a single sample over an area of $1.2\ km^2$ at the equator or $0.62\ km^2$ at a latitude of 45°. On the other hand, elevations (altitude) from Google Maps API have a much higher resolution and accept latitude and longitude with six decimal places (samples over an area $< 0.012m^2$). For a bicycle, an elevation resolution of a $1.0m^2$ is sufficient to accurately compute the slope resistance.

How much power can you generate? There are a range of values for the power that a cyclist can generate. A study [12] has suggested that a non-athletic cyclist could generate 75 watts, a commuter cyclist could generate 150 watts, a healthy touring cyclist may generate 200-250 watts, and a racing cyclist could generate 350-400 watts for an hour or more.

The study collected data from 16 average cyclists for rides of over 100 accumulated minutes during a week. The average speed for this group was about 21 kmph. For every 1% downhill gradient, speeds increased by about 1 kmph., whereas for every 1% uphill gradient, speeds decreased by 1.5 kmph.

The power-to-weight ratio is a common figure to judge if you are a professional, amateur, or recreational cyclist. A professional cyclist may have a power-to-weight ratio of about 6 or more. This means that given a weight of 70 kgs., a professional could generate 70×6 watts for some sustained period (about an hour). An amateur and recreational cyclist may have a power-to-weight ratio of about 3.5 and 2.0 respectively. While it is possible to generate a burst of power 2-3 times your average power, it is not sustainable. This peak power can be used for a minute or more before fatigue sets in.

Increasing power-to-weight ratio is not easy for the average cyclist. You could increase the power generated by training, but at the same time your weight should remain the same. Alternatively, you could reduce your weight and still generate the same power. Finally, you could reduce your weight and simultaneously increase your generated power. All three options are challenging for most average cyclists.

The high power-to-weight ratio explains how professionals can ride uphill at a faster speed than the amateur. With a power-to-weight ratio of 3, an amateur weighing 70 kgs. could generate 210 watts on a 5° slope for a speed of 13.5 mph. A professional with a power-to-weight ratio of 6 with the same weight would generate 420 watts and have a speed of 23 kmph. on the same 5° slope. If the gradient of the slope is doubled to 10°, then the professional will ride at close to double the speed of the amateur.

Real Time Power Analysis In reality, a cyclist does not generate power uniformly. On an uphill, there will be a temptation to increase generated power to maintain a reasonable speed. Likewise on a downhill, the power generated may drop to close to zero, if speed is sufficiently high. In Figure 3.22, the generated power and gradient for a ride of about 2 hours show that power and gradient are weakly related.

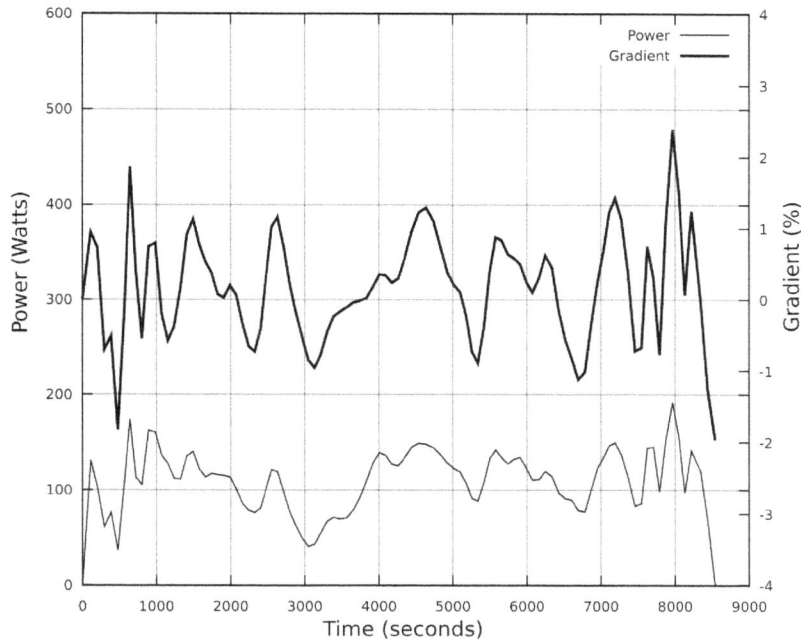

Figure 3.22: Generated Power and Gradient Percent vs. Time in seconds for a ride

Since this data was collected empirically using GPS on a smart phone (see Chapter 7), the accuracy is not very high. Further, the power and gradient data has been smoothed. Still it is apparent that power increases with a gradient and riding with a constant power is not the norm. The generated power fluctuates over an average of 100 watts.

Power Calculators There are several online power calculators to compute power and velocity based on bicycle and rider parameters. The formulas to calculate power or velocity are common and the results from these calculators should be very similar. The typical set of parameters include rider weight, bicycle weight, bicycle velocity, head wind velocity, head wind degree, front area, coefficient of drag, rolling resistance coefficient, temperature, altitude, humidity percentage, and gradient. You can either calculate your expected power or velocity for a given set of variables. The results from different power calculators may not be identical, but should be reasonably close to an expected value.

In Figure 3.23, you can enter the expected power and other fields such as the weight, type of tyres, position on the bicycle, grade, and head wind speed. Clinchers are tyres with inner tubes and are the most common type of tyres with a coefficient of rolling resistance of 0.005. The position on hoods defines the product of the front area and coefficient of drag (0.388). The velocity of 25.45 kmph. is the expected speed for the given power. Similarly, you can set a given velocity and compute the expected power for the same conditions.

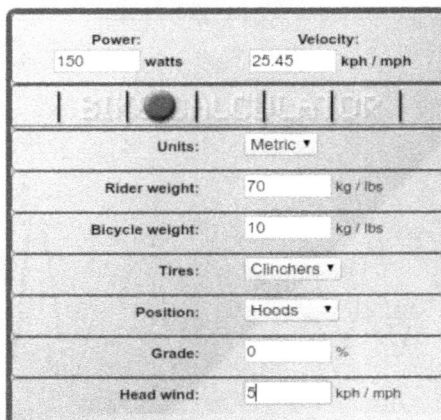

Figure 3.23: Bicycle Power and Velocity Calculator from bikecalculator.com

URL
`http://bikecalculator.com/`
`https://www.cyclingpowerlab.com/PowerComponents.aspx`
`https://www.gribble.org/cycling/power_v_speed.html`
`http://www.kreuzotter.de/english/espeed.htm`
`https://github.com/mkonchady/cycle`
`http://www.tribology-abc.com/calculators/cycling.htm`

Table 3.4: Online Power and Velocity Calculators for Bicyclists

3.2.11 Slope and Head Wind Limits

Although the velocity of a bicycle is directly proportional to the power that you can generate, the relationship depends on the head wind velocity. In the absence of any head wind, power is proportional to the cube of the velocity. If a head wind is included, then power is proportional to the square of the relative velocity. Further on a steep slope, power is proportional to velocity. Since we are of different builds, the maximum power that we can generate has some limit.

Head Wind   When your speed drops to 5 kmph. or below in the face of a strong head wind or a steep slope, you maybe better off walking than riding. This is a head wind or slope fraction limit that you can calculate based on generated power. In Figure 3.24 the plot of head wind (in kmph.) vs. power shows that at 200 watts, it will take a strong head wind of over 75 kmph. to stop your ride. Such extreme events with head winds of over 50 kmph. are rare, but even with 100 watts, you can ride in a head wind of 55 kmph.

Slope   Since slope resistance is dependent on weight, the curves for the maximum slope that you can ride, depend on the combined weight of the bicycle and rider. Logically, it takes less

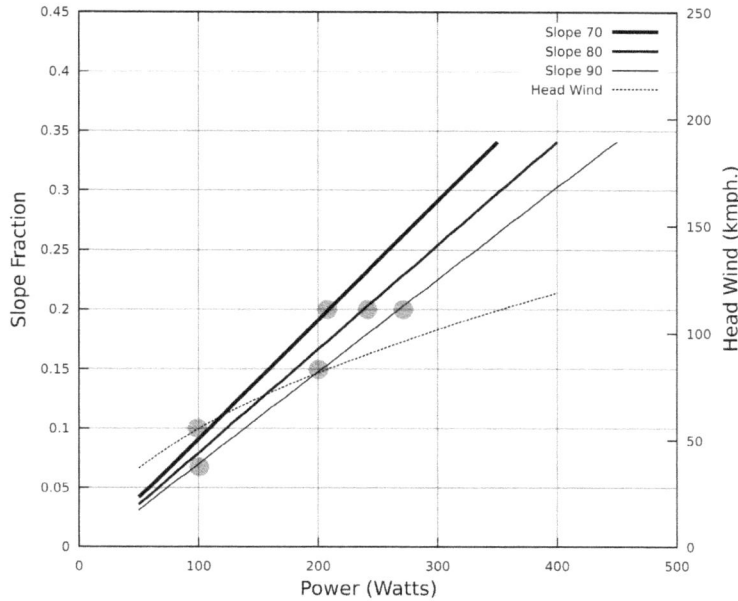

Figure 3.24: Slope Fraction and Head Wind Limits vs. Power for Three weights 70 kg., 80 kg., and 90 kg.

power for riders with lower weights to climb a slope with the same gradient. With a weight of 70 kgs., it takes about 200 watts before reaching the limit with a slope of 0.2 or 20% gradient. When weight is increased to 80 kgs. and 90 kgs., the required power to reach the limit increases to about 240 watts and 270 watts respectively.

The curves for slopes with less weight is steeper since slope is inversely proportional to weight ($s \propto \frac{Power}{mg}$). Therefore for 70 kgs., the slope fraction limit increases by 0.1 from 0.09 to 0.19 when power increases from 100 watts to 200 watts. When the weight is 90 kgs., the slope fraction limit increases by 0.08 from 0.07 to 0.15 for the same increase of power from 100 watts to 200 watts.

3.3   How fast can you coast down a slope?

Riding downhill with little or no effort is a thrill for most bicyclists. Speeds on steep slopes can quickly reach 60+ kmph. and the unwary rider may lose control of the bicycle. Before you ride down a steep slope, you can estimate the highest possible speed that you can reach. Earlier, the slope resistance was computed as $mg \times sin(\theta)$.

When you coast down, this same slope resistance is now a propulsive force and will accelerate the bicycle. The bicycle will stop accelerating when the slope resistance is equal to the sum of the air drag and rolling resistance or when

$$mg \times sin(\theta) = K_a \times v^2 + mg \times cos(\theta) \times C_r$$

111

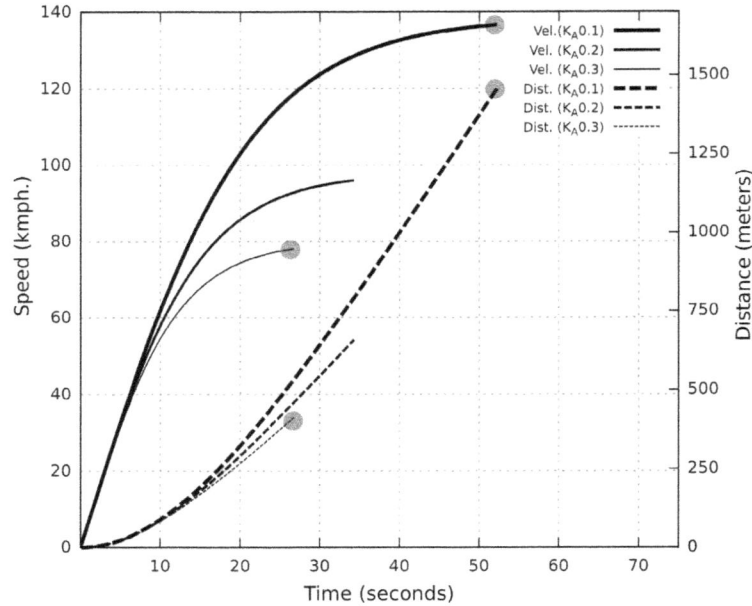

Figure 3.25: Speed and Distance vs. Coast Down Time

From this equation we can calculate the value of v since all the other parameters are known. If the value of $C_r= 0.008$, $K_a= 0.3$, $m = 80$, and θ the gradient is 0.1 (or 5.7°), then the maximum velocity v is 56 kmph. If K_a is reduced to 0.2, the maximum velocity increases to 68 kmph. However, if the gradient is doubled to 0.2 with the same $K_a(0.2)$, the maximum velocity is close to 100 kmph.

In Figure 3.25, the speeds and distances for three different aerodynamic drag (K_a) values and three different gradients shows that maximum speeds and distances differ quite significantly by gradient and K_a. The total weight of the bicycle is the same at 80 kgs. and the coefficient of rolling resistance is also a constant at 0.008. When K_a is 0.1, it takes about 52 seconds and about 1.5 kilometers to reach the maximum speed of 136 kmph. When K_a is 0.3, the peak speed of 78 kmph. is reached in about 26 seconds and a distance of 400 meters.

These results assume ideal conditions where there is no head or tail wind and the surface of the road / tyre is appropriate for the assumed coefficient of rolling resistance. In reality it is hard to achieve these speeds, since downhill roads are typically winding and you need to slow down on a curve. Still, there is a sharp increase in the maximum speed when K_a is lowered and racers go to great lengths to lower K_a through a contorted position. Riders on recumbent bicycles can lower drag in a comfortable position since the center of mass on a recumbent bicycle is lower.

References

[1] Sheldon Brown. Gain ratios: A new way to think about bicycle gears, November 1999. URL https://bit.ly/2LOQ313.

[2] David Gordon Wilson and Jim Papadopoulos. *Bicycling Science*. MIT Press, Third edition, March 2004.

[3] Steven A. Kautz, Michael E. Feltner, Edward F. Coyle, and Ann M . Baylor. The pedaling technique of elite endurance cyclists: Changes with increasing workload at constant cadence. *International Journal of Sport Biomechanics*, 7:29–53, 1991.

[4] Emma Colson. Power to your pedals, August 2002. URL `https://bit.ly/2LFvLLU`.

[5] Bikeradar.com. How to get your bike saddle height right, June 2017. URL `https://bit.ly/2kJDWul`.

[6] Schwalbe Tyres. What exactly is rolling resistance?, May 2018. URL `https://bit.ly/2KF1Imc`.

[7] Pierre Debraux, Bertucci, Aneliya V. Manolova, A. Lodini, and Simon Rogier. New method to estimate the cycling frontal area. *Interntational Journal of Sports Medicine*, 2009.

[8] Bicycling.com. The scientific reason pelotons go so fast, August 2018. URL `https://bit.ly/2M6N7o9`.

[9] Velopress.com. Cheating the wind: The physics of draft-legal racing, January 2018. URL `https://bit.ly/2NHG16u`.

[10] Sjoerd Groeskamp. Translating uphill cycling into a head-wind and vice versa. *Journal of Science and Cycling*, 6(1), 2017.

[11] Chris Barkan. Railroad transportation energy efficiency, November 2009. URL `https://bit.ly/2rVaO6t`.

[12] J. Parkin and J. Rotheram. Design speeds and acceleration characteristics of bicycle traffic for use in planning, design and appraisal. *Transport Policy,*, 17(5), 2010.

4 Tyres

Bicycle wheels are remarkable in that they support about 100 times their weight using a 100+ year old design. While the design with tangential spokes has stayed the same, the size and materials used in a bicycle wheel has changed over the years. When you buy assembled bicycle, you may not be able to select the type of wheels. However, you can change the tyres at a reasonable cost to suit your requirements.

4.1 Dimensions of a Tyre

The diameter of the wheel measured from the bead at 0° to the bead at 180° – called the bead set diameter (see Figure 4.1) is the inner diameter of the tyre that will fit the wheel. The tyre will have a outer diameter larger than the bead set diameter (BSD) that is described on the sidewall of the tyre. If the tyre sidewall reads "$25 - 622, 700C \times 25$", then the tyre will fit a wheel with a BSD of 622 mm. and inflated tyre width of 25 mm. The width of the tyre is important if you will be using an inner tube that must also be of roughly the same width and diameter as the tyre. The term $700C$ is from an older system and means a wheel with BSD of 622 mm.

The range of 1.45 to 2.0 times the inner rim width is a rule of thumb to calculate the minimum and maximum widths of a tyre that will fit on the wheel. So for an inner rim width of 15 mm., the minimum and maximum tyre widths would be 21.75 mm. and 30 mm. respectively. While you could exceed this tyre width range, the diameter of the tyre must fit the BSD.

This is because when the tyre or tube is inflated, the bead of the tyre must sit perfectly in the bead seat. If inner tyre diameter is larger than the BSD, the beading of the tyre may pop out of the rim and it is very likely that you would lose control of the bicycle.

There is more to a bicycle wheel than meets the eye. It appears to have a single function - to roll on some given surface. Other functions including steering control, riding smoothly over rough surfaces, and maintaining stability. Although the function of the wheel is easily understood, there are several factors to consider before buying a wheel and tyre.

- Is reducing the weight of your bicycle a priority for which you would be willing to invest a substantial fraction of the cost of the bicycle on light wheels. Wheels made of low cost material are not necessarily of low quality, however they will tend to be heavier. A large number of spokes and a heavy rim will add to the weight of the wheel.

- Will the surface on which you will ride be rough (trail) or smooth (a city road), or a mix of both. City roads in general are smooth, but a badly maintained road can be almost the same as riding on a trail with loose soil and pebbles.

- Is riding fast or having puncture protection a priority? In general the two requirements are mutually exclusive since a fast tyre is light but may not offer high puncture protection while a sturdy puncture proof tyre will be heavier and correspondingly slower.

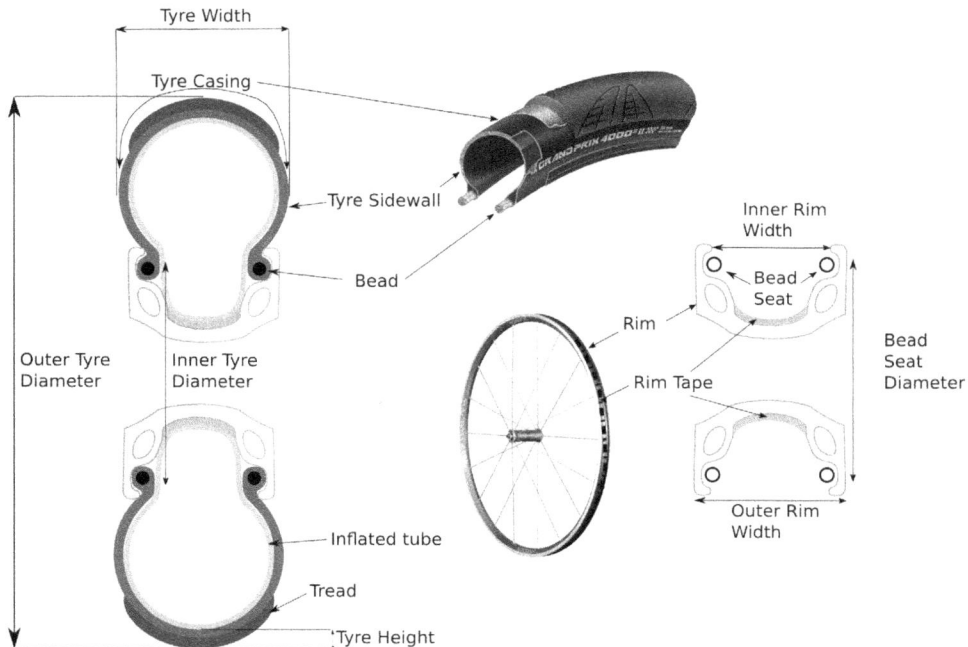

Figure 4.1: Parts of a Tyre and Wheel

Finding the right tyre and wheel for your bicycle can make you more comfortable and even more efficient. You maybe restricted in the choice of tyre based on the frame of the bicycle. Some bicycle frames allow wide tyres while others have narrow forks, limiting the width of the tyre. Narrow tyres are preferred in road bicycles to reduce air drag and weight. However, a narrow tyre is not necessarily faster to ride than a wide tyre. The weight of the tyre is another factor to consider, although a lighter tyre is not always more efficient than a heavier tyre.

Since the beading of the tyre must sit firmly in the bead seat, it is the most rigid part of the tyre. A low cost tyre may use beading made up of woven steel wires. This may add some extra weight compared to a tyre with beading made up of kevlar (a synthetic material lighter than steel, but strong enough to replace steel). A steel beading is rigid and maintains the circular shape of the tyre. Unlike a kevlar based tyre, a tyre made with steel beading cannot be folded for storage. Since a tyre with kevlar beading has some flex, it is easier to remove and install on a wheel than a tyre with a steel beading.

4.2 Choosing a Tyre

In the process of choosing a bicycle, a cyclist tends to spends more time on other components such as the frame, brakes, and so on. However, the tyre is a critical part of a bicycle and a flat tyre will bring your ride to a complete halt. Tyres are the least expensive way to change the way your ride feels. A bigger tyre will give you more cushion, maybe less likely to puncture, and give you more traction. Further, you can lower the pressure of a bigger tyre which will reduce the vibrations of the bicycle on a rough road.

If you are buying a new bicycle, you may have to buy the tyres / wheels that are sold with the bicycle. However given an option to choose a tyre, there are several factors to consider including – tyre width, tyre radius, and tyre surface. The tyre radius and width depend on the wheel.

Often a tyre manufacturer will include the value of coefficent of rolling resistance (C_r) for a particular tyre along with other details. A high C_r for a tyre would likely discourage a buyer unless other factors such as puncture protection and predicted life of the tyre are a priority. The ideal tyre should have a low C_r, puncture protection, and be light weight. However, this is hard to achieve since it is difficult to make a tyre puncture proof without at the same time increasing the weight. The other parameters to consider include - width, diameter, thickness, surface and recommended pressure. Each of these parameters has some effect on the C_r.

The key feature of a tyre for efficiency is the stated coefficient of rolling resistance (C_r). A lower C_r will always reduce the drag due to rolling resistance. Not all tyres are listed with their associated C_r. It is not possible to accurately compute C_r based on the dimensions or materials of the tyre. Often, a test under some standard conditions such as a fixed load and air pressure will give a reasonable estimate of the C_r for the tyre.

4.2.1 What is Rolling Resistance?

The drag due to rolling resistance occurs because the part of the tyre in contact with the road briefly loses its circular shape to become flatter and then regains the original shape when the tyre is no longer in contact with the road. When the tyre is momentarily flattened it gains some energy due to the work done by the road to flatten the tyre. However, when the tyre loses contact with the road and regains its shape due to the tyre air pressure, some energy is lost as heat and not returned to the road. The lost heat energy is responsible for the rolling resistance.

A cyclist must first pedal to generate energy to move the tyre on the road. Since only part of this generated energy is used effectively, the remainder wasted heat energy is called rolling resistance. Reducing this lost heat energy is the key to lowering rolling resistance. Finding the right materials to build a tyre with low rolling resistance is a subject of research for tyre manufacturers.

Rolling resistance depends on factors such as air pressure in a tyre, the surface of the road / tyre, and the dimensions of the tyre. In Chapter 3, the drag due to rolling resistance was computed as the product of the weight with a coefficient of rolling resistance (C_r). This simplification is reasonably accurate, provided the value of C_r is appropriate for the given tyre / road surface and condition.

For a typical mountain bike tyre and a road bike tyre, C_r is estimated at 0.012 and 0.005 respectively. This equivalent to the slope resistance from a uphill gradient of 1.2% to 0.5%. While rolling resistance is not as significant as air drag at higher velocities or a slope resistance from a gradient of 5%, it is a constant drag factor throughout the ride. Additionally C_r is consistent with the real rolling resistance, provided the tyre is inflated, maintained and not wobbling.

Assume a tyre on a bicycle is stationary and is at rest on some surface, then the tyre will be flattened at the area of contact with the surface due to the weight of the tyre and rider. The contact area is exaggerated in the Figure 4.2.

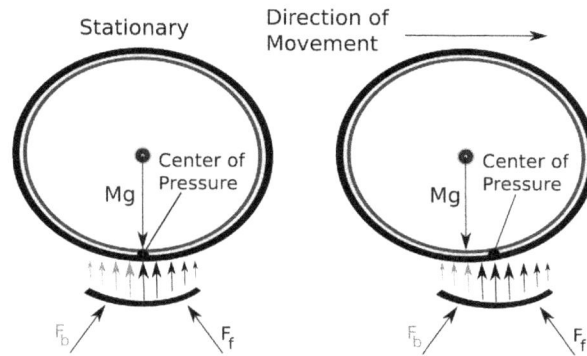

Figure 4.2: Vertical Forces on a Stationary and Rotating Tyre

Since a tyre has a curved surface, the reaction forces of the road are distributed to the left and right of a center of pressure. The forces to the left form a backward force F_b and the forces to the right form a forward force F_f. When the tyre is stationary, both F_b and F_f are equal and the center of pressure is located directly beneath the center of the hub. When the tyre is rotating in a clockwise direction, the force F_b produces a torque to rotate the tyre in clockwise direction, while the force F_f creates a torque to rotate the tyre in an anti-clockwise direction. When the tyre is stationary, the forces to the left and right of the center of pressure are equal, therefore the tyre does not rotate.

In the right of Figure 4.2, the wheel is rotating and moving from left to right. The center of pressure is shifted by a distance a from the center (see Figure 4.3). The force from the hub rotating clockwise is transferred to the wheel and tyre. However, the combined unbalanced forces F_f and F_b causes a resistive counter clockwise torque ($F_f > F_b$).

If $F_f - F_b = F_r$, then F_r acts at the center of pressure to create a counter clockwise torque ($F_r \times a$). The propulsion force F_p acts at a distance r from the axis of rotation. Since the tyre is not moving vertically, the weight of the rider and bicycle (Mg) must be balanced by the normal force of the road. Therefore $F_r = Mg$ and $F_p \times r = k \times F_r \times a$. The torque due to propulsion must overcome other drag forces as well as the rolling resistance and k is a constant that can be estimated based on the contact length and radius of the wheel (see Section 4.3).

In Figure 4.3 if a is 0.40 cms, i.e. the center of pressure F_r is located about a $\frac{1}{16}$ inch ahead of the center of the hub, and the value of $C_r = \frac{F_r \times a}{mg} = 0.004$. If mg, the weight force of the rider and bicycle is fixed, then a smaller a will reduce the rolling resistance since less opposing counter clockwise torque will be generated for the same weight force. Increasing the tyre pressure will reduce the area of the contact patch on the ground which in turn lowers the value of a. Other options include using a wide tyre which increases one dimension of the contact area while reducing the other dimension (a).

117

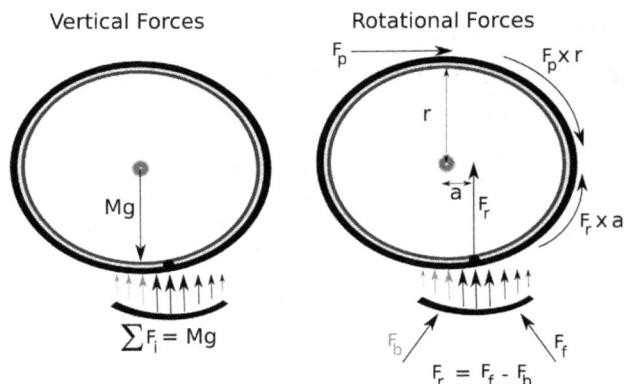

Figure 4.3: Vertical and Rotational Forces on a Rolling Tyre

Measuring Rolling Resistance (1)   The `bicyclerollingresistance.com` website tests the rolling resistance of both road and mountain bicycle tyres. The test simulates a road surface with a rotating drum at a steady speed, and a given load to compute the rolling resistance in watts. The power to overcome rolling resistance can be computed as $v \times Mg \times C_r$, where v is the velocity, Mg is the load, and C_r is the coefficient of rolling resistance. In the test, the load, velocity, and pressure were fixed at 42.5 kg., 28.8 kmph. and 120 psi respectively. If the watts reported for a tyre is 7.7, then the corresponding $C_r = \frac{7.7}{42.5 \times 9.8 \times 28.8 \times 0.27778} = 0.0023$. However if the air pressure dropped from 120 psi to 60 psi, then the watts lost due to rolling resistance increased from 7.7w to 10.9w or an effective C_r of 0.0032. While high air pressure may lower rolling resistance, the loss of power due to vibrations when riding on a more rigid tyre may offset the energy saved.

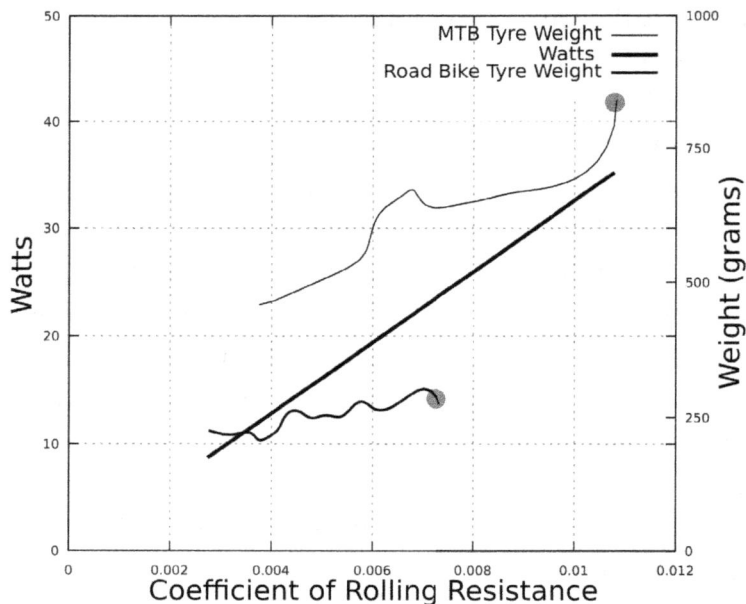

Figure 4.4: Watts and Weight vs. Coefficient of Rolling Resistance at 80 Psi for 92 Mountain and Road Bicycle Tyres

In Figure 4.4, the coefficient of rolling resistance and corresponding weight of tyres in grams is shown for 35 mountain and 57 road bicycle tyres[6]. The smoothed curves for both types of tyres show that the C_r for road bicycle tyres is < 0.008 while the upper limit for C_r of mountain bicycle (MTB) tyres is close to 0.011. The average weights of mountain and road bicycle tyres were 630 and 250 grams respectively.

Clearly, mountain bicycle tyres are heavier due to the larger width of about 2.25 inches compared to a width of 26 mm. (1.02 inches) for road bicycle tyres (the diameter of a MTB tyre was also greater at 29 inches compared a 27-28 inches for a road bicycle tyre). Even though MTB tyres are on average double the weight of road bicycle tyres, the corresponding MTB tyre C_r is not double the C_r for a road bicycle tyre.

From the data given on the website, a heavier MTB tyre did not always imply more watts consumed. A 583 gram tyre consumed more power at 31.4 watts while a heavier 815 gram tyre required just 22.4 watts. The same was true of road bicycle tyres. A 288 gram tyre required 12.3 watts while a 275 gram tyre needed 24.4 watts. Therefore, a heavier tyre does not necessarily imply greater road resistance.

Measuring Rolling Resistance (2) An earlier test in 2010 of rolling resistance conducted by Al Morrison [1] for biketechreview.com evaluated over 140 different tyres under the same conditions - a pressure of 120 psi at 50 kmph. under an identical load. The results did not indicate the weight of each tyre, but mentioned the width of each tyre evaluated.

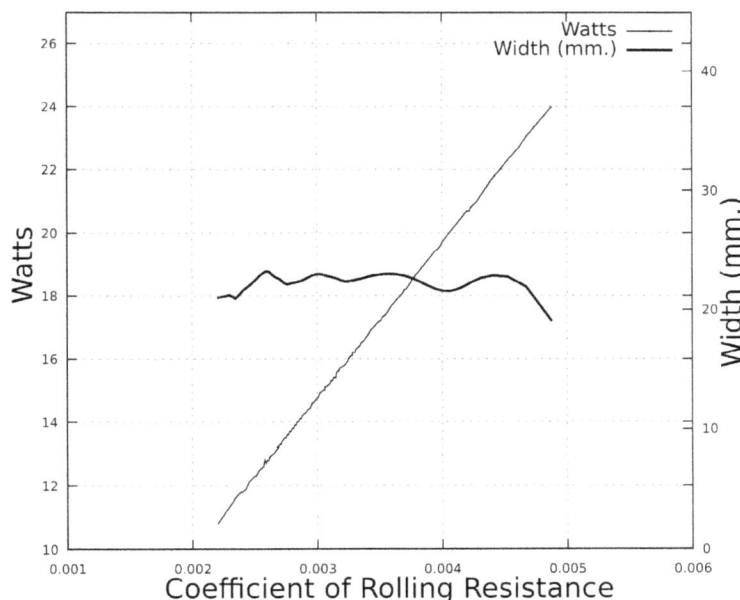

Figure 4.5: Watts and Width vs. Coefficient of Rolling Resistance at 120 Psi for 140 Bicycle Tyres [1]

---

[6]The data shown in the figure was collected from bicyclerollingresistance.com (7/2018).

In Figure 4.5, there is a linear relationship between the coefficient of rolling resistance and watts consumed. This is similar to the results in Figure 4.4 where the relationship between the C_r and watts is linear as well. However, the slope is steeper in the tests from biketechreview.com (52°) compared to tests from bicyclerollingresistance.com (35°).

While there is an obvious relationship between C_r and watts consumed, the relationship between C_r and width of a tyre is not apparent. For example in the test, a tyre of width 19 mm. required 24 watts while a wider tyre (27 mm.) consumed just 13 watts. Since there is no accurate formula to calculate the C_r of a tyre given the weight, width, and other parameters, it is hard to compare the efficiency of tyres without an independent test under similar conditions.

Measuring Rolling Resistance (3)   In another test, Velonews [2] tested 34 tyres at a laboratory in Nastola, Finland to evaluate rolling resistance. The width of the 34 tyres tested ranged from 23 mm. to 28 mm. and included tubeless, tubular, and clincher tyres. The conclusions of the test were that while wide tyres do roll fast, the tread compound used in the tyre was more important.

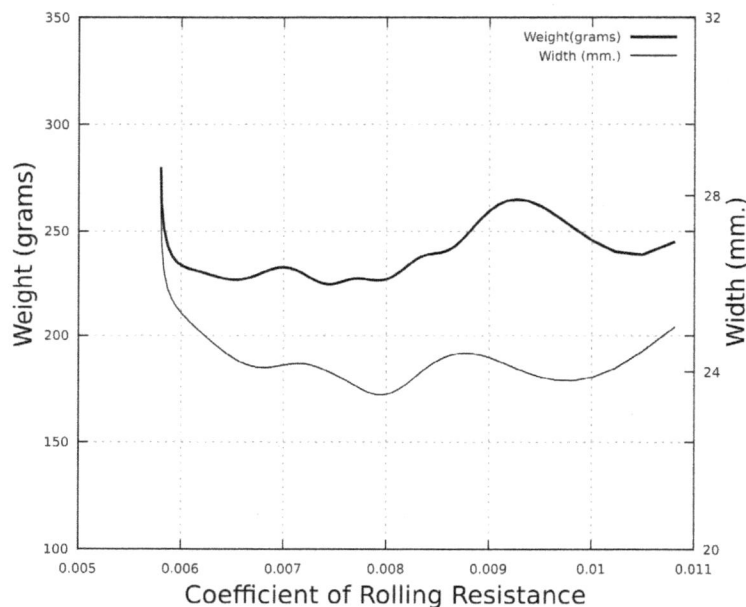

Figure 4.6: Watts and Width vs. Coeffcient of Rolling Resistance at 80 psi for 34 Bicycle Tyres[2]

In Figure 4.6, the coefficient of rolling resistance for 34 tyres shows no relationship with the width of the tyre or the weight of the tyre (the curves are smoothed). The tyre with the least C_r was 28 mm. wide and weighed 280 grams. There was no obvious relationship between C_r and the weight of a tyre. A heavier tyre weighing 318 grams had a C_r of 0.005 while a lighter tyre of weight 210 grams had a C_r of 0.009. The rubber compound used in tyres is a proprietary formula and manufacturers do not reveal the exact mix of compounds to create a tyre with low C_r.

Measuring Rolling Resistance (4)   The previous three methods of measuring rolling resistance were based on formal methods to evaluate the C_r for a tyre under standard conditions in a laboratory. The coast down method is a more approximate method that you can perform on your bicycle with a cyclocomputer.

The coast down method calculates resistance based on the time it takes a car to slow down from some initial velocity V_i to zero on a perfectly flat road. On a bicycle, the same method can be adopted with a modification to calculate the duration of the test. If V_i is 20 kmph., then the test would compute the time to slow down to a final velocity V_f of 5 kmph. on a flat road. The drag forces that slow down the bicycle on a flat road are the air drag and rolling resistance. The deceleration is equal to the sum of the drag forces.

$$m \times (\frac{\Delta V}{\Delta t}) = -(K_a \times V^2 + mg \times C_r)$$

where K_a is the aerodynamic factor (see Chapter 3). With an estimate of 0.3 for K_a, we can solve this equation for C_r. However, since the left hand side includes a differential of V, a numerical solution of the equation with given V_i and V_f values will return the time for the coast down test for a given C_r.

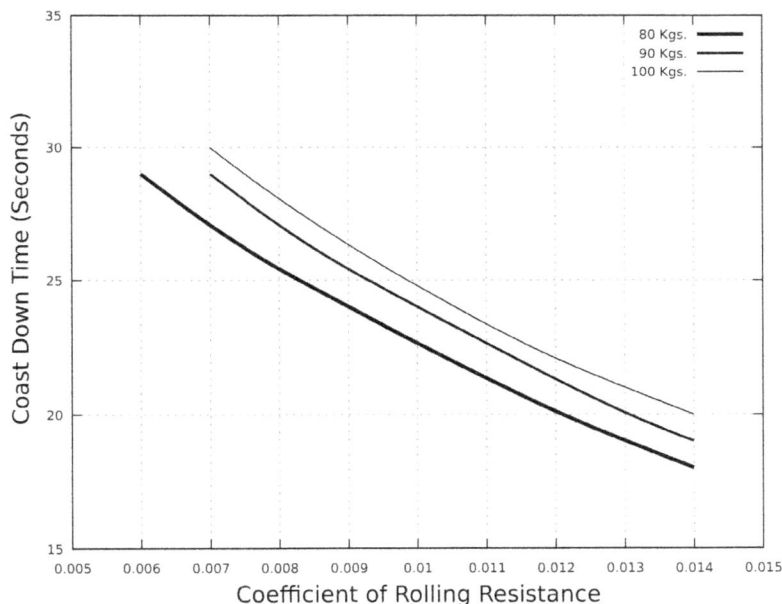

Figure 4.7: Coast Down Time vs. Coefficient of Rolling Resistance for Three Weights

In Figure 4.7 the estimated coefficient of rolling resistance is computed for three different weights - 80 kgs., 90 kgs., and 100 kgs. The weight represents the combined weight m of the bicycle and the rider. A higher C_r in all cases has a smaller coast down time. Logically, a heavier rider has a shorter coast down time compared to a lighter rider.

Compared to the previous tests, this a simple test with just a cyclocomputer and knowledge of the wind speed. In the Figure 4.7, a 5 kmph. head wind is assumed with a start velocity of 20 kmph. and final velocity of 5 kmph. The results can be very different depending on the surface of the road and the type of bicycle for the same weight since the accuracy of the test depends on manual timing results. Still this test should give a rough estimate of the C_r for a given bicycle, rider, and tyre. More sophisticated coast down tests use a variety of sensors to measure C_r with a much higher accuracy. A coast down calculator is available at `https://bit.ly/2KSZcIJ`. The calculator will return an estimated C_r, given parameters such as the wind speed, initial velocity, final velocity, an aerodynamic drag factor (K_a), total weight, and coast down time.

Measuring Rolling Resistance (5)   A website [3] from Engineering Toolbox does provide a formula to calculate the value of C_r from velocity and the pressure of the tyre. However, this formula does not take into account the weight or width of the tyre and may give a rough estimate of C_r.

$$C_r = 0.005 \times \frac{1}{p} \times (0.01 + 0.0095 \times (\frac{v}{100})^2)$$

where p is the pressure in bar and v is the velocity in kmph. Bicycling Science [4] uses a different formula - $0.0046 \times (\frac{100}{p})^{0.44}$ where p is the pressure in psi, gives a lower value for C_r.

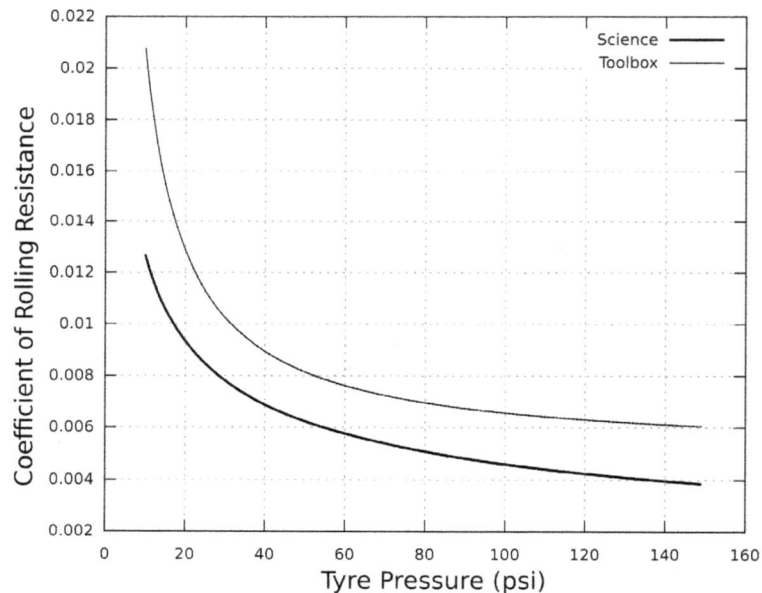

Figure 4.8: Calculating C_r with a formula

In Figure 4.8, both formulas give the same trend of low C_r with increasing pressure. After 80 psi, the reduction in C_r diminishes with still higher pressures. There are a number of other formulas [4] that one could use to calculate C_r and it is not clear which one would be the most accurate. A formula does not take into account the materials used in the tyre and the road conditions. Therefore, experimental results may be more accurate in estimating C_r than a formula.

4.2.2   How much speed do you gain by reducing rolling resistance?

Since tyres with a lower coefficient of rolling resistance will lower the drag due to rolling resistance, speed will increase as well for a given power and fixed drag due to air and slope. Consider a C_r of 0.005 that can be reduced to 0.0045, 0.0040, 0.0035, and 0.0030 with different tyres. In Figure 4.9, the increase in speed is higher when the speeds are lower (curves have been smoothed).

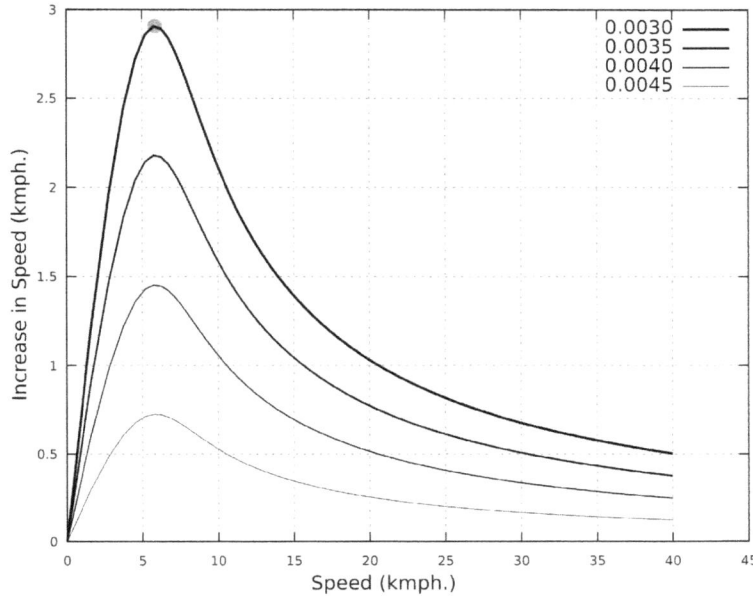

Figure 4.9: Increase in Speed for a given speed and lower coefficients of rolling resistance

At a speed of 5 kmph., using a tyre of $C_r = 0.0030$ instead of $C_r = 0.0050$, increased speed by about 3 kmph. Wilson[4] used the following equation to calculate the increase in speed for a change (reduction) in C_r.

$$\frac{\Delta V}{V} \approx \frac{-\Delta F_r}{3 \times K_a \times V^2}$$

where ΔV is the increase in velocity, V is the initial velocity, $-\Delta F_r$ is the positive change in drag due to rolling resistance from the lower C_r, and K_a is the aerodynamic drag factor (see Chapter 3). In the figure, a K_a of 0.30 is assumed for a commuter with a large front area and the weight of the bicycle and rider is 700 N. If K_a is reduced from a commuter to a racer position (0.11), then the increase in speed will be even larger since in the equation above, the increase in velocity is inversely proportional to K_a.

4.2.3   Minimizing Rolling Resistance

Not all tyre manufacturers list the coefficient of rolling resistance (C_r) when describing a particular tyre. The C_r is also just one of the factors to consider when choosing a tyre. Puncture

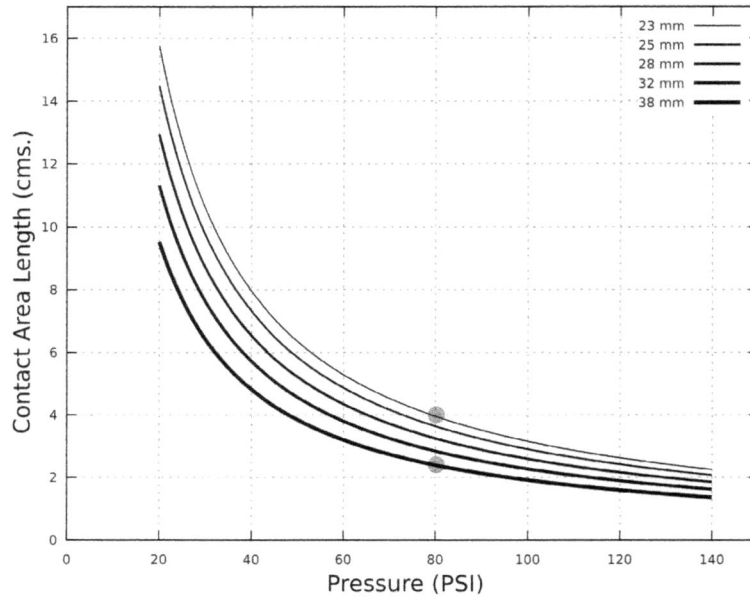

Figure 4.10: Contact Area Length vs. Pressure for Tyre Widths

protection maybe a higher priority than minimizing rolling resistance. A cost effective solution to reduce rolling resistance is to sufficiently inflate tyres. A formula to compute C_r for automobile tyres states that $C_r \propto \frac{1}{p}$ where p is the pressure. Higher air pressure within limits will lower C_r.

Since rolling resistance is proportional to tyre deformation and hence based on the shape of contact area, it would appear that a wider tyre has less rolling resistance. A test of narrow and wide tyres at different pressures has shown that a 60 mm. width tyre at a pressure of 29 psi has the same rolling resistance of a 37 mm. width tyre at 58 psi or twice the pressure [5]. However, most road bikes built for racing still use narrow tyres. This appears contradictory, but other factors besides the width of the tyre come into play when considering rolling resistance.

Road bike manufacturers are constantly looking for ways to reduce a bicycle's weight, since that is one of the primary concerns of many riders when evaluating a bicycle. If tyre width is increased while keeping the circumference the same, the weight of the tyre is certain to increase. While the increase in weight maybe measured in grams per tyre, this additional weight must be saved elsewhere in the bicycle. Narrow tyres will also have slightly less air drag than a wide tyre. However, the primary reasons road bike tyres are still relatively narrow compared to mountain bike tyres is the shape of the frame. A road bike frame is built to minimize air drag and has narrow forks that will only accommodate narrow tyres. A narrow road bike tyre can be inflated to a relatively high pressure to minimize rolling resistance, though the length of contact patch will be larger.

A longer contact patch increases rolling resistance since the elliptical shape of the contact area is less round. Because the length of contact patch of a tyre of width 38 mm. is shorter than the contact patch of a tyre of width 23 mm. at the same pressure, there is less tyre deflection and therefore less rolling resistance (Figure 4.10). At 80 psi and a total load of 80 kgs. or 40 kgs. per tyre, a 23 mm. wide tyre has a contact length of 4 cms., while a 38 mm. wide tyre has a contact length of 2.4 cms (a reduction of 40%). If the area of an elliptical contact patch is defined as $area = \pi \times a \times b$, where a and b are the semi-major and semi-minor axes respectively, then the length of the patch ($2a$) can be calculated from the width b and $area$ of the patch.

For a bicycle of weight 10 kgs. and rider of weight 70 kgs., the force directed down towards the road per tyre is 392 N. The pressure at 70 psi can be expressed as 482,633 N/m^2. The area of the contact patch is the force (392 N) divided by the pressure (482,633 N/m^2) or 8.13 cm^2. The length of the patch on a tyre is $2.0 \times \frac{area}{\pi \times b \times 1.25} = 4.14$ cm. for a 2.5 cm. (diameter) wide tyre.

As the tyre width increases, the contact area length reduces. This benefit does have its limitations. It would not be reasonable to have really wide tyres like the tyres of a formula one car. If the tyres are too wide, then both the weight and the air drag of the bicycle will increase, overriding any benefits from lower rolling resistance.

In Figure 4.11, the pressure of 70 psi is not uniformly distributed across the contact area for two tyres of width 38 m.. and 28 mm. The pressure is maximum at the center of patch and then diminishes based on the distance from the center of the patch. The equation to calculate the pressure at any given point on the patch is $70 \times \sqrt{1 - \frac{x^2}{a^2}}$ where a is the location of center of the patch and x is the distance from the center (Hertz pressure distribution). In reality, the pressure distribution for each tyre, maybe close to the shapes shown in the figure. However, the wider 38 mm. tyre has a more rounded pressure distribution than the shape for the narrower 25 mm. tyre. The pressure along the sidewall of the 38mm. tyre is less than the pressure on the sidewall of the 28 mm. tyre. These differences make it more easier for the wide tyre to roll and therefore rolling resistance is also likely to be less.

Rolling Resistance on a Wet Road   An initial guess would be that rolling resistance on a wet road is less than the rolling resistance on a dry road. This would seem to concur with the larger stopping distances when braking on a wet road (see Chapter 6). On a wet road, the coefficient of friction (μ) between a tyre and the road is about 0.35 compared to 0.7 on a dry road. Therefore, the frictional force to brake on a wet road is reduced by about half compared to the frictional force to stop on a dry road. The rolling resistance drag force is calculated like the frictional force as the product of C_r and the weight.

However, the values of the coefficient of friction and rolling resistance behave differently under wet and dry conditions. On a wet road, the temperature of a tyre can be lower by several degrees centigrade compared to the temperature of the same tyre on a dry road. The formula to calculate C_{r1} at a different temperature t_1 is $C_{r0}[1 + K \times (t_0 - t_1)]$ where C_{ro} is the coefficient of rolling resistance at a temperature t_0. The value of K, the temperature coefficient, depends on the type of tyre and can vary from 0.012 to 0.015 [6].

If the C_r at 25°C is 0.005 (for a clincher tyre) and K is 0.012, then the C_r at 20°C is 0.0053 or an increase of about 6%. However experimental results from a study [6] showed a much higher increase in rolling resistance on a wet road compared to a dry road. One explanation was that the water film thickness on the road had a larger influence on C_r than temperature alone and the increase was more significant at higher speeds. While rolling resistance is higher on a wet road, the drag forces due to air drag and splashing of water along with poor visibility on a wet road are more likely to slow you down.

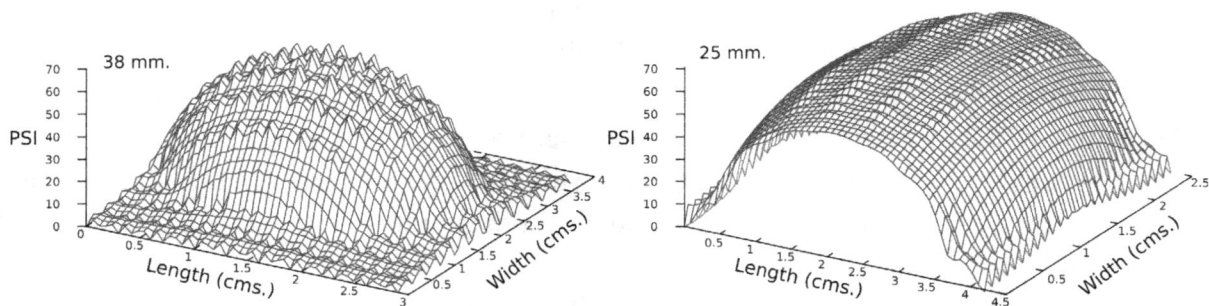

Figure 4.11: Pressure Distribution at Contact Area for Tyres of Width 38 mm. and 25 mm.

4.3 Tyre Width

Consider two tyres, one wide and one narrow, both inflated to the same pressure. Since pressure is computed as $P = \frac{F_W}{area}$ where F_W represents the weight force due to the bicycle and rider. For a given pressure, the size of the contact area with the ground must be identical. However, the shape and length of the contact area will be different (see Figure 4.12).

The length L_n of the contact area for the narrow tyre will be greater than the length L_w of the contact area for the wide tyre. Further, the contact area of the narrow tyre will have an elliptical shape, while for the wide tyre the contact area will be more circular. The deformation of the tyre occurs along the perimeter of the contact surface area. Since a circle has the shortest perimeter for a given area of all shapes, the deformation of the wider tyre will be less than the deformation of the narrow tyre.

You may not always be able to install a tyre of width that you consider optimal since bicycle frames are built with forks that will limit the width of a tyre. In general, road bicycle frames tend to have narrow forks and therefore the maximum width of the tyre that you can install is constrained by the width of the fork. This is one reason to choose a frame *after* you have decided on the type of tyre that you prefer.

There is also no rule that both the front and rear tyres must be of the same width. You could ride with a wider tyre in the rear and narrower tyre in the front as long as the tyres fit in the frame. A really wide tyre may interfere with the movement of the chain, but that would be unusual.

Figure 4.12: Contact Surface Area for a Narrow and Wide Tyre

Contact Area for a Bicycle Tyre and Train Wheel   A single bicycle tyre of dimension 27" x 1.25" with a load of 400 N inflated to 40 PSI creates a contact area of size 14.5 cm^2 and contact length of 1.45 cm. [4]. By contrast a steel train wheel of diameter 89 cms. on a steel track with a load of 27,000 N makes a contact area 80% smaller at 2.9 cm^2.

With this information, the value of the coefficient of rolling resistance C_r for a steel railway wheel can be estimated. The propulsion force F_p must overcome the drag from $\frac{F_r \times k \times a}{r}$ where k is a constant, F_r is the load, a is the distance of the axis of rotation from the center of pressure and r is the radius of the wheel. The expression $\frac{k \times a}{r}$ is the coefficient of rolling resistance. For a train wheel, the values of $k = 0.25$, $r = 0.5m$, and $a = 0.001m$ give a C_r of 0.0005. The C_r of a train wheel is ten times less than the C_r reported for a road bicycle tyre with the least rolling resistance.

4.3.1   Narrow or Wide Tyres?

Earlier recommendations to run a narrow tyre (width of 20 mm.) at the highest possible pressure have been superseded [7]. Tests in a laboratory showed that a narrow tyre was more efficient, but did not take into account the real world conditions where a cyclist would be subject to vibrations leading to a loss of energy and discomfort. If you ride on a tyre with a high PSI, then you will feel every bump which translates to a loss of energy. The force you generate on the pedal is used to vibrate the cycle instead of moving it forward. With a lower PSI, the tyre will deform over the bump and reduce the loss of energy.

Professional cyclists riding on smooth roads use tyres of width 25 mm. or more. However, the benefits of a wider tyre at higher ranges (> 30 mm.) will be reduced by the higher air drag and weight. The increase in front area due to a wider tyre is minimal - the difference in the front area between tyres of 30 mm. and 25 mm. widths for a tyre of diameter 700 mm. is about 0.005 m^2. However, the additional weight of a wider tyre (plus tube for a clincher tyre) will be noticeable.

If the width of a tyre is doubled from 23 mm. to 46 mm. with the same diameter, the material used for the tyre is also roughly doubled. Therefore if the weight of a 23 mm. tyre was 210 grams, then the weight of the 46 mm. tyre would increase to 420 grams. The total increase for two tyres becomes 420 grams. If you are riding at 35 kmph., then this weight corresponds to an increase in rolling resistance by $m \times g \times C_r' \times v$ (less than a watt). The value of C_r' for a heavier tyre is usually greater than the value of C_r for the lighter tyre.

4.3.2 Thick or Thin Inner Tubes

There is nothing more frustrating on a long ride than having a flat tyre after riding for an hour or so. When you would rather use your energy to keep riding, you are forced to spend time (>5 minutes) to remove the wheel/tube, replace the tube, and re-inflate the tube with a small portable pump. Fixing a flat is something a cyclist would prefer doing at home on a lazy Sunday afternoon and not in the middle of a ride.

Although not visible on a bicycle, the inner tube is as important as the tyre. Having an expensive tyre with an inner tube of poor quality will defeat the purpose of using a high quality tyre. A low cost inner tube will be made of cheap synthetic rubber and needs to be thick enough to prevent air leakage. The additional rubber will add to the weight of the inner tube and make it less elastic leading to a higher coefficient of rolling resistance.

The preferred inner tubes are made of butyl rubber and offer better protection than a latex based inner tube. A manufacturer may or may not mention a detailed description of the materials used in the tube. A high quality inner tube will have more elasiticity and therefore support a large range of tyre widths (a range of about 10mm.). Besides having more elasticity, a high quality inner tube will weigh less than its inferior counterpart. The price of inner tubes is roughly inversely proportional the thickness of the inner tube. Very thin tubes can withstand the same pressures as a thicker tube, but must be installed with care.

You would assume that a thick and heavy inner tube will provide the best puncture protection. Between a puncture and a slightly higher coefficient of rolling resistance, a cyclist would possibly prefer the latter. Losing some time on a ride may seem preferrable to having to deal with a puncture. However contrary to common perceptions, a thick inner tube does not offer better puncture protection [8]. Since a thick inner tube has less elasticity, it does not flex around a sharp object that may pierce the outer tyre.

The best inner tube must be lightweight, thin enough to supported high pressure, and elastic as well. While Butyl based inner tubes are lightweight and have high elasticity, some manufacturers use latex based tubes which can be made even thinner than butyl tubes.

4.3.3 How wide should tyres be?

If you can increase your speed by moving from 21 mm. tyres to 25 mm. tyres [7] then if your frame was wide enough, you could keep increasing the width of your tyre as much as possible to get the highest speed possible. Using 25 mm. tyres instead of 21 mm. tyres resulted in a 2.5% increase in speed. While not much, this relatively small gain can accumulate on a long ride of several hours and comes at little to no cost.

However at some point, the benefits of a wider tyre will be lost due to the additional weight. When the width was further increased from 25 mm. to 42 mm. tyres, again there was no noticeable change in speed [7]. Further, 42 mm. tyres were more efficient than 25 mm. tyres in rides on a rough surface. Other reasons to prefer a wide tyre include riding at a lower pressure (more comfortable with less vibrations) and possibly fewer punctures. A wide tyre (42 mm.) with a supple casing which will flex over bumps and potholes, works as well as a narrow tyre (25mm.) [9].

The maximum width of tyres that you can use will be limited by the width of the fork on the frame. Although, tyres upto 90 mm. are available on "fat bicycles", their use is mainly on sand and snow where the depth of the rut created by the tyre should be minimized. A deep rut means that energy is spent sinking vertically into the terrain. This fraction of energy wasted in sinking could be used to generate a forward friction force on a hard surface. Using a fat tyre means that you can ride with air pressure of 5 psi or less and the depth of the rut can be minimized to use energy efficiently.

4.3.4 Riding on Soft Ground

Although you will ride slower on a MTB tyre, the loss of speed is not as much as you would expect. One of the main reasons to prefer a MTB tyre over a road bicycle tyre is if you plan on riding on loose gravel or soft soil. Estimating C_r under such conditions depends on the soil strength. Riding on soft ground with a given weight will create a rut comparable to a briefly flattened tyre on a firm asphalt road. Here the tyre is assumed to be harder than the ground and therefore the wheel sinks into the ground. The estimated C_r under these conditions is the ratio of the area A_2 of the vertical cross section to the horizontal area A_1 [4].

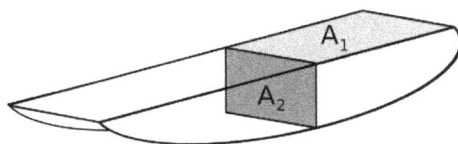

Figure 4.13: A rut formed from a tyre on soft ground

If you can minimize the area A_2, then you can minimize rolling resistance. Essentially, the bicycle should sink as little as possible into the soft ground. The deeper a wheel sinks into the ground, the more energy is lost coming out of the rut. If your tyre is wide, then you can ride at a lower pressure to support the same weight.

If your total weight (rider and bicycle) is 100 kgs. or 980 N, then the length of the rut (contact area length) at 80 psi with a 2 inch wide tyre is 1.11 cm., assuming an elliptical contact patch. For a 1 inch wide tyre, the length doubles to 2.22 cm. If the radius of the tyre is 13 inches (33 cms.) with contact length of 1.11 cms., then the depth of the rut s can be computed from the equation $s^2 + 2 \times 33 \times s - 1.11^2$ giving a depth of 0.0187 cms. When the contact length doubles to 2.22 cms., the rut deepens to 0.0745 cms.

Similarly, increasing the radius of the wheel also reduces the depth of the rut. Under the same conditions with a larger tyre of radius 14.5 inches (36.8 cms) and a width of 1 inch, the depth of rut reduces from 0.0745 cms. to 0.067 cms. Riding on a wider tyre does reduce the depth of the rut more than increasing the radius of the tyre.

Observations

- Race cars have wide tyres to handle the forces needed to corner at high speeds (see Section 5.5). If the contact patch of a race car tyre is not wide enough, then the centripetal forces generated on a curve would drag the car towards the center of the curve. On a bicycle, speeds are usually much lower and therefore there is no necessity to have a really wide tyre.

- You may already have a wheel and therefore the width of a tyre that you can use is constrained by the width of the rim. A general guideline is that tyre width should be between $1.45 \times x$ and $2.0 \times x$ where x is the inner rim width.

4.4 Road Vibrations

An initial guess for tyre pressure would be to inflate the tyre to the maximum allowable pressure. This would seem logical since the contact surface area for a given rider and bicycle weight would be minimized and correspondingly the contact length would also be minimized. In theory this is right, since a smaller contact area implies a smaller deformation of the tyre and therefore less loss of energy.

However with a rigid tyre at high pressure, the rider and bicycle absorb the vibrations from the road. It is rare to find a road that has zero roughness index (measured in millimeters per meter). This value is computed from ratio of the accumulated instantaneous vertical motion to horizontal motion. The vertical motion on a ride uphill takes minutes or longer while the vertical motion from a rough road is instantaneous.

Some of the smoothest surfaces (< 2 mm/m) are airport runways and super highways where speeds of over 100 kmph. must be tolerated with minimal vibrations. Cement and asphalt are the two common materials to make a surface. Depending on the climate, usage, and durability one or the other is used. Bicycle tracks are often located alongside a road for heavier vehicles such as cars and trucks. The same material (asphalt or concrete) is used for bicycle tracks. On a bicycle, asphalt may feel smoother than concrete since there are no joints. Regardless of the material, there will be some roughness on the surface that is more obvious when riding a bicycle than in a heavier vehicle like a car.

Bump Resistance Wilson [4] added a fourth drag force, bump resistance, in addition to the air drag, slope resistance, and rolling resistance (see Figure 4.14). Rumble strips are typically strategically placed to slow down traffic - you are likely to find rumble strips on a downwhill road before it bottoms out. For a larger vehicle like a car, these rumble strips are a gentle vibration that is not very noticeable. But on a bicycle, each rumble strip is jarring and vibrates both the rider and the bicycle. If the height of the strip is over an inch, then it takes some energy to grip the bicycle and retain balance.

Figure 4.14: Bump Resistance

On a rough surface, more energy is used to vibrate the bicycle and the rider, instead of accelerating the bicycle. Apart from the loss of energy due to vibrations, the ride is also uncomfortable. The addition of one or two suspensions (shock absorbers) does make a ride on a rough trail more comfortable, but will not eliminate the loss of energy due to vibrations.

Both bumps and its counterpart, pot holes, will slow you down. A bump of low height (< 1 mm.) will be barely felt on a bicycle with a well inflated tyre and will cause minimal energy loss. Research [4] at the US Army Tank Automotive Center correlated the rate of energy absorption with frequency (f) in hertz and amplitude (d) in millimeters.

Given a frequency (f) and amplitude (d), you can estimate energy loss per second in watts (see Figure 4.15). When frequency was between 1 Hz. and 5.5 Hz., power absorption by the body was estimated at $\frac{f^6 \times d^2}{1000}$ watts, between 5.5 Hz. and 9 Hz. it was $28 \times d^2$ watts, and between 9Hz. and 50 Hz. it was $\frac{f^{2.6} \times d^2}{10.75}$ watts. The watts lost due to vibrations greater than 5 Hz. increased sharply - for a 25 Hz. vibration of amplitude 5 mm., about 10,000 watts were lost. Since average bicyclists ride in a range upto a maximum of 1000 watts, only a limited range of frequencies is shown in the figure 4.15.

Low frequency vibrations have a much lower loss of energy - for a 2 Hz. vibration of amplitude 25 mm., just 40 watts were lost. A cyclist riding over a series of rumbler strips is likely to experience vibrations in these low frequencies and amplitudes. Without any head wind and on a flat, a loss of 40 watts translates to reduction of about 3 kmph. from 28 kmph. to 25 kmph. and another loss of 40 watts further reduces speed to 20 kmph. Therefore within 3-4 seconds, a cyclist riding at 30 kmph. can be slowed down to about half the speed. Given that a rider has no control over the roughness of the road, what is the optimal tyre pressure that would make the ride efficient (limit the loss of energy) and at the same time limit punctures.

Besides rumble strips, there are other objects on a road such as pebbles and debris that you maybe forced to ride over. At a high pressure, the tyre cannot roll over a pebble and instead the pebble gets deflected away from the tyre which makes the ride uncomfortable and bumpy.

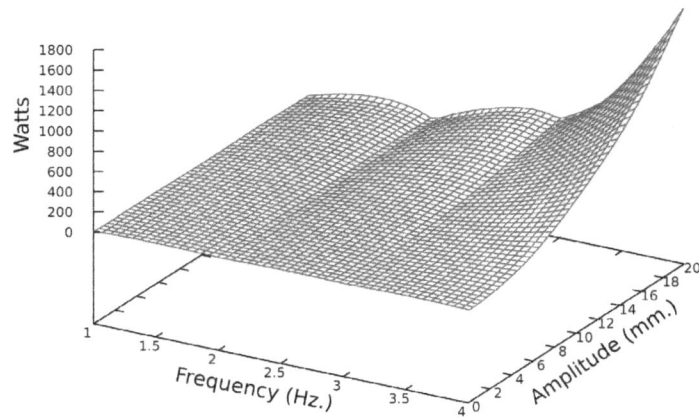

Figure 4.15: Watts Lost due to Vibrations vs. Frequency and Displacement

When the pressure is lower, the tyre rolls over small obstacles and acts like a suspension keeping the ride relatively smooth.

A tyre can be made rigid by either inflating it to a high pressure or if it is made of rubber / tread that does not absorb energy. Narrow road bicycle tyres tend to be stiffer since they have to withstand higher pressures and therefore vibrations from a ride on a rough road will be transmitted to the rider.

Observations

- Wilson [4] recommends a rider stay low before riding over a bump and then rise (accelerate upwards) when the bicycle is on the downslope.

- You will lose more energy if your body is rigid when riding over a bump. A flexible upper body will absorb vibration energy at a lower frequency compared to a stiff body.

- If a bump is steep enough and your speed is high enough, then the wheel may briefly lost contact with the ground. When the wheel regains contact with the ground, the impact will cause vibrations to the wheel that will be transmitted to the rider.

- If you are riding on a rough surface, then you can get out of the saddle temporarily when you ride over a sizeable bump and let the bicycle frame vibrate without your weight on the saddle.

4.5 Tyre Pressure

Pressure[7] is inversely proportional to area and therefore a higher pressure reduces the contact area and correspondingly the rolling resistance. In theory you could increase the pressure such

---

[7]A pressure gauge measures the pressure x inside the tyre relative to air pressure ($14psi$). The absolute pressure in the tyre is $x + 14psi$.

that the contact area becomes a point and rolling resistance is at its minimum. It is not unusual to find tyres and tubes that will handle pressures of 100 psi or more. The material of the tyre and tube must withstand the high forces generated from the air pressure. Pressure of 100 psi is equivalent to being submerged under over 70 meters of water. At about $600\,psi$, a $2\,cm.$ wide tyre would need a contact area of $1.0\,cm.^2$ to support $85\,kgs$. Since this is not practical, what is the optimal pressure for a tyre?

Avoiding Punctures   While you may not gain much by inflating your tyre to the maximum allowable pressure, you stand to lose a lot if you underinflate a tyre. The risk of a puncture is high from an object like a sharp pebble (see Figure 4.16). When the air pressure in the tube of a clincher tyre is low, a sharp rock can press the tube against the rim of the wheel. The tyre with higher pressure does not deform as much as the tyre with low pressure and therefore the tube with high pressure will not press against the rim.

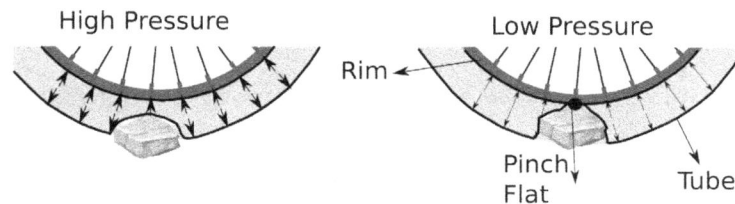

Figure 4.16: A Pinch Flat due to Low Air Pressure

The rim of the wheel has sharp edges and if pressed hard enough against the inner tube will make a hole (a puncture). Since the rim has two edges, you may get two holes side by side. A flat tyre is not what any cyclist would desire since you are forced to stop riding. Therefore maintaining air pressure that is sufficient to avoid pinch flats is what a sensible cyclist should ensure.

4.5.1   Volume of Air in a Tyre

If you change from a narrow tyre to a wider tyre, then you do need to adjust the tyre pressure accordingly. The volume of air in a tyre can increase significantly as the tyre width increases. In Figure 4.17, the volume of air in tyres of three different diameters at widths from 23 mm. to 70 mm. show that the air volume can increase substantially with a wider tyre.

At a width of 25 mm., a 700c tyre has a volume of about 1.0 liter of air. This is the volume when the pressure (about 14.5 psi or 1 bar) inside and outside the tyre is the same. When this width is doubled to 50 mm., the volume of air increases to over 4 liters. The shape of a tyre is assumed to be a torus for which the volume is computed as $2\pi R \times \pi r^2$ where R is the radius computed from the center of the wheel to the inner side of the tyre plus the radius of the tyre and r is the radius of the tyre. Increasing the width from 25 mm. to 36 mm. more than doubles the volume of air. Tyres with smaller diameters will logically have lower volumes of air than a

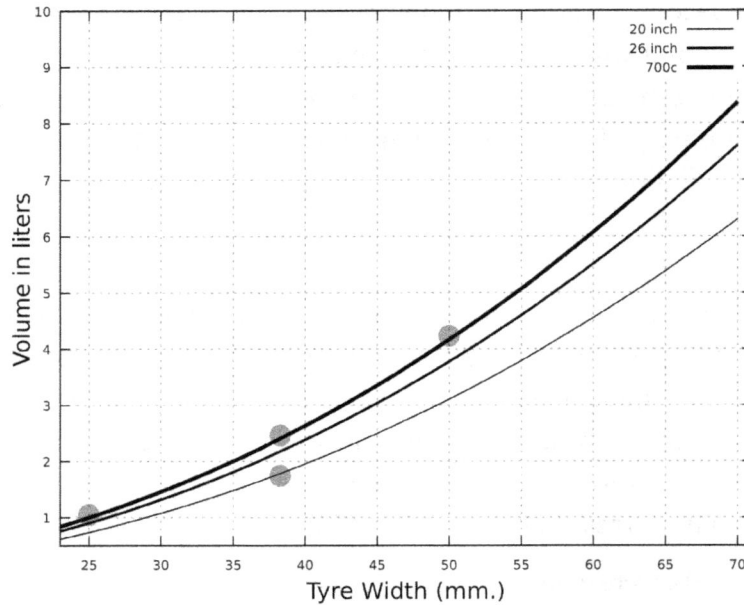

Figure 4.17: Volume in Liters vs. Tyre Width for three tyres of different diameters

larger diameter tyre. For a tyre of width 38 mm., a 20 inch diameter tyre has a volume of 1.75 liters, while a tyre of the same width but with a 29 inch diameter has about 0.6 liters more at 2.35 liters.

4.5.2 Pressure of Air in a Tyre

As you increases the pressure of air in a tyre, you are actually increasing the volume of air in the tyre as well. The pressure increases since a larger volume of air must fill a smaller tyre volume with the higher PSIs. Therefore, you do not need to inflate a wider tyre to the same PSI as a narrower tyre to maintain the same volume of air. If a 700c tyre of width 23 mm. is inflated to 100 psi, then at 25 mm. the effective psi is 84 psi and effective pressure drops rapidly to 36 psi at a width of 38 mm. (see Figure 4.18). Similarly at 70 psi for a tyre width of 23 mm., the effective psi is almost halved when the width is increased to 32 mm.

What is the ideal tyre pressure? On the sidewall of a tyre, you can find maximum air pressure recommended by the manufacturer. However, these values are limits and you don't need to inflate your tyre to the maximum pressure for an efficient ride. In a test to evaluate the difference in energy loss at the maximum recommended air pressure and 2/3 the pressure, for a flexible (supple casing) tyre, the same amount of energy was lost [10]. Three tyres of the same width were as fast at 70 psi as they were at 130 psi.

The ideal tyre pressure lies somewhere between the maximum allowable pressure and the minimum pressure to avoid punctures. This ideal pressure is unique to the combined weight of

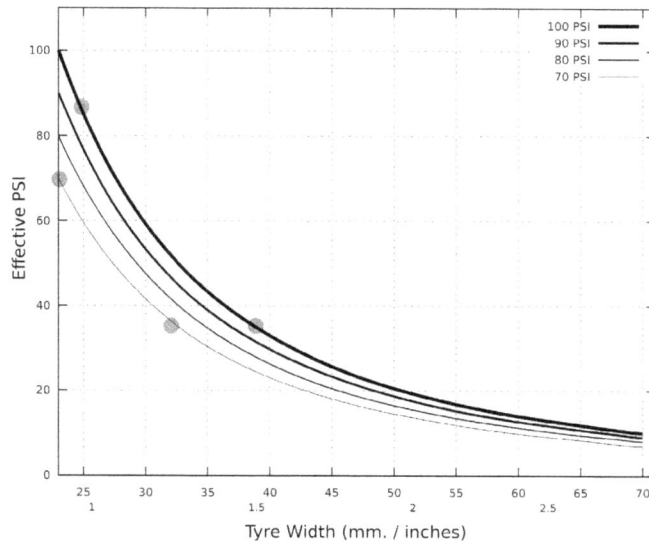

Figure 4.18: Effective PSI vs. Tyre Width for four PSIs on a tyre of width 23 mm.

a cyclist and the bicycle, the type of tyre, and the terrain. A tyre with a rigid casing does not sink as much as a tyre with a more supple casing and will ride with a lower pressure.

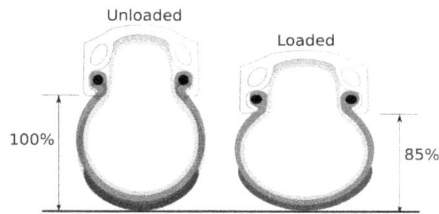

Figure 4.19: Tyre Drop Percent for Loaded and Unloaded Tyre

The recommended air pressure is based on a 15% tyre drop (see Figure 4.19). Tyre drop is the amount the bicycle is lowered when loaded compared to when unloaded due to the tyre deformation. In the Figure 4.19, the tyre drop is 15%. However, accurately measuring tyre drop is hard. Frank Berto measured the optimum pressure for a 15% tyre drop for a 700C wheel with tyres of different widths and weights.

Heine [11] reports that inflating tyres to a pressure that will obtain a 15% drop between a loaded and unloaded tyre is the optimum pressure. An under inflated tyre will have a large contact area and therefore a higher coefficient of rolling resistance (C_r). As a tyre is inflated, this contact area reduces and C_r also reduces. But, the benefits of higher pressure taper off beyond an "optimum pressure" for a given rider and bicycle due to the energy loss from vibrations.

Since your weight is not distributed evenly across the bicycle (more weight on the rear wheel than the front wheel), inflating the rear tyre and the front tyre in the ratio of 60:40 is one possible combination. If the total weight of the bicycle and rider is 100 kgs., then you would assume that 60 kgs. and 40 kgs. are supported by the rear tyre and front tyre respectively. Pressure based on this weight and tyre width from Frank Berto's chart [12] would be a reasonable starting value. However, this would depend on whether you have a load in the front in which case the ratio would be closer to 50:50.

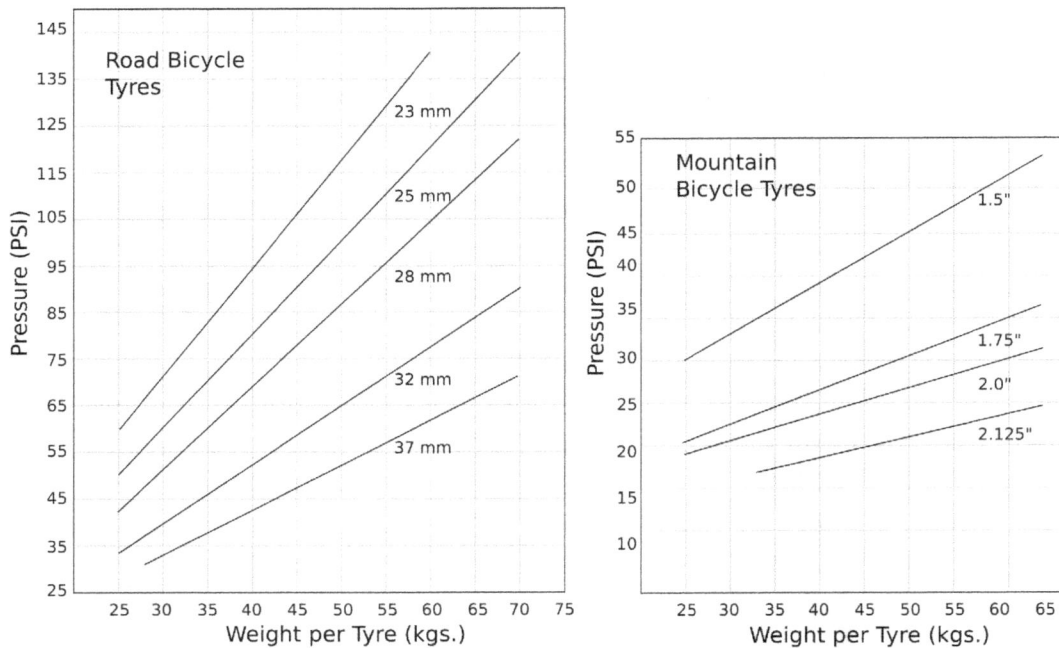

Figure 4.20: Frank Berto's Chart for Tyre Pressure vs. Weight for Road and Mountain Bicycles for tyres of multiple widths

The chart on the left of Figure 4.20 shows the recommended pressure for a 15% tyre drop based on the weight per road bicycle (700c) tyre. If the weight on a tyre is 50 kgs. and the width of the tyre is 32 mm., then the recommended pressure is 65 psi. The chart on the right shows the same data for a mountain bicycle (MTB) tyre of diameter 26 inches and widths from 2.125 inches to 1.5 inches.

However, the chart for the MTB tyres is the *minimum* recommended pressure and not the optimum pressure. Since MTB tyres have knobs on the surface, it is harder to spot the pressure for a precise 15% tyre drop. In both charts the optimum / minimum psi increases more rapidly (higher slope) for narrow tyres than for wide tyres.

In general, there is a tendency to over inflate wide tyres and under inflate narrow tyres. Manufacturers of tyres would like to accommodate riders of a wide range of weights and therefore the maximum pressure written on the sidewall of a tyre is probably higher than what you need, if you are of an average build. Further, recommended values of tyre pressure will differ between manufacturers and the chart in Figure [12]. For a total weight of 85 kgs. on a road bicycle tyre, one manufacturer recommends 95 psi (without considering the width of the tyre) while another recommends 105 psi for a 28 mm. tyre. The chart recommends a lower 75 psi (a weight of 42.5 per tyre and 28 mm. width).

If a tyre is overinflated, then the tread in the center of the tyre that supports most of the load and will wear out sooner. In an underinflated tyre, more of the load will be supported on the sides of the tyre. In both cases of underinflation and overinflation, tyres will wear out faster than if they were correctly inflated. An additional problem with underinflated tyres is that the likelihood of a puncture is higher. An underinflated tube will deform with every bump on the road and may get stuck between the rim and the tyre leading to a "pinch flat".

Finally, most cyclists do not inflate their tyres on every ride. If you inflate your tyre to a recommended pressure once in two weeks, then during the rest of the period, you will be riding on a slightly lower pressure, since all tyres lose pressure over time. For most practical purposes, you want to sufficiently inflate your tyre to avoid a puncture. Common wisdom is that slightly higher or lower than recommended pressure on a smooth road will not make much difference and is a matter of preference. On a rough surface, you can ride more comfortably at a lower pressure, provided the pressure is not low enough to cause punctures. If you are riding uphill, your speed will be relatively low and therefore maintaining traction will not be a significant issue. On a downhill, you can reach high speeds and therefore a lower pressure will give better traction.

Since there is no "ideal" pressure (it depends on the rider and the terrain), Frank Berto's chart is a reference based on your weight and tyre width to get an initial guess and then you could add ±15% of that value to find the value that works for you.

Observations

- If you know that your route will be on a smooth surface, then you could ride on a tyre inflated to a high pressure and not worry about the loss of energy from vibrations. On a rough surface, you would be better off riding at a pressure less than the maximum allowable pressure.

- Air pressure in the tyre increases by about 2 psi for every 5°C increase in temperature. Constant braking on the rim of the wheel can raise the temperature of the rim which in turn increases the temperature of the air in the tyre.

- A heavier person will have to be more conscious of air pressure since more weight on a tyre does lead to greater deformation of the tyre and therefore a greater chance of a puncture.

- More expensive tyres tend to use a flexible lighter material like Kevlar for the beading than a somewhat rigid steel wound cable that keeps the tyre and rim connected. Tyres with Kevlar are lighter and easier to remove / install. The casing of a Kevlar tyre maybe more supple and therefore you would need a higher minimum air pressure to prevent a pinch flat.

4.6 Tyre Diameter

Bicycle wheels and tyres come in a wide range of sizes. The diameter of a bicycle tyre can range from 16 inches to 29 inches. The dimensions of a tyres are typically expressed in inches or millimeters [8]. Naturally, you have to verify that the beading of the tyre will fit in the wheel, since this is critical to keep the tyre in place on the wheel. The tyre diameter in a specification is the outer diameter, while the inner diameter (beading diameter) must perfectly match the diameter of the wheel.

4.6.1 Small Wheels

At first, the argument for using a wheel with a small diameter appears counter intuitive. Conventional wisdom assumes that bigger wheels roll better than smaller wheels and therefore will be faster. However, bicycles with small wheels have large cranks that can easily compensate for the smaller circumference of the wheel. So, a single rotation of the large crank may rotate the smaller wheel three times or more.

Small wheels work very well when the road is smooth - some of the fastest speeds on a bicycle have been set on wheels smaller than the common wheel of diameter 559 mm. Although you can ride at the same speed as a "normal" bicycle, you have to ride at a higher pressure, since the volume of air in a smaller tyre is less than the volume of air in a typical 26 inch tyre. This can mean an uncomfortable ride on a rough road - a good suspension can handle the bumps on the road and will add to the overall weight of the bicycle. The spokes on a smaller wheel are also shorter and therefore may have more tension to handle the weight of a rider.

The other arguments in favor of smaller wheels include easier acceleration, maneuverability, compact size, and less air drag. It takes more power to begin riding a bicycle with a wheel of large diameter. The quick acceleration of a small wheel bicycle is useful when riding in a city where you may need to stop and start many times during a ride. Since the wheel is small, it is easy to make sharp turns. With smaller wheels and a frame that can be folded, it is easier to make a bicycle that can be transported and stored. Riding on smaller wheels will reduce front area since your center of mass maybe lower and therefore air drag will be less. This can make a difference at high speeds.

4.6.2 Large Wheels

On a rough surface with pot holes and bumps, a large wheel will give a smoother ride than a small wheel. While the addition of a suspension can reduce vibrations in a small wheel bicycle, there is a higher likelihood of the bottom bracket or a pedal striking the ground since the clearance at the bottom of the bicycle is low.

---

[8]The units for dimensions are different since bicycles were manufactured worldwide and the local standard unit was used to express the diameter of a wheel or tyre.

For a larger person, riding on a large wheel means a less bumpier ride. At the same pressure, the volume of air in a 29 inch tyre of width 38mm. is roughly 2.35 liters compared to 1.75 liters (25% lower) in a 20 inch tyre of the same width. With a larger volume of air, the pressure can in a large tyre can be reduced for a smoother ride. Riding on a tyre with a lower pressure means that the bumps on the road will not be as noticeable and a suspension may not be necessary.

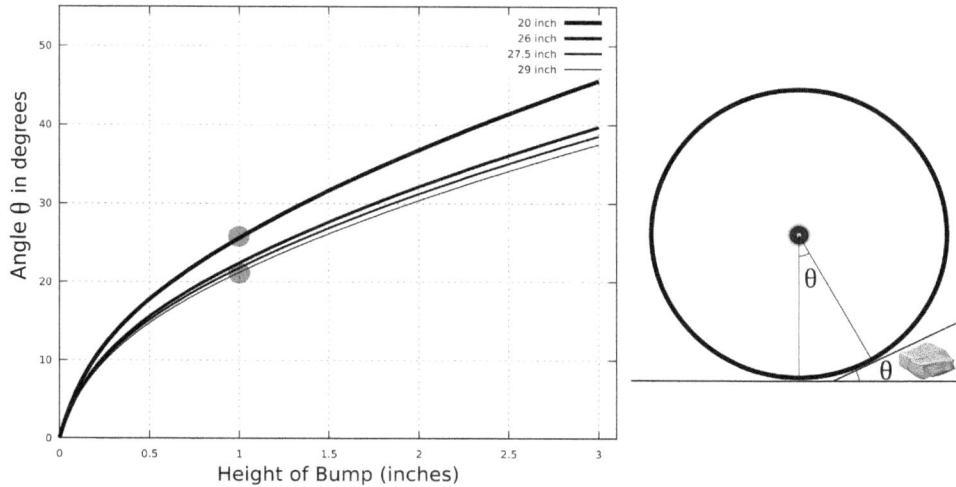

Figure 4.21: Angle of Approach vs. Height of Bump for 20", 26", 27.5", and 29" Tyres

If you ride on a tyre with a large diameter, then you will approach a bump at a lower angle. In Figure 4.21, the angle of approach is computed assuming a rigid tyre. On a 20 inch tyre, you would approach an object of height one inch at about 25° and on a 29 inch tyre, the angle would be about 21°. With higher objects the angle becomes steeper more rapidly on smaller tyres than larger tyres. Bumps in reality are more likely to have rounded than sharp vertical faces and the angles in Figure 4.21 are higher than what you would experience.

Since a small tyre must be inflated to a high pressure to support the weight of the rider, the tyre will not compress much at the point of contact and therefore riding on the bump will lead to a larger vibration than a larger tyre inflated to a lower pressure. Therefore a smaller wheel bicycle like the Moulton uses a full suspension.

Comparing Large and Small Wheels   You would assume that large wheels would weigh more than small wheels. However, there are a wide range of wheels at different price ranges and the more expensive wheels made of carbon fiber may have the same weight as the weight of a small diameter wheel. Similarly, an expensive hub of a wheel can also be practically noiseless and light as well.

Accurately calculating rolling resistance outdoors is hard to accomplish because of the many factors that affect the speed of the bicycle including wind, terrain, and type of tyre. Therefore most tests for rolling resistance are conducted indoors under identical conditions with a rotating drum to simulate the load. John Lafford [13] tested over 200 tyres for wheels of various diameters ranging from 305 mm. to 622 mm (16 inch to 29 inch tyres). He calculated the coefficient of rolling resistance at different pressures and the power consumed at different speeds.

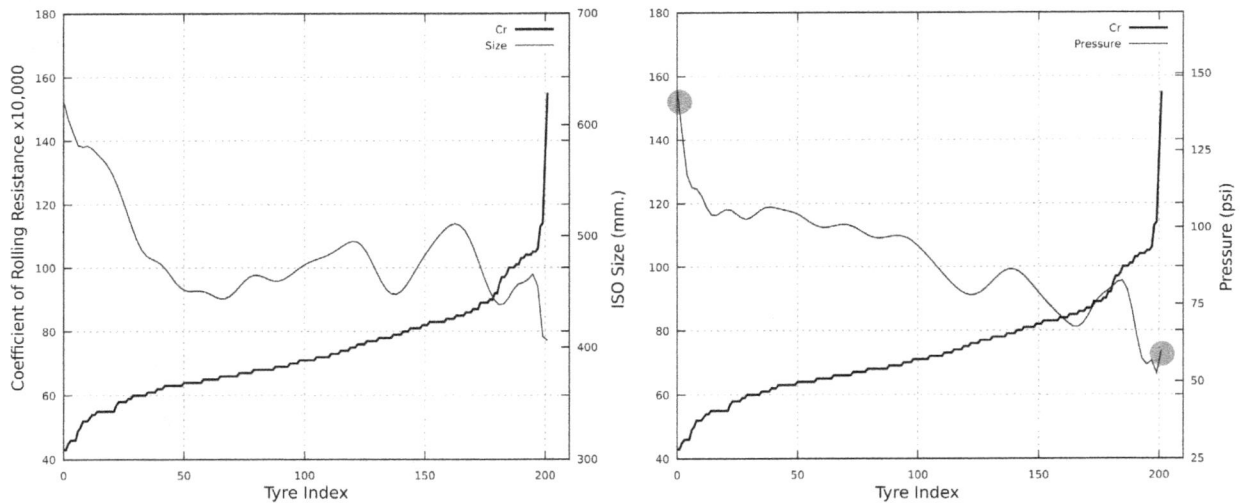

Figure 4.22: $C_r \times 10,000$ for Tyres of different diameters and pressures

In Figure 4.22, the test results for the list of tyres is sorted in ascending order by the coefficient of rolling resistance. As C_r increases from 40×10^{-4} to 150×10^{-4}, the size of the tyres for wheels reduces from 622 mm. to below 400 mm. The data from the tests have been smoothed to show the trend in lower rolling resistance for tyres with larger diameters.

There is a similar trend with air pressure and C_r in Figure 4.22. When the pressure is near 150 psi, the coefficient of rolling resistance is the least. As pressure drops to near 50 psi, the value of C_r increases to near 150×10^{-4}. This is consistent with the theory that on a smooth surface, higher pressures will give lower rolling resistance. Since these tests were conducted in a laboratory, the losses due to vibrations on a real surface with irregularities is not taken into consideration.

However, the trend is not consistent - a tyre inflated to 130 psi had a higher C_r (0.01) compared to the C_r (0.0068) of another tyre inflated to 80 psi. Similarly, the C_r (0.0155) for a tyre on a wheel of diameter 406 mm. had almost twice the C_r for another tyre on a smaller wheel of diameter 305 mm. Despite these anomalies, the trend of lower C_r for larger diameter tyres holds.

Observations

- Interestingly, wheels on a motor cycle may have a smaller diameter ($15 - 17$ inches) compared to the diameter of a bicycle wheel. While the wheel diameter maybe smaller, the width and height of a motor cycle tyre is significantly larger than that of a bicycle tyre to handle the heavier load.

- Getting on and off a bicycle with a large wheel is not difficult since you can tilt the bicycle.

- It is easier to get replacement parts for tyres and wheels of a popular diameter (26 inch).

- If you are forced to ride over an object slightly larger than the width of your tyre, your tyre will ride over a part of the object and the object is likely to fly off in a random direction. This is more likely on a tyre with high pressure. The reaction force may also move your tyre laterally and you may need to steer the bicycle to maintain balance.

4.7 Rough or Slick Tyres

A typical mountain bike tyre has knobs (protrusions) of various shapes along the surface of the tyre. These knobs interlock with irregular features on a trail. Knobs come in a number of shapes and sizes for different conditions [14] - the knobs for a muddy trail may not be appropriate for a gravel trail.

If you know that you will only be riding on a relatively smooth surface like asphalt, then slick tyres will give better traction. Although, these tyres have a smooth surface without any treads, they have more traction than a treaded tyre. This appears counter intuitive, because a smooth tyre surface may imply that your tyre is more likely to slip due to lack of friction with the road. However having more of the tyre surface in contact with the road actually improves grip. An average cyclist riding on a straight path cannot generate the power required to make a slick tyre slip on a smooth road.

Contact Area   A slick and treaded tyre of the same width, diameter, pressure, and rubber compound will make contact areas of about the same size. Although the shape of the contact areas may not be identical, traction is greater with a slick tyre. At a microscopic level, an asphalt road has small irregularities that are not obvious. So, the entire contact area of the tyre and the road are in not in contact.

In Section 4.3, a wider tyre had less deformation because of a more roundish contact area compared to a narrow tyre. The smaller perimeter of the round shape represents the deformation on the edges of the contact area. Within the inner part of the contact area, there are differences in how a slick and knobby tyre make contact with a surface.

In Figure 4.23, the shapes of the contact areas for slick and knobby tyres are not identical on both smooth and rough surfaces, even though the width and pressure of the tyres maybe identical. The slick tyre has a longer contact area on smooth and rough surfaces (the figure is simply a representation of how a tyre may make contact at a magnified level). The longer contact area of the slick tyre means that the vertical deflection of the tyre will be less than the deflection of a knobby tyre with the same pressure. Since rolling resistance is proportional to the degree of tyre deformation, the slick tyre will have less rolling resistance than the knobby tyre.

A slick tyre is often mistaken for a *bald tyre*. Any tyre that has been used for long will lose rubber on the surface due to wear and tear. Once a tyre has lost tread and knobs have all but disappeared, the thickness of the rubber in the tyre is less than that of a new tyre. With less

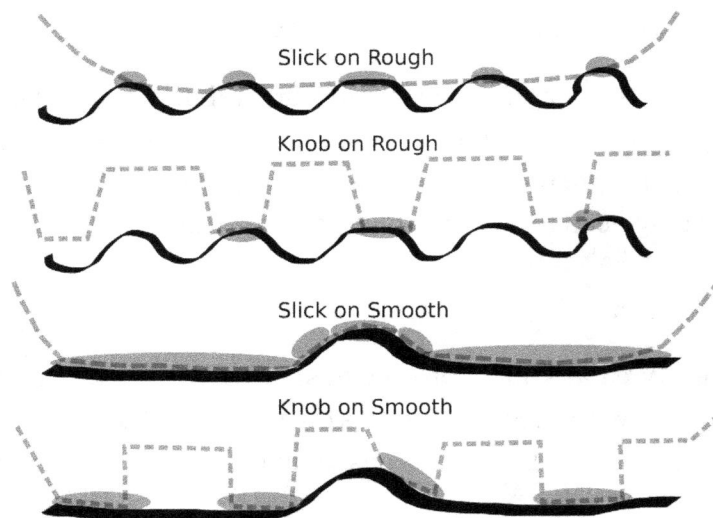

Figure 4.23: Contact Areas for Slick / Knobby Tyres on Smooth / Rough Surfaces

rubber to protect an inner tube, you are more likely to have a puncture with a bald tyre. A new slick tyre has a smooth surface, but the amount of rubber in the tyre is the same or more than the amount of rubber in a similar treaded tyre. Therefore, you are *not* more likely to have a puncture with a slick tyre, at least till the tyre has been used for some time.

If knobs do not make as much contact with a rough or smooth surface, then there would appear to be no reason to use them on a tyre. Knobs on a tyre do give the appearance of ruggedness and durability, which may encourage a cyclist to ride on any surface. Beyond the psychological motivation, appropriately shaped knobs do help to dig into an irregular surface [14] and finding a tyre with right knobs is a subject of research for the rider. Since large knobs do not give as much contact, the surface of a cross trail or hybrid bicycle tyre may have smaller knobs to ride at a reasonable speed on both a smooth and rough surface.

The tread on a hybrid tyre maybe smooth in the center of the tyre that makes contact with the surface of the road when riding straight. On either side of the central strip, there maybe other strips with fine ribs to interlock with irregularities on the road when cornering. The tread on the edge of the tyre that does not make contact with the road maybe used to protect the tyre from objects on the road.

Observations

- Using slick tyres in wet weather may appear risky, but in reality a bicyclist cannot generate the power to hydroplane, which occurs when the tyre is sliding on a film of water with loss of traction and control. However, if you are using rim brakes, then wet weather will limit braking torque due to a wet rim and therefore is a greater risk than skidding due a wet slick tyre. Knobs do not help braking in wet weather since the squishing of the knobs when braking on a wet surface do not give greater traction than a slick tyre.

- On a wet surface with mud, knobbed tyres can out perform slicks. Matej [15] has summarized his experiences with riding on several slicks and knobby tyres on a variety of surfaces.

- Rotating tyres, i.e. exchanging the front and rear tyres to maintain uniform wear and tear is fairly common in automobiles. On a bicycle, the rear tyre will wear out sooner than the front tyre, since the rear tyre bears a larger fraction of the load. One reason not to rotate bicycle tyres is to maintain a front tyre that is in good condition to brake and corner safely.

- If you observe your tyre rotating and you see the surface of the tyre bobbing up and down, you may have uneven wear / tear. You may need to remove the tyre and check if the inner tube is seated properly or if any objects are between the tube and tyre.

4.8 Types of Tyres

The ideal tyre should be light, have a low coefficient of rolling resistance, and be puncture resistant. However, finding a tyre with all these features is difficult since having a puncture resistant tyre may mean making the tyre thicker and therefore heavier. A tyre with a lining made up of kevlar or some other similar material can alleviate this problem, but may also add to the cost of the tyre.

The most common of tyre is the clincher – an inner tube is inflated within a tyre that is attached to a wheel with rims to keep the beading of the tyre in place. Clinchers are popular since they are easy to maintain. You can fix a puncture by simply replacing the inner tube or patching the existing tube.

Tubeless tyres are similar to clinchers except that air in the tyre is inflated instead of the inner tube. Without an inner tube, the sides of the tube and wheel must be sealed perfectly to avoid any leakage of air. The sealant must be checked for leakages over time and fixing a puncture on a tubeless tyre is more consuming than on a clincher tyre.

Although tubeless tyres are increasing in popularity, they have yet to replace the clincher tyre. The major benefits [16] of a tubeless tyre are

- Less rolling resistance: The friction between the surface of the inner tube and the inner surface of the tyre is absent.

- Less weight: The tubeless tyre will be weigh slightly less than a clincher tyre with an inner tube.

- More puncture protection: A clincher tyre can be made more puncture proof with a thicker inner tube, tire liner, or a protection belt within the tyre. All these modifications add to the weight of the tyre. Tubeless tyres have puncture protection liquid to seal a hole and do not lose air in a sudden burst.

The third type of tyre is the tubular tyre that is more common on bicycles used in races. It has a lower coefficient of rolling resistance but is more complicated to install than a clincher tyre. The tyre is glued on to the rim of the bicycle and installation is more tedious.

The best tyre for you is based on your requirements and a clincher tyre is a safe initial tyre. How much do you gain from the additional complexity of a tubular tyre? Curt Austin's bicycle calculator assigns default C_r values of 0.005, 0.004, and 0.012 for clincher, tubular, and mountain bicycle tyres respectively. Other web sites give different values for C_r depending on the surface (concrete, paved, or off trail).

4.8.1 How much will the use of tire liners slow you down?

If you ride on roads that are pot holed and have debris, then there is a strong likelihood of ending up with a puncture. As any bicyclist will tell you, a flat tyre is the end of the ride and can be frustrating when you don't have a spare tube, pump, or cartridge. Adding a tire liner is a precaution that can prevent punctures. A tire liner is made up of polyurethane [17] that keeps debris from puncturing the inner tube of a tyre. The tube is installed on the inside surface of the tyre and the inner tube is then installed inside the tyre (see Figure 4.24).

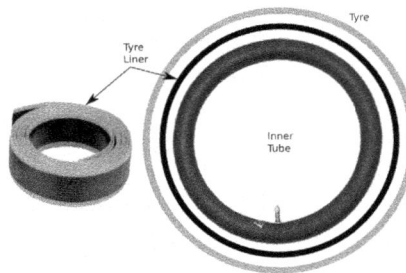

Figure 4.24: A Tire liner between Inner Tube and Tyre

Assume a tire liner weighs 50 grams and you add two tire liners for a total of 100 grams. The additional weight of tire liners alone is negligible compared to the combined weight of a bicycle and rider (say 80,000 grams) and therefore adds very little to slope resistance. Since air drag is not based on the weight of the bicycle and the tire liner is within the tyre, using a tire liner does not change the front area. Therefore, only rolling resistance will increase with the installation of a tire liner.

The tire liner is usually slightly longer than the circumference of the tyre and there is some overlap. The pressure of the inner tube keeps the tire liner in place on the tyre. Over time, the tire liner sticks to the outer surface of the inner tube and acts like an inner tube that is fatter on one side. If we take an average case with a head wind of 5 kmph., on a slope of gradient 0.01, and a total weight of 80 kgs. we can compute the loss in velocity due to the addition of a tire liner for different values of the coefficient of rolling resistance (C_r) at different speeds (see Figure 4.25).

When the C_r increases by 50% to $1.5 \times C_r$, the speed lost due to the higher rolling resistance is limited to a range of $0.5 - 1.0$ kmph. The loss of speed is higher when the bicycle speed is less than 20 kmph. and at higher speeds air drag becomes dominant. The loss of speed due to higher rolling resistance is not as significant as the loss of speed due to air drag. As C_r increases

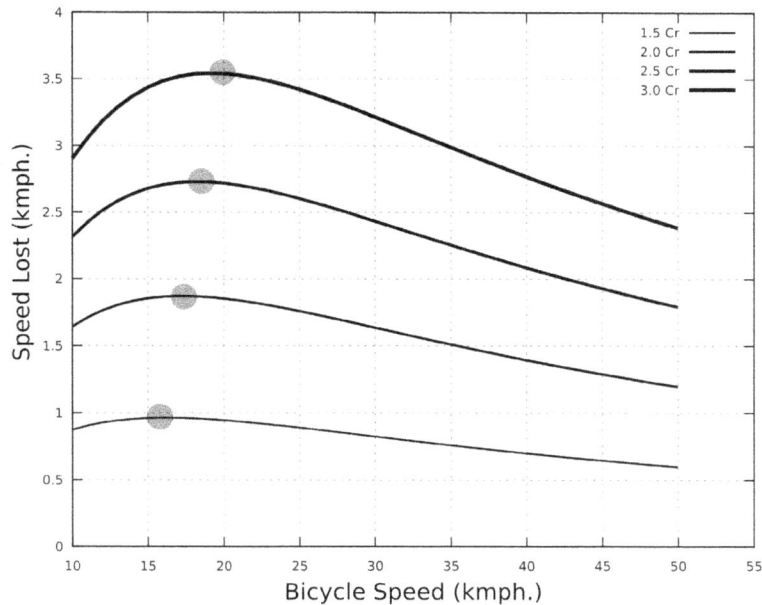

Figure 4.25: Speed Lost vs. Bicycle Speed for Multiples of Coefficient of Rolling Resistance

further to $3.0 \times C_r$, the peak speed lost moves closer to 20 kmph. and then decreases as air drag increases.

Adding a tire liner will increase C_r, but without actually measuring the increase in C_r, the speed lost due to the higher coefficient of rolling can only be estimated. If you assume your C_r doubles with the addition of a tire liner, then your maximum speed loss is limited to 2 kmph. The use of a tire liner maybe acceptable if you are willing to sacrifice some speed in exchange for the higher protection from a puncture.

The alternative to using a tire liner is to carry a spare tube, a portable pump, and tyre levers. All three items will add some minimal weight to the total weight of the bicycle. See Chapter 3 to estimate by how much will additional weight slow you down.

References

[1] BikeTechReview.com. AFM Tire esting, 2010. URL `https://bit.ly/2uRnooM`.

[2] Velonews. Where the rubber meets the road: What makes cycling tires fast?, October 2016. URL `https://bit.ly/2nA5UKu`.

[3] Engineering ToolBox. Rolling resistance, 2008. URL `https://bit.ly/2B8sj72`.

[4] David Gordon Wilson and Jim Papadopoulos. *Bicycling Science*. MIT Press, Third edition, March 2004.

[5] Schwalbe Tyres. What exactly is rolling resistance?, May 2018. URL `https://bit.ly/2KF1Imc`.

[6] Jerzy Ejsmont, Leif Sjogren, Beata Swieczko-Zurek, and Grzegorz Ronowski. Influence of road wetness on tire-pavement rolling resistance. *Journal of Civil Engineering and Architecture*, 9:1302–1310, 2015.

[7] Jan Heine. Tires: How wide is too wide?, January 2014. URL `https://bit.ly/K6B060`.

[8] Cycling.co.uk. Inner tubes can be performance upgrades too, January 2015. URL `http://bit.ly/2Q6Br6o`.

[9] Jan Heine. 12 myths in cycling : Wider tires are slower, January 2018. URL `https://bit.ly/2uTajeT`.

[10] Jan Heine. Tire pressure take-home, March 2016. URL `https://bit.ly/2MON2PH`.

[11] Jan Heine. Optimizing your tire pressure for your weight. *Bicycle Quarterly*, 5(4), 2006.

[12] Frank Berto. All about tire inflation, 2006. URL `https://bit.ly/2vfEmOL`.

[13] John Lafford. John Lafford's Tyre Rolling Resistance Data, 2002. URL `https://bit.ly/2Oip8zH`.

[14] Christophe Noel. Mountain bike tire design basics: Know your knobs, February 2018. URL `https://bit.ly/2Mu4V9F`.

[15] Matej Gorsic. Gravel cyclist's tire dilemma: Knobbies or slicks?, November 2017. URL `https://bit.ly/2On8crE`.

[16] Schwalbe Tyres. Tyre types, January 2018. URL `https://bit.ly/2odOV2K`.

[17] mrtuffy.com. What is a Mr. Tuffy anyway?, 2018. URL `https://bit.ly/2M50YLo`.

5 Wheels and Balancing

If the tyre is like the muscle of a bicycle, then the wheel is like the skeleton that bears the load of a much heavier rider. It is remarkable that thin tensioned spokes attached to the rim of a wheel can support a load of a few hundred kilos [1]. A wheel of a mountain bicycle is subject to even higher forces when riders jump from obstacles a few meters above the ground.

5.1 Wheels

The secret behind a wheel that appears flimsy but has great strength is the tensioned (stretched) spoke or wires. The physics behind the construction of bicycle wheels has not changed in many decades because it works very well. Materials used in the rims, hubs, and spokes have changed to reduce weight and improve strength, but other aspects of wheel construction have remained the same. The ideal wheel should be strong (to support a load), rigid (not have lateral deflection on a load), and light (for easy acceleration).

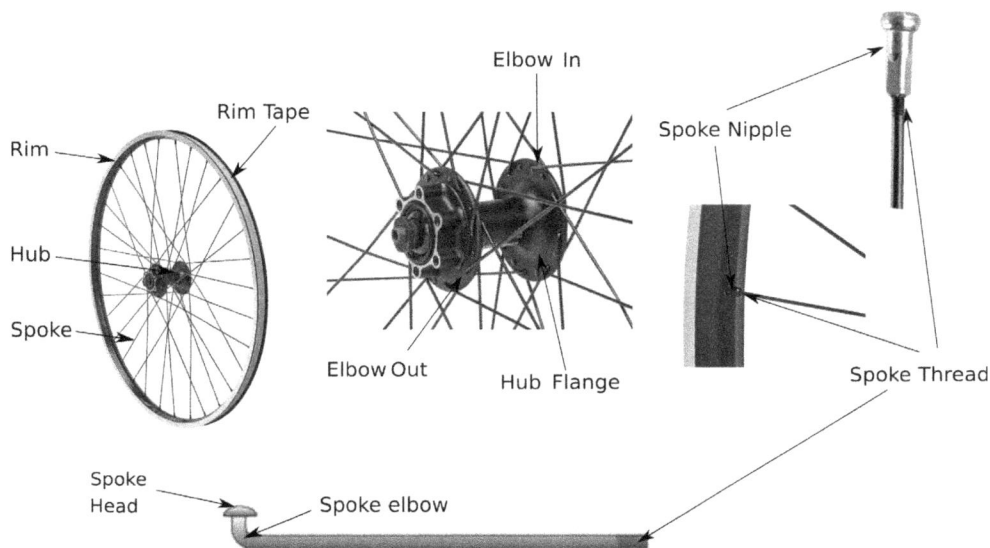

Figure 5.1: Parts of a Wheel (courtesy Specialized and Madegood.org)

5.1.1 Parts of a wheel

A bicycle wheel has relatively few parts - the hub, rim, spokes, and spoke nipples. The pressure of the air in the tyres make a ride smooth but are not necessary to support a load. Most rims are made of steel, an aluminium alloy, or carbon fiber. The rim is designed for the three tyre types - clinchers, tubulars, and tubeless. A rim for a clincher tyre will have rim tape to protect the inner tube from punctures when the head of the spoke nipple juts out of the rim. Rims come in a number of designs and manufacturers will display a cross section of a rim to illustrate the rigidity of the rim. A rigid wheel that does not bend laterally is important for performance.

The spokes connect the rim to the hub. A spoke has a head at one end and a thread at the other end. The spoke head is placed in the hole of the hub flange. Typically, the number of spokes in a wheel is even. A wheel with 32 spokes will have 16 spokes on one side of the hub and another 16 on the other side. Some wheels do have an uneven distribution of spokes on either side. A wheel may have more spokes on the drive (gear) side than on the other side. This maybe to maintain the same spoke tension in all spokes in the wheel (see Section 5.1.4).

The sum of the number of holes in both hub flanges should match the number of holes in the rim. A spoke nipple sits firmly in an embedded hole of the rim beneath the rim tape. The thread of the spoke is fitted into the hollow inner thread of the spoke nipple that projects inwards towards the hub. The length of the spoke thread is barely a few millimeters, but is sufficient to create a tensioned spoke that will support a relatively large load. You can adjust the spoke tension by turning the lower part of the nipple close to the rim with a spoke wrench.

The hub is attached to the frame of the bicycle through the chain stay and seat stays. Some hubs are fastened with a solid threaded axle or with a skewer (quick release). The rotating parts of the hub are the flanges to which the spokes are attached. Sealed inside the hub are ball bearings to keep the rotation smooth.

5.1.2 How does a wheel support a load?

The spokes of a wheel play a central part in the support of a load. While supporting a load, the spokes must maintain the circular shape of the rim and support the weight of the rider. The use of spokes in a wheel has worked for centuries to keep the cost and weight of the wheel low, while still being rigid enough to support relatively large loads. In Figure 5.2, a loaded tandem bicycle for three riders will have to support 200 kilos or more on just two wheels with relatively narrow rims and tyres.

Figure 5.2: A Loaded Bicycle with Cargo and a Tandem Bicycle (courtesy Rediff.com and tandem-central.com)

Spokes Spokes are a frequently overlooked component when considering the performance of a bicycle. They are thin stretched wires a few millimeters wide, often made of steel, and transmit force from the hub to the rim. The braking and accelerating forces are transmitted by spokes to and from the hub. A load on a normal spoke will compress the spoke, which may buckle if the force of the load is excessive. The set of tensioned spokes in a wheel with a hub as an anchor prevent the rim from buckling under a load. The most common material for a spoke is steel for several reasons -

- It has high tensile strength, a single spoke of diameter 5 mm can handle a tension of 1000 N and is relatively lightweight (5 - 6 grams).

- You can create the required tension with threading on a small length (8-10 mm.) of the spoke to keep it firmly attached to the rim.

- It is more affordable compared to other materials such as carbon fiber and is corrosion resistant.

The basic spoke has a uniform diameter from the head to the thread. The average diameter of a spoke is about 2 $mm.$ with a minimum of 1.5 $mm.$ and a maximum of 3.0 $mm.$ or more. A spoke with a larger diameter can handle a larger load. Some spokes are butted – thinner (1.8 $mm.$) in the mid-section and fatter (2.0 $mm.$) at the ends. Even though butted spokes are thinner in the middle, they can still handle a tension [9] of 1300 N/mm^2 [2]. One of the advantages of a butted spoke is less weight due to a thinner cross section. Another possible advantage is more flex at the center of the spoke when the load is high and a lower likelihood of snapping in the middle. A third advantage is that the spokes are wider at the ends that connect to the rim and better able to handle the transfer of torque. Butted spokes handle loads in the mid-section leaving the threads and elbows less stressed [1]. Spoke failure often occurs at the bottom of the thread where the spoke is the thinnest.

How do spokes support a rider's weight? While you could tighten a spoke to the maximum allowable tension, you can support a weight of 80 kgs. with a lower tension of about 500 N/mm^2. If the rear wheel supports 50 kgs. and the front wheel supports the remaining 30 kgs., then the rear axle will press down towards the ground with a weight of 50 kgs or about 500 N. Three spokes close to the ground contact area support this downward force at approximate fractions of 25%, 50%, and 25% with the central spoke supporting the maximum load. If the central spoke is tensioned to 500 N/mm^2, then the tension will reduce to 250 N/mm^2 when it is part of the contact area and increase back to the original tension of 500 N/mm^2 once the wheel rotates further. This cycle usually takes a fraction of second.

In a study [3] with a test rig to analyze how a spoke wheel supports a load, the rim was found to carry very little of the load and its purpose was mainly to transfer load forces to the spokes. The rim also distributes the load to multiple spokes. The study confirmed that the few spokes near the ground handle almost all the load by losing tension and the remaining spokes have a slight increase in tension. The maximum load that a wheel can handle is the load at which the spokes become lose (zero tension).

---

[9]Although tension is a force, the amount of compression and elongation depends on the force per unit area.

The fraction of the load carried by a spoke is in direct proportion to the tension [1]. Since spokes on the gear side are at a higher tension than the spokes on the non-gear side, they carry a larger load and therefore are more likely to fail sooner.

The study also found that pressure of air in a tyre does not apply a force on the rim when a load is handled. The pressure of air in an inner tube directed towards the rim is balanced by an upward pressure on the bead of the tyre [1]. Therefore air pressure acts equally in both directions and does not contribute to the net force on the rim.

Spoke Fatigue   It is amazing that the cycle of losing tension and regaining tension in a spoke can happen a million times or more before a spoke finally gives way. At 20 kmph., on a 32 spoke 26 inch wheel, a single spoke has to handle the load for just a hundredth of a second. Even though a spoke must support a load for a fraction of a second, a one hour ride at this speed translates to about 10,000 revolutions or 10,000 times the spoke has to briefly support the load.

Spokes that are poorly manufactured will have flaws that gradually grow bigger with many such cycles and eventually become obvious to the rider. One common area where a spoke fails is at the thread, where the spoke attaches to a nipple. The length of the spoke thread is limited and over time the thread may be stripped. The spoke nipple sits perpendicular to the rim while the spoke thread and nipple will rarely be perfectly aligned. Instead the spoke comes at an angle to the rim with a sharper angle on the geared side of the wheel (see Figure 5.8). This slightly skewed entry of the thread on the rim adds to the stress on the thread.

How does the hub stay centered?   If you removed the spokes in the upper half of an unloaded wheel, the hub will still be centered in the wheel. The tension of the spokes in the lower half of the wheel are sufficient to keep the hub in place. The hub is attached to the frame of the bicycle with either a threaded solid axle or a skewer (quick release) to the seat stay and chain stay tubes. You still do need the spokes in the upper half of the wheel to transmit the weight force to the hub. When the wheel rotates, these same spokes are now in the lower half of the wheel to support the normal force from the ground.

A Stiff Wheel   The stiffness of a wheel is the resistance of the wheel to a sideways (lateral) force. A stiff wheel is preferred to a less stiff wheel since less energy is lost due to lateral flex when pedalling or cornering. The *stiffness* of a rim is measured from the amount of deflection as a lateral load is increased unlike the *strength* of the rim that depends on the maximum sustainable load before the rim gives way.

Increasing the tension of spokes on the wheel did not make the wheel more rigid [4]. A spoke will lose the same amount of tension under a load if it is near the maximum allowable tension or within some range. However increasing the thickness of spokes did increase the rigidity of a wheel as did a heavier rim. Wider hub flanges do make a wheel more rigid but spoke tension will need to be higher to support a vertical load and it takes more effort to rotate.

Instead of the thin rim in Figure 5.1, a deep rim with shorter spokes is assumed to be more rigid. Manufacturers also claim that deep rims are more aerodynamic and therefore you need to use less power at higher speeds.

5.1.3 Spoke Patterns

The spoke patterns of wheels has evolved over many decades and the patterns that "work" are well known. The most common spoke patterns in Figure 5.3 are the five patterns from 0 cross or radial to 4 cross. The number 3 in the 3 cross means that when a wheel is viewed from the side, a spoke will intersect with 3 other spokes. The intersection of spokes is not physical, but instead in the line of sight.

If a rim has 36 holes, then a hub must have 18 holes on the left and right flanges respectively. The holes in a rim may not be centered, instead rim holes are shifted alternately closer or away from one edge of the rim. In all patterns half the number of spokes are installed on the left side of the hub and the other half on the right side of the hub. The number of spokes or the pattern do not have any effect on the comfort of the ride [5], which largely depends on the width and pressure of the tyre. A touring bicycle that will need to support more than the average weight will have wheels with 36 spokes. For most other bicycles, 32 spokes is sufficient to handle the weight of an average rider and some cargo.

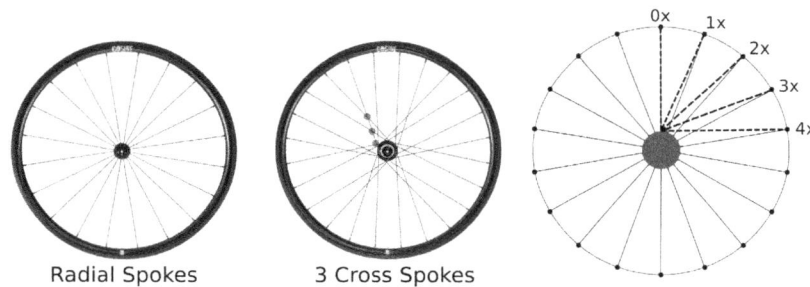

Figure 5.3: Different Spoke Patterns (wheels courtesy Wiggle Ltd.)

There are several different spoke patterns that can be categorized by the number of crossings. In the 3 cross spoke pattern in the center of Figure 5.3, every spoke intersects three other spokes. The right wheel shows how spokes for 4 patterns would be placed. As the number of crossings increases, the angle that the spoke departs from the hub becomes more tangential.

A 0x pattern is the same as the radial pattern on the left of the figure. Each spoke is threaded to the rim hole that is the shortest distance from the hub flange hole. This also means that spokes in the radial pattern are the shortest and the longest spokes are found in the 4x pattern.

Why do you need a crossed pattern? The radial pattern is often found on a front wheel with rim based brakes. It is a simple straightforward pattern and works well when you do not need to transfer torque from the hub to the rim. The frame and fork are driven forward from the rear

151

to rotate the front wheel in the same way a wheelbarrow moves when pushed from behind. A radially spoked wheel is slightly stiffer than a cross (tangentially) spoked wheel.

The remaining spoke patterns from 1x through 4x in Figure 5.3 are used when torque must be transferred from the hub to the rim or in the case of disc brakes, from the disc to the rim. A wheel can even have radial spokes on the non-drive side and crossed spokes on the drive side.

When spokes cross or intersect each other they provide a brace for the spoke whose tension is changing. The tension does change fairly sharply and quickly when a spoke is directly above the ground contact area. The 3 cross pattern is the most common pattern for a 32 spoke wheel. Two cross patterns are found on wheels with fewer (24-28) spokes, each of which is tensioned higher than the spoke tension on a wheel with more spokes.

The spokes with higher crossings are slightly longer - a 3x crossed pattern will use spokes that are about 5 mm. longer than a radially spoked road bicycle wheel. The exact length of a spoke depends on several parameters including the rim diameter and hub dimensions (see Section 5.1.5). The weight penalty due to longer spokes in a crossed pattern is measured in grams and will not be noticed by a rider.

The process of building a wheel begins with spoke lacing which is essentially a pattern of connecting the hub to the rim with a set of spokes. While the description is simple, it is important to get the pattern right on the drive side. Every spoke is either a *trailing* or a *leading* (see Figure 5.4) spoke. In a 36 spoke wheel, 18 spokes are used on the left ride and the remaining 18 on the right side. The figure shows only one side for simplicity.

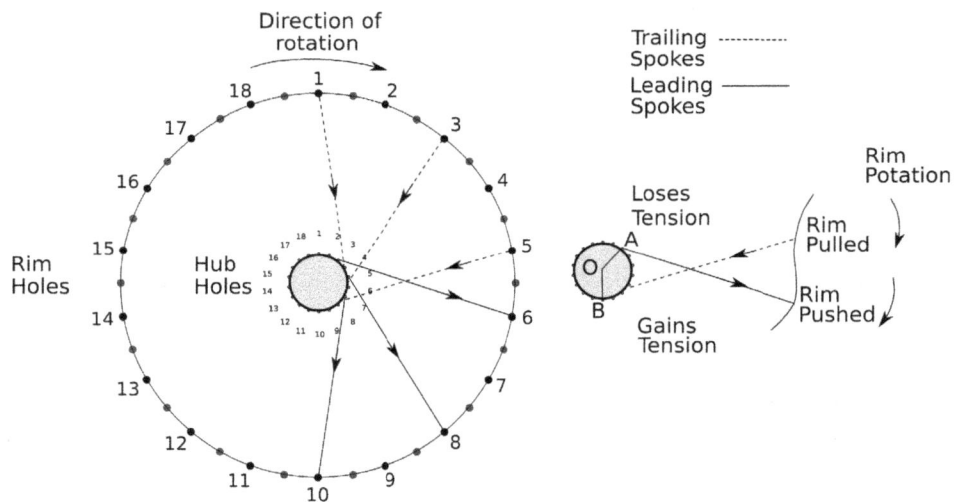

Figure 5.4: Spoke tension changes in trailing and leading spokes on a rotating wheel

Installing spokes in a rim follows a repetitive pattern. Starting in the right side from hole 10 on the rim (near the valve hole) and hub hole 7, moving counter-clockwise you skip 3 holes on the rim and 1 hole on the hub, to link hub hole 5 with rim hole 8 (elbow out), hub hole 3 with rim hole 6, hub hole 1 with rim hole 4, and so on till all rim holes on the right side are connected with a spoke to a hub hole.

152

This completes 9 spokes on the right side and another 9 spokes are linked in the same manner on the left side. These 18 spokes are the leading spokes and are laced in the direction of rotation of the wheel. The hub holes on the flanges of both sides may not line up and the matching hub hole on the opposite side is the first hub hole moving anti-clockwise. Leading spokes are connected to the hub flange with the elbow out (see Figure 5.1).

The first trailing spoke on the right side begins at hub hole 11 and is linked to rim hole 5 crossing 3 leading spokes (7-10, 5-8, and 3-6) with elbow in. The first two crossings are inside (under) the spoke and the third crossing is outside (over) the spoke. The next two trailing spokes connect hub hole 9 to rim hole 3 and hub hole 7 to rim hole 1 respectively. The remaining 6 trailing spokes on the right side are similarly linked moving in the anti-clockwise direction. The last remaining 9 of the 36 spokes are also linked on the left side. A number of good references [1, 6, 5] explain step by step how to build a wheel.

The purpose of a spoke pattern is to transfer torque from the hub to the rim. Since spokes are fastened to the rim and cannot rotate they can only transfer torque by gaining or losing tension. Brandt [1] calls the leading and trailing spokes, pushing and pulling spokes respectively. A pushing spoke pushes the rim outwards by losing tension and a pulling spoke pulls the rim inwards by gaining tension (the flex in the rim is exagerrated in the figure).

If you are pedalling, then torque is transferred from the hub to the rim, however if you are using a disc brake, then torque is transferred from the disc to the rim. All the spokes participate in the transfer of torque. The torque for a given spoke is the product of the change in tension ($\Delta tension_{leading}$ or $\Delta tension_{trailing}$) and the hub flange radius or lever arm (OA for leading spokes and OB for trailing spokes). Although leading spokes lose tension, the loss of tension generates a torque in the same direction as a trailing spoke that gains tension. The total torque on the wheel is sum of all the individual spoke torques. The torques from the change in tension in both the leading and trailing spokes are in the same direction (clockwise). The use of trailing and leading spokes makes the wheel more rigid when accelerating or braking instead of all spokes in one direction relative to the hub.

Since torque is the force driving the rim and therefore the bicycle, it would appear that increasing hub flange radius would increase torque for a given change in spoke tension. However, the cyclist would have to generate this additional torque to rotate a hub with a large flange radius. The use of gears to increase the number of wheel rotations with a small radius hub works well and can be more finely adjusted based on the rider's power.

Which is the best cross pattern? In a finite element analysis study [7] by *williamscycling.com* of spoke lacing patterns, the spoke tensions for a model 28 spoke wheel were calculated when loaded and unloaded using standard lacing patterns. In a 3x/3x pattern the spoke tensions were 650 N/mm^2 and 390 N/mm^2 on the drive and non-drive sides respectively or about 40% less on the non-drive side. With the rider on the bicycle the spoke tensions near the bottom of the wheel decreased while the spoke tensions in the rest of wheel increased slightly. When torque was applied (pedalling), the maximum tension was 747 N/mm^2 in a trailing spoke on the drive side and the least tension was 248 N/mm^2 in a leading spoke on the non-drive side.

The 2x/2x pattern also had similar results with slightly higher spoke tensions. Both these results indicated that these two patterns are structurally stable and should not result in spoke failure, if sufficient tension is maintained. Ideally, the spoke tension range for a particular spoke should not be too large (> 200N) and the maximum tension should be within the tensile strength of the spoke. Of all the patterns tested, the 3x/3x pattern had the least tension range and slightly lower tensions than other patterns.

The study also evaluated the loss of energy due to lateral flex for different cross patterns. A wheel which flexes more than another wheel will not be as efficient in transferring torque from the hub to the rim. Some energy is wasted when the wheel flexes and is dissipated in the form of heat. The 3x/3x pattern had the least power loss due to wheel flex. Finally, the quality of a wheel may depend more on how well it has been built than on any particular spoke pattern and the quality of the hub, rim, and spokes.

The radial pattern has been used on the drive side as well as the non-drive side. On the drive side, the torque of the hub will generate a high spoke tension due to the near perpendicular angle of the spoke to the hub's tangent (small lever arm). On the non-drive side, the spoke tensions are lower. Therefore some wheel builders have used the radial pattern on the non-drive side and a crossed pattern on the drive side. Spoke patterns alone do not have any effect on the stiffness of a wheel [1].

The differences between spoke lengths of individual patterns is less than 3% – a longer spoke will generate more torque on the rim. Increasing the diameter of the wheel is a better option to use longer spokes than with an alternate spoke pattern. A radial pattern generates the highest stress at the hub flange hole compared to cross patterns. Some manufacturers used a thicker hub flange to prevent a crack in the hub due to spoke tension.

5.1.4 Spoke Tension

A bicycle spoke is a stretched or tensioned wire. When a spoke is tensioned, a load applied to the wire reduces the tension in the wire. When the tension in a spoke is zero, the length of the spoke is its natural length without the application of any force, either a compressing or stretching force. Since spokes on a wheel do not twist or bend on a normal ride, torque is transmitted by changes in spoke tension which in turn deform the rim. The changes in tension are not sufficient to make visible changes in the shape of the rim.

Though all the spokes in a wheel are tensioned, only a few spokes support a load at any given time. A few spokes to the left and to the right of the ground contact area support the load. These support spokes lose tension when they are located near the ground contact area, yet are still rigid enough to maintain the structure of the wheel. Once these support spokes move away from the ground contact area, other spokes handle the load and the previous support spokes regain the original tension. This process of spokes losing and regaining tension is repeated many times when the wheel rotates. Even though only a few spokes near the ground contact area support the load, the remaining spokes in the rest of the wheel are needed to maintain the shape of the wheel and rim.

The tension in a spoke depends on the tension in all other spokes and the forces on a wheel can be analyzed by looking at the tension in all other spokes simultaneously. If the tension in all spokes is high enough and evenly distributed, then a wheel will be more durable. When the tension is not high, a spoke may become lose and will rattle when the wheel rotates. Evenly distributed tension makes it less likely that one or more spokes will be handling a larger share of the load.

The tension for a single spoke can range from 930N (for a light rim) to 1250N [5]. For a wheel with 36 spokes and individual spoke tension of 1000N (T_{spoke}), the weight force on a section $\frac{2\pi}{36}$ (S_{spoke}) of the rim for a single spoke is over half a ton ($\frac{T_{spoke}}{S_{spoke}} = \frac{36 \times 1000}{2\pi} = 5730N$ or $573kgs$.). Although this compression force is distributed across the circumference of the rim, the material of the rim needs to be strong enough to withstand this force.

What is the correct spoke tension? If the spoke tension is too high, then the force on the hub flange hole will be excessive leading to a failure or crack in the flange. The rim hole for the spoke also may crack due to similar reasons. The flat edge of the spoke used to increase tension with a spoke wrench may also become rounded, making it difficult to adjust spoke tension.

When spoke tension is too low, the rim will have to handle the load that the spoke should have supported. Other spokes that are properly tensioned can only handle a load when they are over the ground contact area. A loose spoke (low spoke tension) in a wheel will rattle noticeably. Also a wheel with spokes that are not uniformly tensioned will be more likely to be out of true. When you ride with spokes of unequal tension on the same side of the wheel, there will be a tendency to equalize tension by deforming the rim.

You can measure tension with a spoke tension meter. The meter gives a deflection reading that can be translated into a tension value from a table. A deflection reading of 22 [8] may translate to a tension of 860 N for a round steel spoke of diameter 2.0 mm. If the diameter of the spoke is 1.5 mm, then the same reading of 22 indicates a higher tension of 1370 N. The tension depends on the diameter as well as the spoke material (steel, aluminium, or carbon). Two spoke manufacturers may have different tension tables for a spoke of the same diameter and material.

How is spoke tension distributed in a wheel? In rear wheels, the spokes on the gear side or right side will have more tension than the spokes on the left side of the wheel. Similarly, in a front wheel with a disc brake, the spokes on the side of the disc will have more tension. The spokes on the drive side will be a few millimeters shorter than the spokes on the other side. Spokes that are almost parallel to the plane of the rim will transmit torque efficiently whereas a spoke that attaches to the rim at angle will have parallel as well as perpendicular components of force and therefore are less efficient. The spoke tension on the drive side of the wheel can be as much as $1\frac{1}{2}$ times the spoke tension on the non-drive side. The spoke tension in the rear wheel is higher than the spoke tension in the front wheel, since the rear wheel will carry a larger fraction of the load.

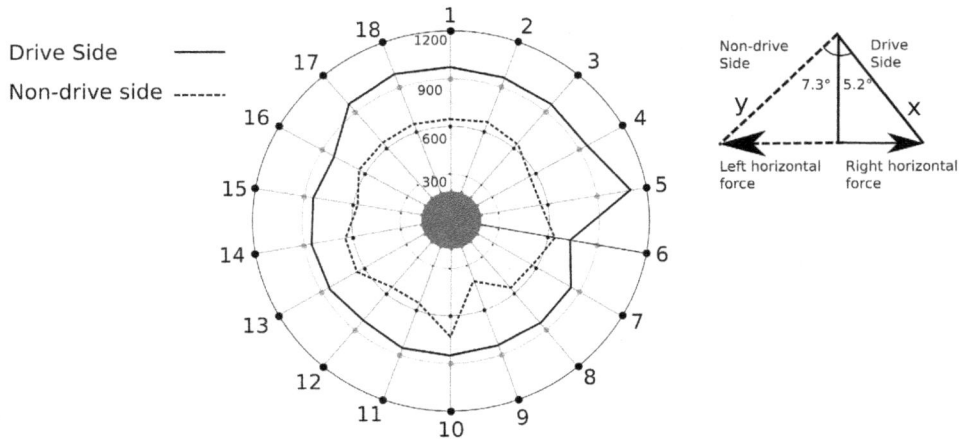

Figure 5.5: Circular Plot of Distribution of Spoke Tension on the Drive side and Non-drive side of a Wheel

A graphic representation of spoke tension in a 36 spoke wheel in Figure 5.5 shows how spoke tension can be corrected. The drive and non-drive side have average tensions of about 1000 N/mm^2 and 600 N/mm^2 respectively. Spoke 5 on the drive side has a higher than average tension while spoke 6 has a lower than average tension. Similarly on the non-drive side, the tension of spoke 9 should be increased and the tension of spoke 10 decreased to the average. The main goal of visualizing spoke tension is to ensure that spoke tensions are roughly evenly distributed on either side and the tension on the drive side is higher than the non-drive side by about 40% while at the same time maintaining the trueness of the wheel.

In Figure 5.5 the left horizontal and right horizontal forces should be equal for the wheel to be aligned along the plane of the frame. If the tension on the drive side is x and the tension on the non-drive side is y, then $x \times sin(5.2) = y \times sin(7.3)$ where 7.3° and 5.2° are the angles the non-drive and drive spokes make with the vertical respectively. The ratio of the tension x on the drive side to the tension y on the drive side is $\frac{sin(7.3)}{sin(5.2)} = 1.4$ or about 40% more on the drive side.

Do you need a tension meter? If your wheel is in true and the spokes are reasonably tensioned, it is remarkable that you can ride on a wheel for many thousands of kilometers without ever measuring the tension on a spoke. The range of spoke tensions on which you can safely ride is fairly broad and if your tension is within that range and the wheel is close to true, you could ride without accurately measuring spoke tension. A Bontrager rim may have a range from 500-1320 N while a Mavic rim may have a range of 1000-1300 N [8].

A spoke thread may have 56 threads per 2.5 cms. and therefore one turn of the spoke thread moves the spoke by about 0.045 cms. or 0.45 mm. The length of the spoke thread may range from

8-10 mm. or about 20 threads. Each full turn of the spoke can increase the tension by about 50 N.

Visualizing Tension   A common assumption is that once a wheel is trued (without any wobble), the tension of spokes on each side should be relatively the same. However, this not necessarily the case [9]. Although, the wheel may spin without any wobble, the tension in the spokes on one side may not be uniform. Compared to the average tension on one side, the tension of some spokes maybe loser or tighter by a relatively large margin.

While this may not be a problem initially, over time a spoke that is loose will tend to become more loose leading to a wheel that is out of true. A spoke that is too tight will have high stresses near the threads at the rim or near the hub which can lead to cracks in the nipple hole or hub flange hole.

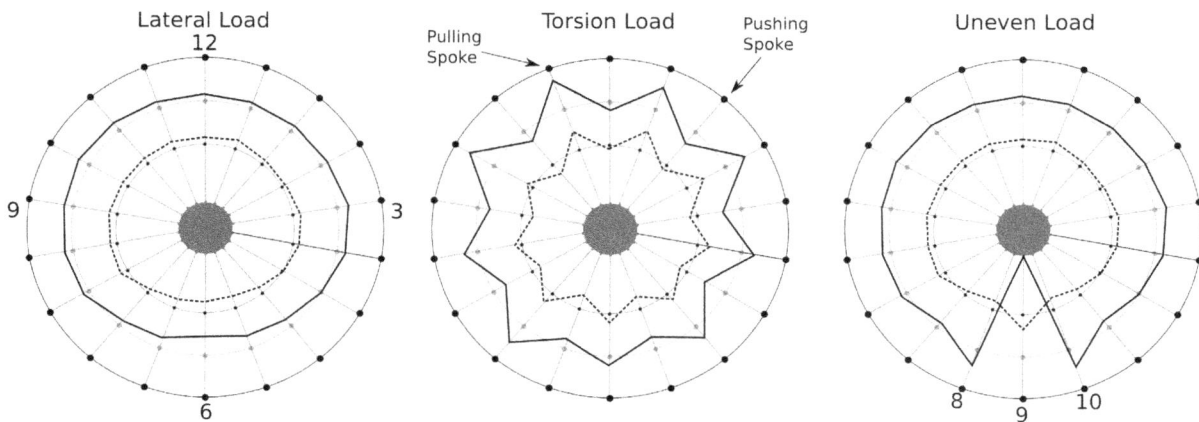

Figure 5.6: Visualizing Spoke Tension under three different loads (Radial, Torque, and a Missing Spoke)

In Figure 5.6, spoke tension for three different scenarios shows that individual tension can change substantially. In the left most figure, the effects of a radial load or a rider sitting on the bicycle show that the tension near the ground contact or at the 6 o'clock position lose tension. This is what you would expect since the few spokes near the ground contact area support the entire load. The spokes at the 3 o'clock and 9 o'clock positions have a slightly higher tension. The rim which is flattened near the ground bulges slightly on the sides and therefore the rim is pushed slightly outward at the 3 o'clock and 9 o'clock positions. At the 12 o'clock position, there is no change in the tension.

When you pedal, you are applying a torque to the wheel. The pulling or trailing spokes gain tension and pull the rim inwards. The pushing spokes or leading spokes lose tension and push the rim outwards. The changes in the shape of the rim are minute and not obviously visible. The same type of load is applied when using a disc brake except the pulling spokes lose tension and the pushing spokes gain tension.

The right most figure shows tension when a spoke (no. 9) is missing or without any tension (lose). The immediate neighbouring spokes 8 and 10 increase in tension to compensate for the loss of tension. The rim section at spoke 9 will be pushed outwards and therefore the rim sections at 8 and 10 will be pulled inward. The tension on the spoke opposite side of spoke 9 will increase in tension since a rim section is pushed laterally to one side.

Tyre air pressure also influences the tension in a spoke. When there is no air pressure on a wheel, the tension in spokes is higher than with an inflated tyre. The air pressure in an inflated tyre pushes the rim inward and therefore ironically decreases tension. The decrease in tension will be noticeable when the air pressure is 100 psi or more.

Modelling Spoke Tension Ian Smith [10] published a model of the change in spoke tension using finite element analysis when a 100 kg. load (or 1000 N) is applied to a whell. He modelled a 36 spoke wheel in the 3 cross pattern for a 600 mm. diameter wheel. The normal force at the bottom of the wheel was distributed over three spokes. The spoke (19) above the center of the ground contact patch supported a load of 350 N and the two neighbouring spokes (18 and 20) each supported 250 N.

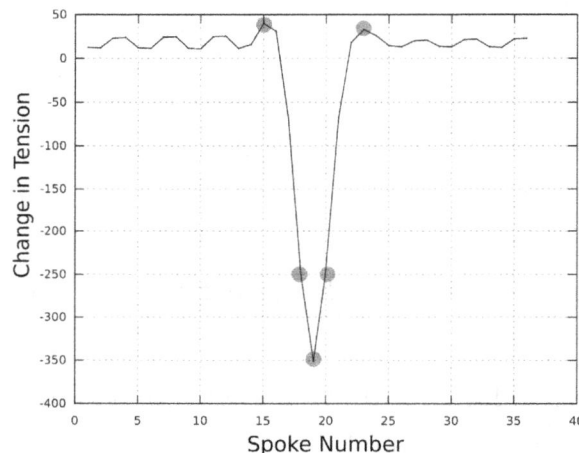

Figure 5.7: Change in Spoke Tension under a lateral load vs. Spoke Number

In Figure 5.7, the change in the spoke tension vs. individual spokes shows that there is a sharp reduction in the tension of the spoke that is directly above the ground contact patch. Five spokes out of the 36 spokes lost tension while the remaining 31 spokes gained some tension with the sharpest increase at spokes 23 and 15. The remaining spokes gained less tension.

The analysis showed that a single spoke handles a large part (50%) of the load for a brief period. While a missing spoke is not catastrophic, the neighbouring spokes must handle the load that would have been supported by the missing spoke. This will lead to more stress on such spokes over time.

5.1.5 Spoke Length

There are many spoke length calculators on the Web that you can use to precisely calculate the spoke length for wheel. Of course, the calculator will only give results as precise as the input data that you provide. It is important that spoke lengths be accurate upto a millimeter. If the length of a spoke is too short for the wheel, then you may not be able to stretch it to generate enough tension. On the other hand, if a spoke is too long, then it may force the nipple above the edge of the rim and puncture an inner tube.

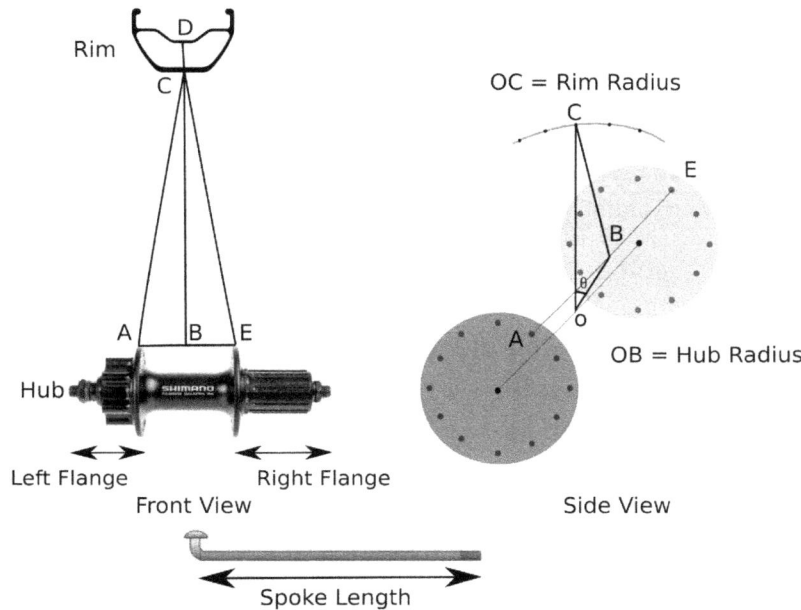

Figure 5.8: Spoke Length Calculation (hub courtesy Shimano)

Input Parameters and Calculation A geared hub will have flanges that are not equi-distant from the center of the hub. The flange of the geared side of the hub maybe slightly closer to the center of the hub than the flange of the non-geared side. Three measurements on the hub viz. the total length of the hub (h), the distance of the geared side (right) flange to the edge (h_g), and the distance of the non-geared side (left) flange to the other edge (h_{ng}) are sufficient to calculate AB and BE in Figure 5.8. If h is 120 mm., h_{ng} is 16 mm., and h_g is 30 mm., then AB and BE are 44 mm. and 30 mm. respectively. The spoke lengths for the geared side and non-geared side must be calculated separately. The spoke length for the non-geared side is the length $AC + \frac{CD}{2}$.

We can calculate $AC = \sqrt{(AB^2 + BC^2)}$, however BC is still unknown. The point B is located at some imaginary location on the line joining the hub holes at the same location on both flanges. The distance OB is the distance from the center of the hub axle to any hub hole and OC is the rim radius of the wheel. Both of these distances are usually specified by the manufacturer. The value of BC^2 by the cosine theorem for a triangle is $OB^2 + OC^2 - 2 \times OB \times OC \times cos(\theta)$. The value of θ is unknown and depends on the number of crossings in the spoke pattern. For a zero crossing or radial pattern, the value of θ is 0 and $BC^2 = (OC - OB)^2$.

For a non-zero crossing X, the value of θ is calculated as X times the angle between a hub flange hole. If the total number of rim holes is 36 and each hub flange has 18 holes, then $\theta = X \times \frac{360}{18}$ or in general for h rim holes $X \times \frac{720}{h}$. There are several spoke length calculators on the Web that accept a general set of input parameters and use the same formula to calculate a spoke length.

URL
https://www.prowheelbuilder.com/spokelengthcalculator
http://www.sapim.be/spoke-calculator
https://spokes-calculator.dtswiss.com/en/
https://www.wheelpro.co.uk/spokecalc/
https://github.com/mkonchady/cycle/

Table 5.1: Online Spoke Length Calculators

The length of a drive side spoke is shorter by about a few millimeters compared to the length of a non-drive spoke. The calculation for CE in Figure 5.8 is identical except that instead of AB the distance BE is used in the calculation. In most hubs, the radius of a hub flange hole OB is the same for both the left and right flanges.

Number of Spokes It may appear that having fewer spokes means a lighter wheel and therefore less drag force. The standard number of spokes on a wheel ranges from 32 to 36, but wheels with as few as 16 spokes are also in use. With fewer spokes, you may need a thicker rim since load is distributed at fewer locations on the circumference of the rim. The spokes also may need to be thicker in diameter and have a higher tension.

Both of these changes to the wheel will increase the weight of the wheel which negates the benefits of fewer spokes. However, materials to build the rim that are light and rigid (like carbon fiber[10]) can be used without the additional weight. Still, replacing all steel spokes in a wheel with carbon fiber spokes will reduce the weight of both wheels by about 200 grams. This will make a minimal difference in the speed that you can ride for a given power (see Section 5.3.1).

Another problem with fewer spokes is that one broken spoke will increase the load that the remaining spokes must carry. On a normal wheel with 32 spokes, you can ride safely with 1-2 missing spokes. The number of spokes is usually the same in both the rear and front wheels, but sometimes a rear wheel may have a few more spokes to support the additional weight in the rear of the bicycle. The rear wheel must also support additional forces such as the torque from the rotation of the hub (see Section 5.2). The additional weight of spokes in a wheel is offset by the additional strength and durability of the wheel. A wheel with more spokes is easier to true and maintain spoke tension.

---

[10]A carbon fiber spoke can weigh as little as 3 grams compared to about 6 grams for a steel spoke.

Life of a Spoke For every revolution of a wheel, a spoke that supports the load near the ground contact area will undergo some compression and will return back to its tensioned length once the load is absent. This cycle is repeated many times during the life of a spoke. The fatigue life (number of revolutions) for a spoke can range from 700,000 to 1.5 million (1400 kms. to 3000 kms. for a 26 inch wheel with circumference of 200 cms.). While this is a manufacturer's estimate of the life of a spoke, in reality spokes can last for upto 10,000 kms. (5 million revolutions) or more. The life of a spoke depends on the loads it handles and heavy loads can reduce the spoke life to the manufacturer's limit.

5.2 Forces on a wheel

The ideal wheel should be light but at the same time strong enough to withstand the forces applied while riding. This is a common requirement in most vehicles since a heavier vehicle will reduce efficiency. To date, there is no ideal material that is relatively cheap to manufacture, light and yet strong enough to withstand the loads a wheel must handle. Steel is the most common material for building spokes and the rim while lighter composites such as carbon fiber have also been used. In general, the lighter the material, the more expensive the components. Further, the individual components of a wheel - the rim, hub, and spokes should be compatible. A light rim will not work with thick spokes. The tension that you could generate when tightening a thick spoke may damage a light rim.

The strength of a wheel is the maximum load it can support without collapsing. The stiffness or rigidity of a wheel is the deflection of the wheel under a given load. The more stiff a wheel, the less it deflects. In a rigid rim, the section that supports a load is longer and therefore more spokes loose tension. The tension lost per spoke is less and therefore stresses in the spoke elbows and threads is also less which extends the life of the wheel.

5.2.1 Load on a wheel

Brandt [1] defines two types of loads - static and dynamic, each of which has radial (parallel to the wheel plane), lateral (perpendicular to the wheel plane) and torsional (twisted) components. The rider's weight is a dynamic load since the wheel's rotation changes the load on the spokes of the wheel. The static loads are the tensioned spokes that are needed to keep the structure of the wheel. The rider changes the load when accelerating, braking, or riding over bumps.

The load on a tandem bicycle is substantially higher than on a normal bicycle. A tandem bicycle may weigh as much as 20 kgs. due to the larger frame and components such as the crank, chain, and handle bars. The total weight of two average riders maybe 130 kgs. Therefore, each wheel must support 75 kgs. and this weight excludes any cargo. It is rare to find more than 36 spokes on the wheel of a normal bicycle. However, wheels of a tandem bicycle may have 40 or more spokes to handle the additional weight.

The best wheel is a wheel with a rigid rim and many spokes [1] – the load is spread over many spokes. The number of spokes in a wheel is typically proportional to the diameter of the wheel. Large wheels like the ones used in the high wheel bicycle had 80 spokes while smaller wheels like the 16 inch wheel can use 20 spokes or less. Folding bicycles that use small wheels still use 32 spokes since most hub manufacturers build hubs with 32 holes in the flanges.

A rim that is radially more stiff is preferable to less stiff rim, since the load affected zone is distributed across a larger section of the rim. Longer tensioned spokes are stiffer than shorter tensioned spokes. Since the difference in spoke length between spoke patterns is a few percent, all spoke patterns have about the same effect on rigidity. Smaller wheels are less stiff than larger wheels, but do not affect the smoothness of the ride which is dependent on air pressure.

The lateral stiffness of a wheel is the resistance to sideways deflection of the rim. A rear wheel with gears will have more spoke tension on the drive side than on the non-drive side. The greater the difference in spoke tension, the more easily the wheel will deflect towards the non-drive side.

Torsional stiffness is the resistance offered by the wheel when a torque is applied to the hub. Thickness of spokes has a greater effect on torsional stiffness than the rim. Using spokes with 2X diameter results in a wheel four time more rigid than a wheel with 1X diameter spokes.

Inner tube pressure presses down on the rim of a wheel. The bead of the tyre generates an opposing upward force and is locked into the rim by inflation pressure. Therefore, it is important to verify that the bead of a tyre is seated tightly on the rim before riding. Air pressure of 80 psi or more can violently lift an improperly seated tyre off the rim.

5.2.2   Truing a Wheel

In the ideal case, the shape of the rim of the wheel will be a perfect circle. Since air pressure is distributed uniformly in a tyre and absent any particles in the tyre or tube, the wheel and tyre will have a circular shape. The aim of truing a wheel is to get a rounded wheel rim that is aligned with the plane of the bicycle frame. You can iteratively true a wheel by iteratively tightening and loosening spokes.

A rim can be deformed from the ideal shape along the $x, y,$ and z axes (see Figure 5.9). Lateral truing corrects deformation in the x axis while radial truing corrects deformation in the y and z axes. The corrections to create a true wheel are not straightforward since you can only tighten or loosen spokes to make corrections in any direction. Also since a wheel with spokes is tensioned, loosening the tension in one spoke can change the tension in other spokes.

Small errors (0.5 mm. or less) are corrected a quarter turn and large errors may need two or more turns. You may need to correct errors in both the radial and lateral directions. One option is to start with the direction with the largest error and alternately make adjustments. For simple truing, you can place the bicycle upside down and tighten or loosen spokes by observation alone. This is obviously not as accurate as truing with a stand.

You can either tighten or loosen a spoke depending on the condition of the neighbouring spokes. A single tight spoke in the neighbourhood of loose spokes should be loosened and vice

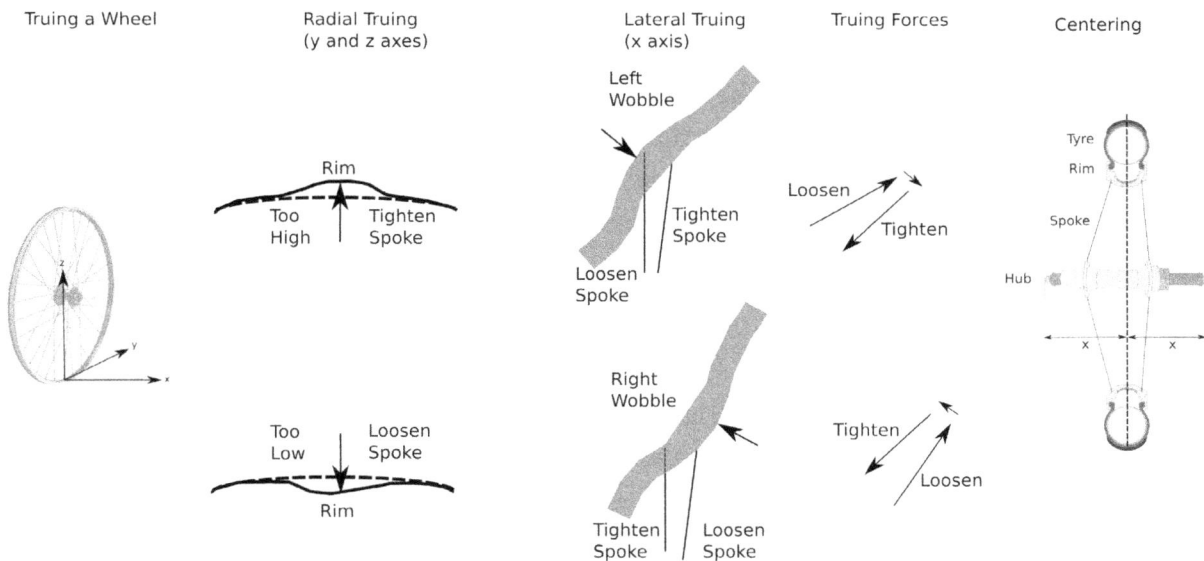

Figure 5.9: Truing a Wheel by Tightening and Loosening Spokes

versa to minimize the number of corrections. Brandt [1] recommends working with the largest errors first and then make finer adjustments. He also suggests taking care to keep tensions within a range (not too tight or too loose).

Radial Truing A wheel rim with radial errors will appear to visibly bounce up and down while rotating. Any bounce up (a high spot) or bounce down (a low spot) must be corrected to make a circular shape. To detect the high and low spots, you will need a "reference probe" like a scale or some other gadget. The spoke that is centered on the rim section that is projected upwards is tightened to pull the rim inwards and likewise the spoke on the rim section is projected inwards is loosened to push the rim outwards.

Lateral Truing The purpose of lateral truing is to remove any wobbles and keep the wheel in a single plane. The reference probe must touch the side of the rim when you rotate the wheel. Unlike radial truing where individual spokes are tightened or loosened, in lateral truing, spokes adjusted in pairs. To adjust a right wobble, loosen the spoke at the center of the wobble and tighten the spoke directly opposite on the other side of the wheel.

Centering The purpose of centering is to ensure that the frame, front wheel, and rear wheel lie within the same plane. The rim of a centered wheel will be mid-way between the lock nuts of the hub. A truing stand will have a reference probe to detect centering errors. If the rim is off center to the left, then you would loosen the spokes on the left side of the wheel and tighten the spokes on the right side. Brandt [1] recommends half turns and quarter turns for large and fine adjustments respectively. Centering a wheel is obviously more tedious work since all spokes are adjusted and will take more time.

163

5.3　Wheel weight

The weight of a wheel is the sum of the weights of the rim, spokes, and hub. It is often quoted that next to the frame, the wheels are the most important component for performance (speed). If you have a carbon fibre frame, then you may assume that replacing a standard wheelset with a carbon fibre wheelset will improve speed. The cost of a carbon fibre road wheelset is relatively high and although the benefits are not easily computed, we can estimate the gain in speed.

5.3.1　How much faster is a light wheel?

A common assumption is that more weight on the wheels is roughly twice as worse as added weight on the frame [11]. The rotating weight on the rims is counted twice as much as the same weight on the frame, while the total weight of all spokes is counted $1\frac{1}{2}$ times and the weight on the hub is counted the same as weight on the frame. We first consider the three drag forces.

Since aerodynamic drag depends on the shape, whether the additional weight is on the frame or wheels will not matter. Similarly, slope resistance is the same regardless of the location of additional weight. The effect of heavier wheels on rolling resistance and bump resistance is tiny, since the additional weight is distributed evenly along the circumference of the wheel. In Chapter 4, heavier tyres did not always have a higher rolling resistance.

A heavier wheel will slow you down when you accelerate. It takes more kinetic energy to generate a higher speed (accelerate) when weight is added to the wheels compared to the frame. The kinetic energy of the bicycle can be divided into translational (or linear) kinetic energy and rotational kinetic energy. The translational kinetic energy is defined as $\frac{1}{2}m_b v^2$ where m_b is the total mass of the bicycle including the wheels and v is the velocity of the bicycle. The rotational kinetic energy is defined as $\frac{1}{2}m_w R^2 \omega^2$ where m_w is the weight of the wheel alone, R is the radius of the wheel, and ω is the angular velocity. The product $m_w R^2$ is the rotational analog of mass or moment of inertia for a wheel (or a hoop) and w is $\frac{v}{R}$. The total kinetic energy of a bicycle is

$$\frac{1}{2}m_b v^2 + \frac{1}{2}m_w v^2$$

Therefore adding weight to the wheel of a bicycle will add to both the translational and rotational kinetic energies for a given velocity v, while additional weight on the frame alone, will only increase the required translational kinetic energy. Fortunately, this additional energy only comes into the equation when you are accelerating. If you are riding at a steady pace, i.e. generating enough power to overcome drag forces alone, then the additional wheel weight will only add to the slope and rolling resistances.

Power evaluation with a heavy and light wheel　With a few assumptions, it is possible to get the difference in power required to accelerate with lighter wheels. If you assume that you are riding on flat terrain, with zero head wind, an aerodynamic factor of 0.23 (hands on the hoods), and a rolling resistance of 0.005, then you can compute the power required to accelerate with wheels weighing a total of 1.0 kgs. or 2.0 kgs. for a bicycle weighing 8 kgs. (without wheels) and a rider of weight 70 kgs.

Based on the equation for the total kinetic energy of the bicycle, the 1.0 kg. or 2.0 kgs. will contribute to both the translational and rotational kinetic energies. The total power required for three different acceleration rates of 1.0 kmph./second, 0.5 kmph./second, and 0.25 kmph./second for a fixed duration of 10 seconds with the two wheels shows that the difference due to wheel weights is minimal (see Figure 5.10 - six curves are shown which appear as three pairs with very little separation in a pair).

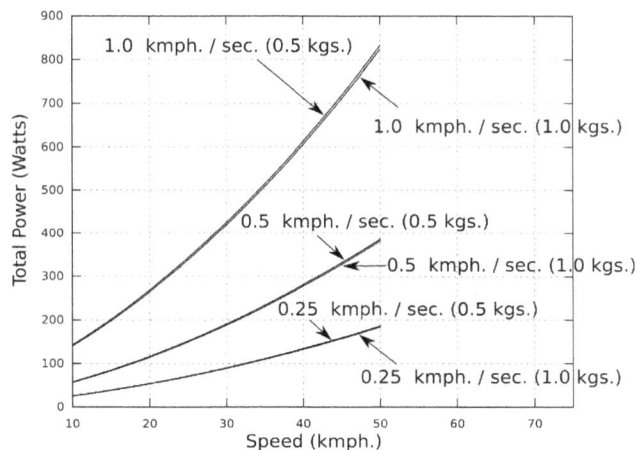

Figure 5.10: Total power required to accelerate at three different rates for two wheels of weight 0.5 kgs. and 1.0 kgs. vs. speed

If you are riding at 20 kmph and need to accelerate to 25 kmph. in 10 seconds, then your acceleration rate is $\frac{5}{10}$ or 0.5 kmph. / second. At 20 kmph. you will be generating about 64 watts under the assumptions given above. This is the power required to overcome all the three drag forces alone at 20 kmph. The power required to ride at 25 kmph. under the same conditions is about 109 watts, a difference of 45 watts. The kinetic energies for speeds of 20 kmph. and 25 kpmh. in Table 5.2 show that although the rotational KE for a 0.5 kg. wheels is half the rotational KE for a 1.0 kg. wheel, the difference in the total KE in both cases is not substantially different.

The translational KE is over 95% of the total KE for both light and heavy wheels and the KE due to the wheels in the translational KE is also minimal. The bulk of the KE in a bicycle comes from the relatively heavy rider (70 kgs.). The additional power needed for acceleration is the difference between the total KE at 20 kmph. and 25 kmph. divided by the period (10 seconds).

Speed (kmph.)	Wheel wt. (kgs.)	Rotational KE (J)	Translational KE (J)	Total KE (J)
20	0.5	15	1219	1234
20	1.0	30	1234	1264
25	0..5	24	1905	1934
25	1.0	48	1929	1977

Table 5.2: Kinetic Energies for Speeds of 20 kmph. and 25 kmph. with wheels weighing 0.5 kgs. and 1.0 kgs.

With 0.5 kg. wheels, it takes 1934 - 1234 J or 70 watts to increase the KE by 700 J in 10 seconds and with 1.0 kg. wheels it takes 1977 - 1264 J or 71 watts to increase the KE by 713 J in 10 seconds. Therefore, the total power required to accelerate from 20 kmph. to 25 kmph. is 45 + 70 = 115 watts and 45 + 71 = 116 watts for 0.5 kg. and 1.0 kg. wheels. In Figure 5.10, the benefits of low wheel weight are only noticeable at high speeds and high accelerations.

Observations

- When considering weight alone, there does not appear to be much benefit from a lighter wheelset. But the more expensive wheelsets come with other components such as the rim, hub, and spokes that have a higher quality than a standard wheelset.

- Typically the more expensive rims are more durable and stay true longer. A stiffer wheel may increase efficiency when applying force to the ground.

- There are fewer spokes in a more expensive wheel that may improve aerodynamics and the quality of the spokes will be better.

5.4   Steering

When a bicycle is stationary, it appears to be unstable and easily topples over. A beginner cyclist may attach trainer wheels to the rear wheel to keep the bicycle stable, when it is not in motion. A bicycle has only two points of contact with the ground and its base of support (the line joining the ground contact points) forms a line and not a polygon. The base of support for a tricycle or a recumbent bicycle is a triangle.

The center of mass or center of gravity is the average of all masses factored by the distance from a reference point. For equal weights attached to the ends of a line, the center of mass is the center of the line. In a bicycle, the center of mass changes depending on the position of the rider and the bicycle.

A bicycle becomes unstable when the center of mass is not over the base of support. While a stationary bicycle is unstable, bicycles are designed to be stable when in motion. Even a riderless bicycle with some initial velocity will be balanced for a short period.

The mathematics to describe bicycle motion, stability, and steering is complex and a few papers describing the reasons for the stability of a bicycle in motion have been published. Frame designers and bicycle manufacturers have experimented with tubes of different sizes and angles to optimize a bicycle for speed, handling, and comfort.

While speed is largely determined by aerodynamics and weight, handling depends on the geometry and distribution of mass on the bicycle. Part of the reason that it is difficult to mathematically describe the motion of a bicycle is that a rider can shift to different positions on the bicycle to change the direction or speed. Unlike other vehicles, the rider is four or five times heavier than the bicycle and can control its motion.

5.4.1 Rake and Trail

Bicycles and other two wheelers have been designed keeping in mind that stability is critical to keep riders safe. The *rake* is defined as the perpendicular distance between the hub and the steering axis (see Figure 5.11) and the *trail* is the distance between the projection of the steering axis to the ground and the contact point between the ground and tyre.

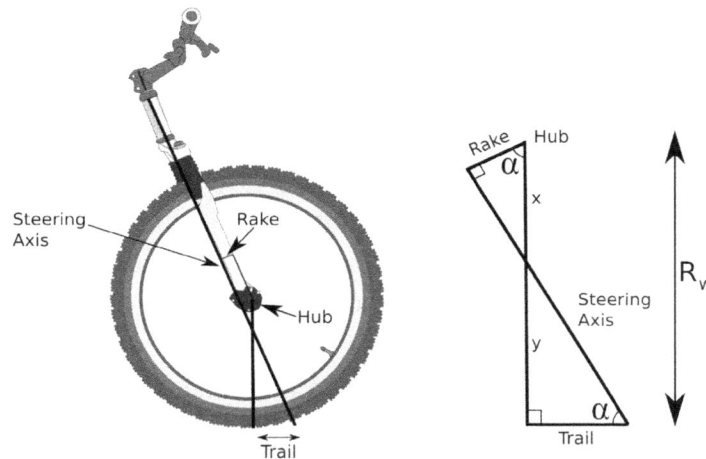

Figure 5.11: Rake and Trail of a Bicycle Wheel

For any given head angle α that the steering axis makes with the ground and a given rake, we can calculate the trail. The radius of the wheel R_w is the distance from the hub to the ground contact point or $x+y$ in Figure 5.11. Therefore $R_w = x+y$ where $x = \frac{Rake}{cos(\alpha)}$ and $y = Trail \times tan(\alpha)$. Solving the equation, we get $Trail = \frac{R_w cos(\alpha) - Rake}{sin(\alpha)}$

If you change the angle α, the rake, or the radius of the wheel, you can change the value of the trail. Small wheels will have a small radius R_w and therefore the trail will be smaller than the trail of a larger wheel. A steep angle between the steering axis and the ground will lead to less trail and vice versa (see Figure 5.12). A large α in the formula for trail will increase $sin(\alpha)$ and reduce $cos(\alpha)$.

Similarly a large rake will position the hub forward relative to the steering axis and therefore trail is small. The numerator in the formula for trail will be smaller given that all other values are constant. A small rake will position the hub relatively close to the steering axis and therefore trail will be larger. To keep the trail length within a range, a shallow head angle maybe combined with large rake and similarly a steep head angle will be combined with a small rake.

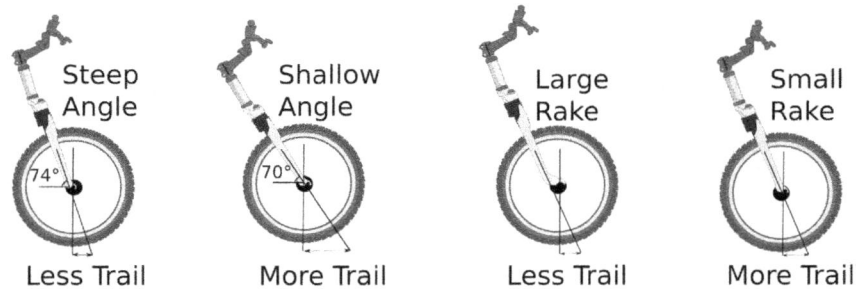

Figure 5.12: Changing Trail with Head Angle and Rake

What trail length is the best? With the three parameters wheel size, steering axis angle (head angle), and rake, you can change the length of the trail. Is a long or short trail better? A short trail is associated with better handling but less stability. You can make sharp turns at a moderate speed without losing your balance.

Early bicycles had longer trails to ride relatively smoothly on rough roads. A bicycle with more trail will follow the road closely and turn with a slight lean. The steering axis had a shallow angle such that the load would be distributed along the axis of the fork. With a steep angle, more of the load is supported at the fork crown which can lead to a failure of the fork after repeated vertical shocks. Once roads became smooth, steeper head angles and the associated smaller trail were more feasible. Forks comes in different shapes and different angles on bicycles (see Figure 5.13). A bicycle with a curved fork like the left bicycle in Figure 5.13 need not always have a larger rake than a bicycle with a straight fork. The hub can be positioned ahead of a straight fork to increase rake.

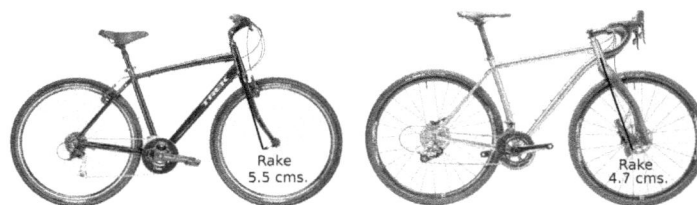

Figure 5.13: Forks and rake (courtesy Trek and Whycycles)

You can also adjust trail by changing the radius of the wheel. A small wheel will have a shorter trail than a large wheel if the other two parameters - the rake and head angle are the same. Small wheels are easier to turn and more responsive, but a ride on small wheels on a pot holed road will be rough.

The wheel base or distance between the centers of the small front and rear wheels will be small and therefore the base of support is also small implying less stability. Folding bicycles with small wheels tend to have short trails and therefore the steering is more "twitchy". Increasing the trail by decreasing the head angle would change the appearance of the bicycle and therefore is less common.

The third method to adjust trail is by changing the rake or offset. The fork ends at the hub and can be placed at some perpendicular distance x from the steering axis. The end of the fork is either curved, angled relative to the head tube, or the fork ends with a sharp angle at the hub. A large x means that the hub is more forward relative to the steering axis and therefore rake is larger.

In a shopping cart the forks are bent backwards to flip the wheels when moving forward. This allows you to steer the cart from the back. The wheels of a shopping cart or bicycle "trail" point towards a pivot. The pivot is where the steering axis meets the ground and in a bicycle it is ahead of the ground contact point. In a shopping cart, the pivot is behind the ground contact point and therefore the wheels point backwards.

Name	Head Angle	Rake	Trail	Mech. Trail	Wheel Flop
MTB Trek Superfly 26"	70°	4.1 cms.	7.4 cms.	7.0 cms.	2.4 cms.
Trek Emonda 47 cms.	71.2°	4.5 cms.	6.7 cms.	6.3 cms.	2.0 cms.
Trek Emonda 62 cms.	73.9°	4.0 cms.	5.5 cms.	5.3 cms.	1.5 cms.

Table 5.3: Head Angle, Rake, Trail, Mechanical Trail, and Wheel Flop for 3 Trek Bicycles

In Table 5.3, the head angle, rake, trail, mechanical trail, and wheel flop for three bicycles from Trek show that trail can change by almost two centimeters between bicycles (see Section 5.4.2 for mechanical trail and wheel flop). Since the radius of the wheel depends on the width of the tyre, the trail maybe marginally larger for wider tyres. In Table 5.3, a 44 mm. tyre width is assumed for the MTB and 25 mm. tyre width for the two road bicycles.

The smaller 47 cms. frame road bicycle has a smaller wheelbase by about 4 cms. compared to the larger 62 cms. frame road bicycle. The trail of the smaller road bicycle is slightly larger to provide more stability with a shallower head angle (71.2°). The difference in the head angle accounts for a trail longer by 1.7 cms., but the longer rake reduces the trail by about 0.5 cms. The MTB has the longest trail to provide more stability with slower steering.

On a bicycle with a small frame for short riders, the wheelbase may not be long enough to prevent toe clip. When you turn a bicycle with a small frame, the rear of the front tyre may clip your foot on the pedal. This is a problem if you want to make a quick turn on a road. Smaller frames tend to have more shallow angles to prevent toe clip.

Clearly, there is no ideal trail length and it depends on the type of bicycle, frame size, and terrain. Most bicycles have a trail length between the mid-50s to low 60s (mms.) [12]. This mid trail range is a compromise between speed and stability. Motorcycles may have much longer trails since stability at high speeds is more important compared to the trail of a slower bicycle.

5.4.2 Mechanical Trail

Wilson [13] defines the mechanical trail as the perpendicular distance from the ground point of contact to the steering axis. In Figure 5.14, the mechanical trail is the length of AB. Since we know the value of trail is $\frac{R_w cos(\alpha) - Rake}{sin(\alpha)}$ and $AB = AC \times sin(\alpha)$, the mechanical trail AB or L_{mt} is $R_w cos(\alpha) - Rake$.

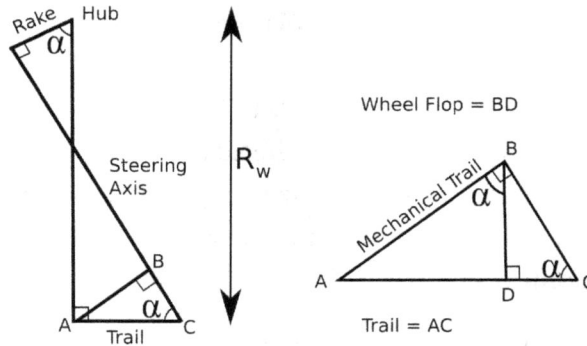

Figure 5.14: Mechanical Trail and Trail

You could also consider the mechanical trail as a projection of the trail at an angle of $(90 - \alpha)$ along an axis parallel to the rake. The length of the mechanical trail AB in Figure 5.14 has a large effect on the handling and stability of a bicycle. This is because the mechanical trail is a lever arm that acts along the steering axis. A perpendicular force acting at the point A will cause the steering axis to rotate to the left or right. The point A is where ground forces act on the tyre.

Wheel Flop When you turn the handle bars of your bicycle, you are losing a small amount of potential energy since the bicycle and rider are slight closer to the ground than before the turn. This loss of potential energy ends when the turn no longer reduces the height of the bicycle. As a result of this tendency to gravitate to the lowest height of the bicycle, a slight turn of the handle bars is accelerated by the bicycle to a position with the least potential energy.

The wheel flop BD in Figure 5.14 is the product of L_{mt}, the length of the mechanical trail, and $cos(\alpha)$. It reflects the amount by which the which the front wheel dips when turned to an equilibrium point (when L_{mt} or the trail becomes zero). As you turn the handle bars of the bicycle, the front end drops and the distance between where the steering axis meets the ground C and the ground contact point A reduces. At some steering angle θ, the distance AC is zero and therefore trail, mechanical trail, and wheel flop are also zero.

In Table 5.3, the wheel flop for the MTB is larger than the wheel flops for the two road bicycles. The wheel flop length is proportional to the mechanical trail which is in turn based on the trail length. A bicycle with a large trail will also have a large wheel flop which means that the bicycle will turn more than you would expect when making a turn. This does help make a smooth turn on a curve of small radius, but may give the rider less control on a turn of large radius.

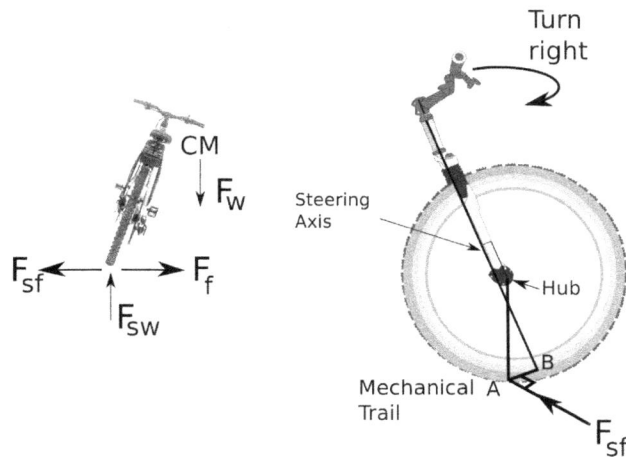

Figure 5.15: Mechanical Trail acts as a lever arm to turn handle bars to the right

In Figure 5.15 the forces acting on a bicycle leaning to the right will generate a reactionary force perpendicular to the mechanical trail. A bicycle will tend to the lean to the right if a rider moves slightly to the left to keep the center of mass (CM) positioned over the base of support joining the two ground contact points.

The two forces F_w and F_f are the weight force and the centripetal force respectively. The reactionary forces are F_{sw} and F_{sf}, the former counters the weight force while latter acts perpendicular to the wheel in the rider's frame of reference. Even as the wheel tilts to the right, the force F_{sf} has a component perpendicular to the mechanical trail. The force F_{sf} effectively generates a torque along the steering axis to turn the handle bars to the right.

Since torque is the product of the lever arm (mechanical trail) and the force perpendicular to the lever arm, a longer mechanical trail will generate more torque on the handle bars. This means that the front wheel will turn more sharply to the right when the frame leans to the right [14]. Although, this may not appear to add stability to the bicycle, the turn to the right is actually to counter the leftward force (see Figure 5.16).

In Chapter 3, the effective force moving the bicycle forward was shown as the sum of the two friction forces on the front and rear tyre. The friction force on the forward tyre acts in the reverse direction of motion and the friction force on the rear tyre acts in the direction of motion. When the bicycle is aligned on a straight path, the effective force is the difference between the magnitudes of the force.

When the front tyre is turned to the right, the friction force $F_{friction\_f}$ acts in the reverse direction of motion. The rear tyre is still moving in the same direction as earlier and therefore $F_{friction\_r}$ has the same direction as before. Now, combining the two friction forces generates an effective force $F_{effective}$ towards the left. To counter this leftward force, the frame will lean to the right. When the turn is complete, the rear wheel will be aligned with the front wheel and the leftward force will vanish. If you do continue to keep the front tyre turned to the right with the handlebars, then the bicycle will move in a circular path around the center of rotation.

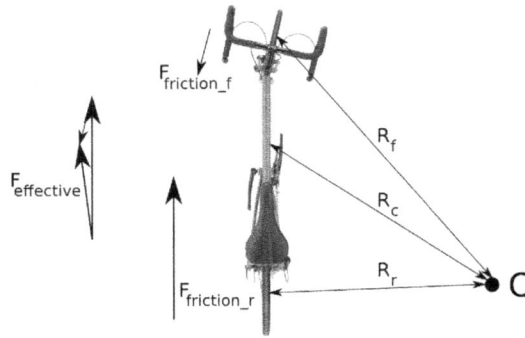

Figure 5.16: Right turn generates a leftward force

The rear wheel is closer to the center of rotation than the front wheel ($R_r < R_f$) and the radius of the cyclist R_c is greater than R_r but less than R_f. All three speeds of the front wheel, rear wheel, and rider are different around the common rotation point O. The differences in speed mean that the rider is no longer balanced over the front and rear wheel due to the induced acceleration.

5.4.3 Turning and Leaning

You can steer a bicycle by either turning the handle bars to turn the wheel or lean into a turn. The lean by itself will not turn the bicycle, but its effects create a lever force on the steering axis to turn the handle bars (see Figure 5.15).

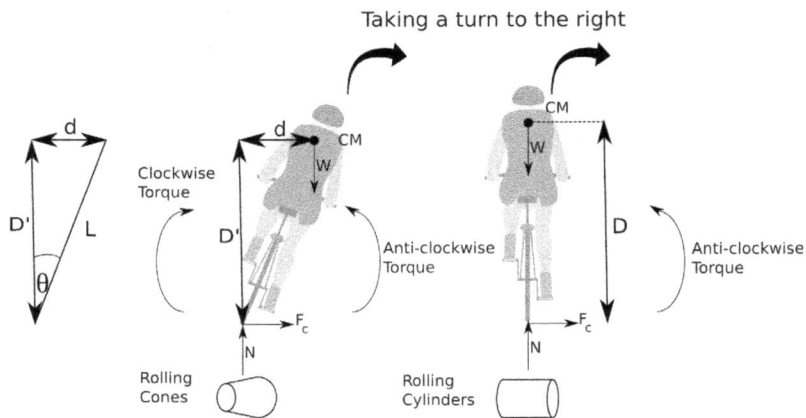

Figure 5.17: Rotational and linear forces when leaning and turning

In Figure 5.17, the rotational forces (torques) are balanced when a rider is leaning and turning but unbalanced when the rider is sitting upright. For an upright rider making a turn to the right in an inertial frame of reference, the centripetal force (F_C) creates a anti-clockwise torque with the pivot at the center of mass (CM). The remaining two forces, the weight (W) and the normal force (N) act directly through the pivot point CM when the bicycle is not leaning and therefore cannot generate any torque. Since there is an anti-clockwise torque ($F_c \times D$) on the bicycle, the rider will find it hard to stay balanced when turning and sitting upright.

In the center of Figure 5.17, the rider is leaning into the turn and therefore the normal force N has a component that will generate a clockwise torque ($N \times d$). Both the anti-clockwise and clockwise torques must be balanced for the rider to stay stable.

When the bicycle is leaning, the anti-clockwise torque is defined as $F_c \times D'$ or $\frac{mv^2}{r} \times L \times cos(\theta)$ where θ is the angle of lean to the right and r is the radius of the turn from the center of the mass. The clockwise torque is $N \times d$ or $mg \times L \times sin(\theta)$. If both anti-clockwise and clockwise torques are equal, then $tan(\theta) = \frac{v^2}{rg}$. This is the same formula to calculate the angle θ of a banked curve for a given velocity and radius (see section 5.5).

If you increase the lean angle θ, then d the perpendicular distance from the line of action of N increases and therefore the clockwise torque will also increase. However, if the lean angle is large then the torque due to centripetal force may not be sufficient to counter the clockwise torque and you are likely to fall over.

Likewise, you can increase the centripetal force by taking a turn at high velocity or reducing the radius of the turn. A large centripetal force means a large anti-clockwise torque. You may not be able to lean at a sharp angle to generate enough clockwise torque to counter the ant-clockwise torque.

Figure 5.18: Leaning into a Turn (Photo by Simon Connellan on Unsplash)

You would assume that by just leaning the bicycle alone, you could make a turn. However it turns out that although the tyre may appear to be rolling in the form of truncated cones and therefore should put the bicycle into a turn, two attached rolling cones do not take a curved path [15]. A single rotating cone does follow a circular path, but two connected cones follow a linear path just like two rolling cylinders.

Although, you cannot make a turn by leaning alone, you do need to lean into a turn to generate a centripetal force on the rear tyre. As the bicycle leans, the center of mass falls closer to the ground. The center of mass is at the highest position when the rider is sitting upright and therefore the potential energy of the rider and bicycle is the maximum [13]. The loss in potential energy is converted to a torque to turn the handle bars.

Common wisdom also recommends not pedalling while turning since there is a possibility of striking a pedal to the ground at the 180° position where the pedal is closest to the ground. Although grounding the pedal is not harmful in most cases, repeated contact of the pedal and the road surface will damage the pedal over time. With a higher bottom bracket, you can avoid this problem.

5.5 Cornering Forces

Taking a corner at a reasonable speed is one of the many pleasures of riding a bicycle. Even on an inexpensive bicycle, you can take a turn at a relatively high speed, albeit with some caution. The area of the contact patch for a combined (bicycle + rider) weight of 40 kgs. per tyre on a 2.5 cms. wide tyre at 70 psi was computed as 8.13 cm^2 (see Chapter 4). If the contact patch of a tyre is divided longitudnally into two halves, then the area of the contact patch on the side that the cyclist leans into, will grow as the lean angle increases and the area on the opposite side will reduce (see Figure 5.19).

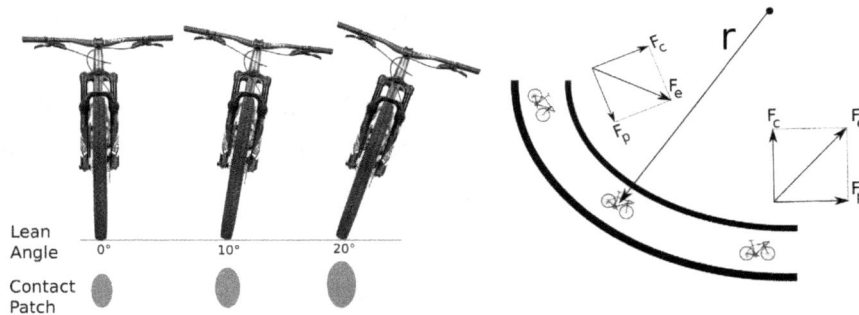

Figure 5.19: Contact Patch Areas for Cornering

The area of the patch on the left side of Figure 5.19 roughly reduces with $tan(\theta)$ where θ is the lean angle. At 45°, the entire contact area of the patch is on the right side of the tyre. As the area of the contact patch on the right side grows, the length and width of the contact patch grows as well.

In theory you could keep increasing the lean angle to increase the contact area, but at some point (> 45°), you may lose your balance [16]. As you tilt the bicycle to the right, your path will also veer to the right. The greater the tilt, the larger the rate of rotation, leading to a sharp turn.

5.5.1 How fast can you turn on a flat and banked road?

Centripetal force F_c is defined as $\frac{mv^2}{r}$ where m is the combined mass of the rider and the bicycle, v is the tangential velocity of the bicycle and the r is the radius of the circular path. Since $F_c \propto v^2$, centripetal force increases rapidly with increase in velocity. The length of the trail (see Section 5.4.1) will limit the turning radius r. A bicycle with a long trail is more stable, but requires a larger r to make a turn compared to a bicycle with a shorter trail. Therefore, not all values of r will be feasible on a bicycle with large trail. The effective force (F_e) on the bicycle is the sum of the centripetal and propelling forces, F_c and F_p. Often, it is difficult to generate both forces F_c and

F_p through a turn, since the bicycle leans to one side, there is a possibility of striking the lower pedal on the ground.

Turning on a Flat Road   For a bicycle tyre to roll on a circular path, there should be enough static friction (F_s) between the tyre and the road surface to generate the required centripetal force. The maximum value of F_s is defined as $\mu_s N$ where μ_s is the coefficient of static friction and N is the normal force. On a dry asphalt road, if μ_s is 0.75 and the combined weight of the rider and bicycle is 80 kgs., then the maximum static friction force $F_s = 0.75 \times 80 \times 9.8 = 588N$. If the required centripetal force F_c exceeds $588N$, then the tyres are likely to slip since the maximum static frictional force F_s is insufficient to provide the required F_c.

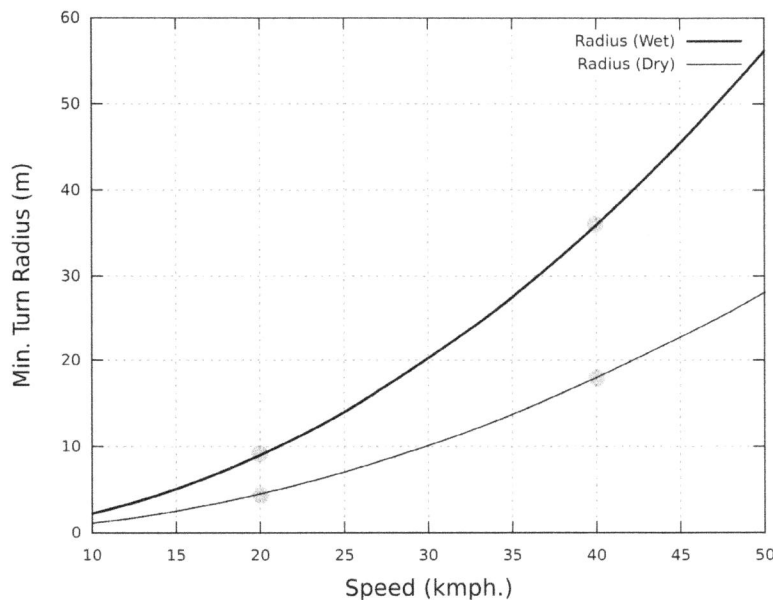

Figure 5.20: Minimum Turn Radius in meters vs. Speed in kmph.

When the radius (r) of the path is not large enough, F_c will be greater than F_s. At 30 kmph., if r is 7 meters and m is 80 kgs., then the required centripetal force is $\frac{80 \times (30 \times 0.27778)^2}{7} = 793N$. This exceeds $588N$ and therefore the bicycle will slip. As speed increases to 50 kmph. (see Figure 5.20), you need a larger radius to limit the magnitude of the required centripetal force. On a wet road if μ_s is 0.35, then the static friction force F_s is lower, while the required F_c to make a turn at a given speed does not change on a wet or dry road. Therefore, the radius for a turn on a wet road to maintain traction (no slippage) can almost double. Although the mass was used in the calculation of the centripetal and static friction forces, the turn radius is not dependent on the mass. The turn radius is defined as $\frac{v^2}{\mu_s \times g}$. If v increases, then the radius must increase as well. If the radius is small, then μ_s must be large enough to provide the centripetal force.

On a dry road at speeds of 20 kmph. and 40 kmph., you would need a minimum turn radius of 4.5 meters and 18 meters respectively (see Figure 5.20). The minimum turn radius increases with the square of the velocity and therefore is four times as much, when the speed is doubled. The minimum turn radius is inversely proportional to the coefficient of static friction μ_s. On a wet road at 40 kmph., the minimum turn radius increases to 36 meters.

As you lean into a turn, static friction force F_s of the tyre on the road resists the force on the tyre away from the turn. The sharper the angle of lean, the greater the friction force. The maximum lean angle θ for a dry and wet road with $\mu_s = 0.70$ and 0.35 are about 35° and 20° respectively using the formula $tan\theta = \mu_s$.

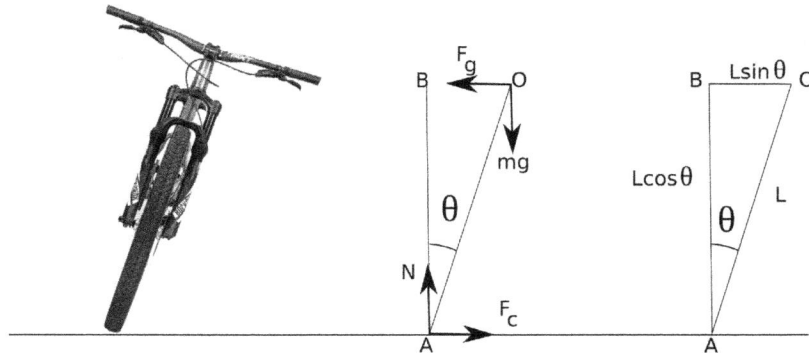

Figure 5.21: Forces on a leaning bicycle making a turn

In Figure 5.21 a bicycle leaning at angle θ making a turn will have forces acting at points O and A. The locations O and A represent the center of mass and the ground contact point respectively. The vertical forces are N and mg while the horizontal forces are F_c the centripetal force and F_g the centrifugal force in the cyclist's frame of reference. The distance L is the distance from O to A. Calculating torque around the point A, the forces $F_g \times Lcos\theta$ and $mg \times Lsin\theta$ must be equal if the bicycle is not tilting further and moving in a steady arc about the turn. Since F_g must equal F_c to prevent travelling sideways, therefore $F_g = \mu_s mg$ or $\mu_s mg \times Lcos\theta = mg \times Lsin\theta$ and therefore the steepest lean angle $\theta = tan^{-1}(\mu_s)$.

The calculation of the minimum turn radius from American Association of State Highway and Transportation Officials (AASHTO) uses a formula, $\frac{0.0079 \times V^2}{tan\theta}$, based on the lean angle θ from the vertical and the speed V in kmph [17]. Using the same equation to equate the torque around point A, we get

$$\frac{mv^2}{r} \times Lcos\theta = mg \times Lsin\theta \ or \ \frac{v^2}{gtan\theta} = r$$

where v is the speed in meters per second. At 35° and 20° lean angles and a speed of 20 kmph. the minimum turn radius is 4.5 meters and 8.7 meters respectively. The pedals are likely to strike the ground when lean angles are greater than 20°, although you could take a turn at higher lean angles without pedalling and maintaining the pedal on the inner side of the turn at the 0° position.

Turning on a Banked Road   On a banked road turn, you can can ride a little faster since some part of the centripetal force F_c is derived from the normal force (see Figure 5.22). The normal force N acts perpendicular to the surface of the road and therefore has components in the x and y directions - $NSin\theta$ and $NCos\theta$ respectively. The x component of the normal force N acts in the same direction as the centripetal force F_c towards the center of the turn. All or part of the centripetal force F_c maybe derived from the normal force depending on whether the bicycle is leaning or not. At the rated speed v of the banked curve when all of the centripetal force is derived from the normal force, $F_c = NSin\theta$. When your velocity exceeds v the rated speed of the curve, the frictional force adds to the normal force to generate enough centripetal force to keep the bicycle on a track without slipping. Without the frictional force, the bicycle would move upwards away from the turn to increase the radius of the turn.

The frictional force F_{sf} that is generated by leaning acts parallel to the road surface and opposes the tendency of the bicycle to head straight. Like the normal force, the frictional force also has two components in the x and y directions - $F_{sf}Cos\theta$ and $F_{sf}Sin\theta$ respectively (see Figure 5.22). Unlike on a flat road (where $\theta = 0$), the centripetal force on a banked road includes a part of the normal force - $F_c = NSin\theta + F_{sf}Cos\theta$. If $F_{sf} = \mu_s N$ the maximum frictional force, then the centripetal force $F_c = N(Sin\theta + \mu_s Cos\theta)$.

The angle θ of a banked curve maybe computed assuming a safe velocity v such that all of the centripetal force is generated from the normal force alone, i.e. frictionless or $\mu_s = 0$. Therefore the centripetal force $\frac{mv^2}{r} = Nsin\theta = \frac{mgsin\theta}{cos\theta} = mgTan\theta$ or $\theta = Tan^{-1}(\frac{v^2}{rg})$. If the safe or rated velocity is 25 kmph. and the radius of the turn is 12 meters, then the angle θ of the banked curve should be 22°. Increasing the radius of the turn from 12 meters to 15 meters will lower the angle θ to 18° and vice versa, decreasing the radius to 9 meters increases the angle θ to 28°. Realistic angles for a banked curve are much lower to avoid the risk of a vehicle sliding down the curve.

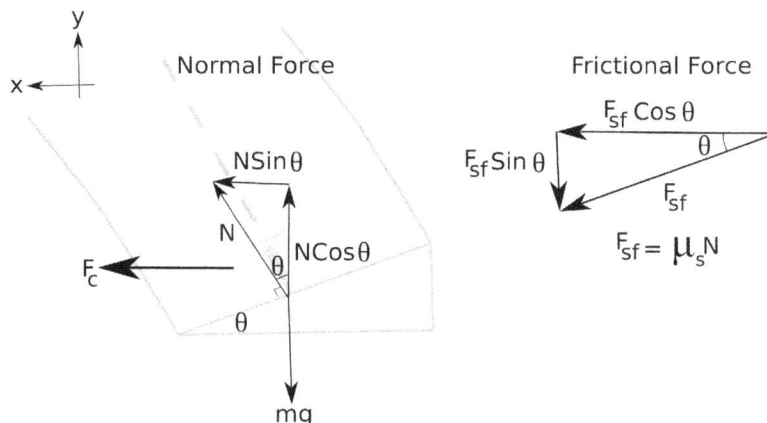

Figure 5.22: Normal and Frictional Forces on a Banked Turn

The maximum velocity on a banked curve, where the centripetal force is derived from both the normal and frictional forces can be computed by summing the forces in the x and y directions.

$$\Sigma F_x = F_c = \frac{mv^2}{r} = NSin\theta + \mu_s NCos\theta$$

$$\Sigma F_y = 0 = NCos\theta - mg - \mu_s NSin\theta$$

From the forces in the y direction, $N = \frac{mg}{Cos\theta - \mu_s Sin\theta}$. Replacing N in the sum of the forces in the x direction, the maximum velocity is

$$v = \sqrt{\frac{gr(Sin\theta + \mu_s Cos\theta)}{Cos\theta - \mu_s Sin\theta}}$$

On a flat road ($\theta = 0°$), the maximum velocity for a radius r of 9 meters is 28 kmph, where all of the centripetal force is derived from the static frictional force. If theta is increased to 20°, the maximum velocity increases by over 40% to about 40 kmph. On a banked curve, the additional of the horizontal component of the normal force to the frictional force, makes it possible to generate a higher centripetal force than with just the frictional force alone.

In theory you could increase the steepness of a banked curve to allow turns at high speed, but a slow rider would find it difficult to stay on a track without sliding down. Besides, you may also need to stop on a banked curve for whatever reason. Therefore, banked curves for average cyclists may have conservative angles between 1° to 6°.

The AASHTO formula for the minimum turn radius on a banked turn is $\frac{V^2}{127 \times (e + \mu_s)}$ where V is the velocity in kmph., e is the elevation fraction, and μ_s is the coefficient of friction. If the centripetal force is generated from the sum of the frictional and normal forces (see Figure 5.22), then the turn radius r is

$$\frac{mv^2}{N(sin\theta + \mu_s cos\theta)} \simeq \frac{v^2}{g(\theta + \mu_s)} = \frac{V^2}{127 \times (\theta + \mu_s)}$$

where V is the velocity in kmph. and θ is relatively small ($< 10°$). On a dry road, the minimum turn radius at 20 kmph., a 20% elevation, and a μ_s of 0.70, would be 3.5 meters. With a more common 2% elevation, the minimum radius increases to 4.4 meters at the same speed. Doubling the speed to 40 kmph. increases the minimum turn radius by almost four times to 17.5 meters.

Banked Road vs. Flat Road Maximum Speed A banked road will let you turn faster than on a flat road. The difference is not substantial, but is in the range of 5%-17% depending on the condition of the road, tyre, and slope (see Table 5.4). On a dry road, you can ride roughly 5% and 9% faster when the slope is 4% and 8% respectively. When the road is wet, you can ride at even higher speeds on a banked road, since the contribution of the normal force N to the centripetal force depends on the weight force and the angle θ of the slope, but not the coefficient of static friction.

Slope %	+Dry Speed %	+Wet Speed %
4	5	9
8	9	17

Table 5.4: Maximum Speed Increase in % for Two Slopes on a Dry and Wet Surface

Leaning on a Flat Road Leaning on a turn is one of the thrills of riding a bicycle. On a bicycle, you can make suprisingly sharp turns compared to other two wheelers. Leaning too far is a painful event that most riders will not encounter, since with a little experience you can sense a critical lean angle at which point you will lose control if you lean even further. The critical angle is estimated at 45° [16]. On a flat road when you lean, you will stay in a stable position, if the anti-clockwise and clockwise torques are balanced (see Figure 5.23).

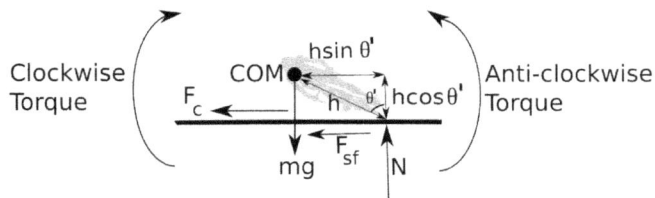

The clockwise torque about the COM in Figure 5.23 for a lean on a flat road is $F_{sf} \times h\cos\theta'$, where h is the distance of the COM to the ground contact point, must equal the anti-clockwise torque $N \times h\sin\theta'$ to maintain a lean angle of θ'. As the angle θ' increases, the distance $h\sin\theta'$ also increases while $h\cos\theta'$ decreases, making it increasingly difficult to balance both torques.

The static friction force $F_{sf} = N \times \mu_s = mg \times \tan\theta'$ will increase with larger θ', but when F_{sf} exceeds $\mu_s mg$, the tyre will begin to slip to the right in Figure 5.23. The maximum angle θ' will depend on the position of the COM and the value of μ_s which depends on the condition and surface of the road. When the bicycle is in equilibrium, $F_c = \tan\theta' \times mg$ which must be $\leqq \mu_s \times mg$. If $\tan\theta'$ exceeds μ_s, the tyre will begin to slip.

Figure 5.23: Leaning on a Flat and Banked Road

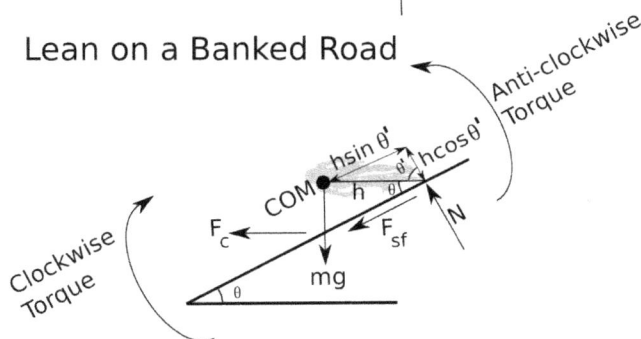

Leaning on a Banked Road On a banked road, the part of the centripetal force, $N Sin\theta$, due to the normal force, appears to remain the same whether you lean the bicycle or not (see Table 5.5). However, the normal force N is slightly higher with a lean compared to the N without a lean. The maximum normal force $N = \frac{mg}{Cos\theta - \mu_s Sin\theta}$ is greater than mg, since it must not only prevent vertical motion but also provide a part of the centripetal force.

Centripetal Force	Flat Road		Banked Road	
	No Lean	Lean	No Lean	Lean
Normal	0	0	$N Sin\theta$	$N Sin\theta$
Frictional	0	$mg \times tan\theta'$	0	$N tan\theta' \times Cos\theta$

Table 5.5: Normal and Frictional Forces on a Flat / Banked Road with and without Lean

Without any lean on a banked road, $N = \frac{mg}{cos\theta}$ and the contribution of the normal force to the centripetal force is $mg \times tan\theta$. When the lean angle is θ', the normal force $N = \frac{mg}{Cos\theta - tan\theta' Sin\theta}$ and a larger θ' will increase N which in turn increases both the normal and frictional components of the centripetal force. The maximum velocity is calculated in the same way as before by summing the forces in the x and y directions and substituting the normal force N as the square root of $\frac{gr(Sin\theta + tan\theta' Cos\theta)}{Cos\theta - tan\theta' Sin\theta}$.

In Figure 5.24, the maximum velocity for several lean angles and three banked turns shows that on sharply banked turns you can make a turn at a rather high velocity without slipping. The radius of the turn is 10 meters in Figure 5.24 (a larger radius would imply higher maximum velocities). A banked turn of 36° would be typically found in a velodrome while much smaller angles of 4° or 8° are more common on an urban road.

With a lean angle of 10°, the maximum velocity on a 8° and 36° banked turns is about 20 kmph and 36 kmph. respectively. If the lean angle is increased to 30°, the same maximum velocities increase to about 30 and 51 kmph. For banked turns of less than 10°, every additional degree of lean increases the maximum velocity by about 0.5 kmph. Clearly, the steeply banked turns in a velodrome allow for high speed turns that would not be possible on urban roads. Further, some lean angles may not be possible without skidding on a wet road due to the low coefficient of static friction.

In reality there are many other factors to consider including wind, type of tyre, and road conditions. However, even a simplified analysis of cornering shows the risks in taking turns at high speeds and how a wet road can change the safe speeds and lean angles for a turn. Once a tyre starts slipping, your weight pushes the wheel away from the turn and you are likely to fall.

Observations

- To stay in equlibrium, you can apply force on the handle bar away from the turn to reduce torque due to the normal force.

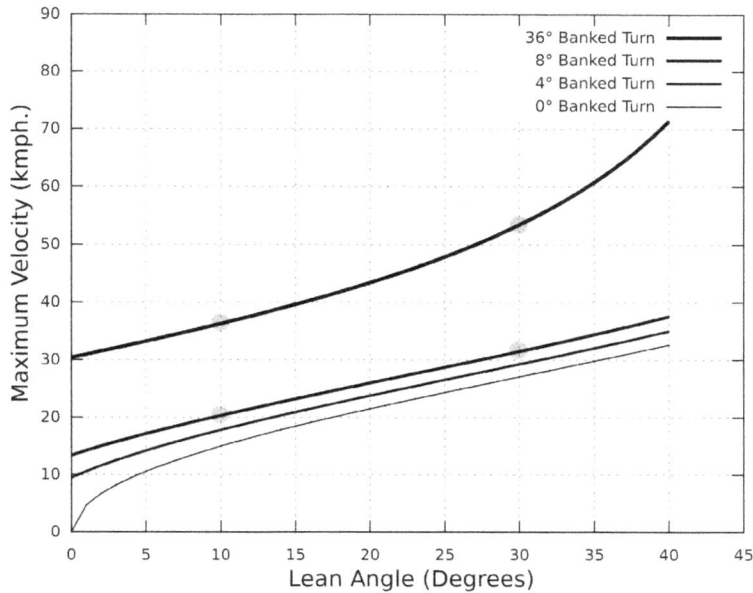

Figure 5.24: Maximum Velocity in kmph. vs. Lean Angle in degrees for three banked turns

- On a banked curve, without a frictional force component of F_c, speed must be maintained at the rated velocity to ride on a path following the turn without veering to the inner or outer edges of the turn.

- The normal force $N = \frac{mg}{cos\theta}$ is higher on a banked road of angle θ than the normal force mg on a flat road and even higher for a lean angle of θ' on a banked curve at $\frac{mg}{cos\theta - tan\theta' sin\theta}$.

5.6   Balancing

As long as the center of mass remains over the line joining the two ground contact points, a bicycle will stay balanced. The center of mass moves to the left or right if you lean in one direction or other. The support line joining the ground contact points moves when you steer the bicycle. Therefore balancing is a combination of steering and leaning to keep the center of mass over the support line. When a bicycle is in motion, you can steer it left or right to keep the center of mass over the contact points line. This is hard to accomplish in a stationary bicycle since a slightly off center bicycle will quickly lose equilibrium.

Track Stand   Some riders use a technique called a "track stand" to keep an almost stationary bicycle stable. In a track stand, the handle bars are turned about 45° which turns the front wheel (see Figure 5.25) and therefore the line of support moves to one side of the frame. When the bicycle moves back and forth, the front contact point moves laterally. The frame leans to one side to stay balanced and the center of mass (CM) stays over the line of support. This technique requires some minimal movement back and forth. On a fixed gear bicycle, you can simply rotate the pedals back and forth to move backwards and forwards. In other bicycles, you may need a slight incline to move backward.

Fear of falling One of the biggest fears of a beginner cyclist is losing balance and falling over. While this is a serious problem on a busy road, with a little practice it is possible to stay balanced. The reason a bicycle falls over is because the center of mass is not located over the line of support. The center of mass may move if you shift your position on the bicycle or when you steer the bicycle. A load in the back of the bicycle may shift during a ride and as a result the center of mass will also shift as well.

When a bicycle is initially leaning to the left, you would assume that steering to the right would be the best way to restore balance. Instead if you steer in the same direction (left), you are actually generating a force to move the frame from left to right which restores balance. Small turns of the front wheel generate a much higher centripetal force to keep a bicycle on track. The centripetal force and its counterpart in the rider's frame of reference, the centrifugal force, are proportional to the square of the velocity.

Turning smoothly is another problem that a beginner cyclist faces. To turn right, you would assume that you should simply turn the handle bars right and apply a correction torque on the handle bars to keep the bicycle along a steady curved path. Once you begin applying a torque to turn the front wheel, the angle of the turn will continue to increase unless you apply a restraining torque on the handle bars. Counter steering is a method to turn using the lean and steering of the bicycle.

5.6.1 Counter Steering

The main purpose of counter steering is to create a lean in the direction of the turn. If you want to turn right, then you initially turn left to create a lean to the right (see Figure 5.25). In the figure, F_{cg}, F_{cp}, and CM are the centrifugal force, the centripetal force, and center of mass respectively.

The initial turn to the left is brief and creates a centrifugal force in the rider's frame of reference to lean the frame of the bicycle to the right. Following the lean, the handle bars will also turn to the right (see Figure 5.15) to steer the bicycle to the right. A rider then simply needs to allow the bicycle to turn right at the appropriate radius by applying a controlling torque at the handle bars. The total time to counter steer is relatively short and cyclists do it almost unconsciously.

Riding too close to the curb edge may not give you enough room to make a turn. If the edge is to your left, then a right turn will require an initial left turn. If your front tyre hits the curb edge, you are likely to stop or even worse lose your balance. If you ride into a rut or a narrow pot hole in the road, you cannot steer the bicycle and you will lose your balance. Bicycle tyres are narrow compared to the tyres of motorized two wheelers and staying balanced is harder on a bicycle on a pot holed road.

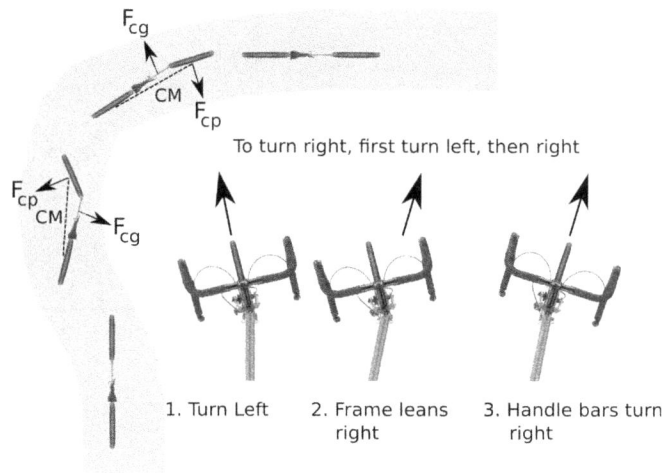

Figure 5.25: Counter steering to make a right turn

Is it possible to steer without counter steering? You could simply turn the handle bars in the direction of the turn. To turn right, you would pull the right handle bar closer to you and push the left handle bar further away. Now, the front wheel points to the right, but the front ground contact point is to the left of the frame. Therefore the frame will lean to the left to keep the center of mass over the line of support. To correct the outward lean and stay balanced, you would need to move your body to the right. This is more likely to take more effort than counter steering.

5.6.2 Wobbling

You may not need to always make a sharp turn. If you have to avoid an object or puddle in the road, then you may have to quickly steer to the left and right. The front wheel will travel a more wavier path than the path of the rear wheel (see Figure 5.26).

The wavelength is the distance between two consecutive waves. The amplitude of the wave is the distance from the mean path to either the left or right extreme. Following the initial wave, the amplitude of successive waves should diminish till the cyclist resumes riding on a path with no waves.

In a steady turn, the rear wheels will follow the same path as the front wheels. A wobble is a more irregular turn and the path of the rear wheel will lag behind the path of the front wheel. Since the front wheel makes a wavier path, the centripetal acceleration will be greater for the front wheel. Centripetal acceleration is defined as $\frac{v^2}{r}$ or equivalently $v \times \omega$ where ω is the angular velocity expressed in radians per second. Steering in the front gives you more control to quickly accelerate the front wheel.

The delayed centripetal acceleration of the rear wheel means that you have more control over a mass located on the front wheel. Often cargo is transported over the rear wheel and the lateral movement of the cargo during a ride can make it difficult to stay balanced. In the front

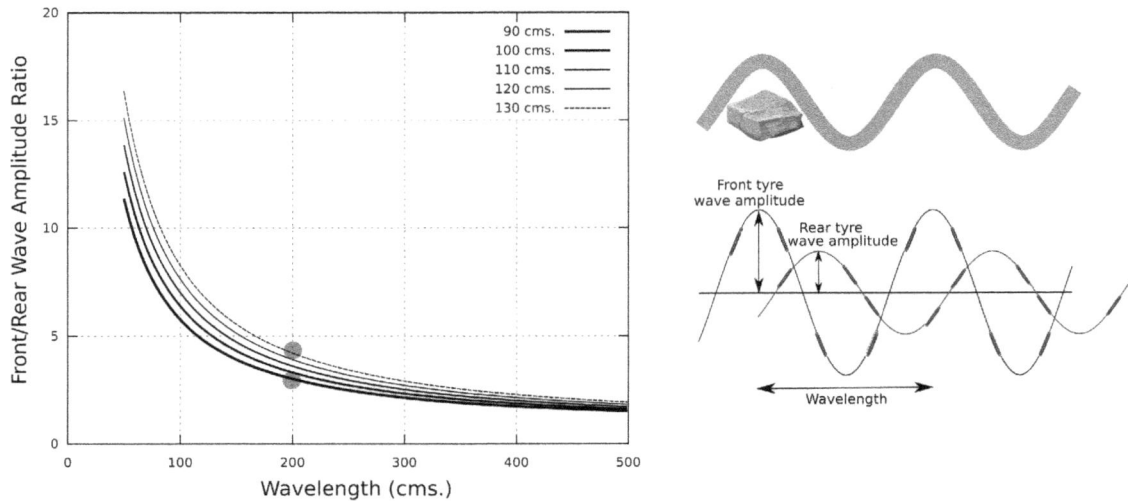

Figure 5.26: Front / Rear Wave Amplitude Ratio vs. Wavelength for different wheelbases

you have more control since you can quickly accelerate left or right. Still loads on the rear wheel are more common - a rear rack with panniers is convenient for large loads.

The formula to calculate the amplitude of the rear wheel A_r for a given wavelength and wheel base W is defined as

$$A_r = A_f \times \frac{1.0}{\sqrt{1 + (\frac{2\pi \times W}{\lambda})^2}}$$

where A_f is the amplitude of the front wheel and λ is the wavelength [13]. In Figure 5.26 the ratio $\frac{A_f}{A_r}$ decreases rapidly as the λ increases. The longer the wavelength, the more time it takes to complete a cycle and the more closely a rear wheel will follow the path of the front wheel.

The wheel base W also affects the front / rear wave amplitude ratio. For a short wheel base W of 90 cms., a wavelength λ of 200 cms., and a front wheel wave amplitude A_f of 20 cms., the rear wheel wave amplitude A_r is 6.66 cms. or a ratio $\frac{20}{6.66} = 3.0$. When the wheel base is increased from 90 cms. to 130 cms. the ratio increases from 3.0 to 4.2. The rear wheel is slower to follow the path of the front wheel when the distance between the front and rear wheels is larger.

5.6.3 Shimmy

A shimmy is an uncontrollable wobble at some frequency that occurs unexpectedly and typically at high speeds. In a shimmy, the handle bars oscillate laterally and a rider may panic and lose control of the bicycle. It is defined as a 3 hz. to 10 hz. steering oscillation of the fork, front wheel, and handlebar. A first guess would be that the wheel is not aligned (or in true) and therefore the wheel wobbles on a ride. A non-aligned wheel will not roll smoothly and should cause a disturbance felt on the handle bars for every revolution of the wheel. However, this type of wobble would happen on every ride and at all speeds and is also relatively easy to fix unlike a shimmy that is triggered at some given velocity.

The oscillation depends on the exchange of energy between the frame and the wheels. Normally, energy flows from the frame to the front wheel. The frame effectively trails the front wheel and is pushing it forward. When the energy flow is reversed, the wheel transfers energy to the frame which leads to the oscillation felt by the rider. A trailer behind a vehicle may fish tail when it is being towed – in this case energy is flowing from the vehicle to the trailer. If the trailer is propelling the vehicle forward by fishtailing, then energy flows from the trailer to the vehicle.

The two suspected triggers for shimmy are low torsional stiffness of the frame and high speed. Since the frame of the bicycle is made up of tubes that can stretch or bend, a frame can store and release energy. The period at which energy is absorbed and released depends on the torsional stiffness of the frame. At high speeds a tyre may generate a lateral force due to the angle around the steering axis which is transmitted to the frame.

The polar mass moment of inertia I_{zz} defines the the resistance to torsional deformation. For a wheel, it is defined as $\frac{1}{2} \times m_w \times r_w^2$ where m_w is the mass of the wheel and r_w is the radius of the wheel . A wheel with mass of 1 kg. and radius of 0.35 meters would have a I_{zz} of $0.06\ kg - m^2$. The larger the radius of the wheel, the higher the polar moment of inertia and therefore the velocity to trigger a shimmy is higher. Wilson [13] defined the trigger velocity for a shimmy as

$$V = \frac{k \times I_{zz}}{m_h \times c \times L_{mt}}$$

where k represents the composite stiffness or torque per unit area, m_h is the mass of the head tube, c is a constant, and L_{mt} is the mechanical trail. A high stiffness (high k) and high polar moment of inertia implies that the velocity to trigger a shimmy will be high. The value of k is used to define the stiffness of a spring in Hooke's formula $F = kx$ where x is the displacement of the spring. A stiff spring compared to a less rigid spring needs more force for the same displacement. For a k value of 1000 N/m, I_{zz} of $0.06\ kg - m^2$, m_h of 1.0 kg., c of 70 kg/s, and L_{mt} of 40 mm, the trigger velocity is 79 kmph. The constant c is the stiffness between the rider and the frame measured in kg/second. You can increase the trigger velocity by using

- a wheel of large diameter which will have large polar moment of inertia since more of the mass is distributed further from the axis of rotation

- a stiff fork and head tube will have a higher $k,$ but a very rigid bicycle may not be comfortable on a rough road

- a large L_{mt} or mechanical trail increases stability, but also increases the lever force of the wheel and therefore reduces the trigger velocity

- a small c means that the rider is not sitting firmly on the saddle and will dampen oscillations compared to a rider who is sitting firmly on the saddle.

Because a shimmy can start unexpectedly, a rider will find it unnerving and disturbing. Since the cause of the shimmy and the trigger speed cannot be pinned down, a rider will hesitate to ride fast on the same bicycle following the experience. The common advice is to stay calm, slow down, and ride just above the saddle to dampen oscillations.

5.6.4 Gyroscopic Effects

If you remove a wheel from a bicycle and tilt it to the left or right, you will find that there is little resistance to the movement. However if the wheel is rotating, you will find that there is more resistance to a tilt to the left or right.

To steer the bicycle, the force due to the gyroscopic effect must apply a torque at a distance equal to the height of the center of mass. The mass of the rotating wheel is much less than the combined weight of the bicycle and rider and torques due to the gyroscopic effects alone are not sufficient to keep a bicycle balanced. A study [18] with a custom bicycle has shown that gyroscopic effects from wheel rotation are not necessary for stability of a bicycle.

In an experiment to show that gyroscopic effects are not needed to balance a bicycle, David E. H. Jones [19] added a second front wheel and spun it in the opposite direction of the front wheel. This effectively nullified the gyroscopic effects from the front wheel alone. He was able to ride the bicycle despite the absence of gyroscopic effects. He created four unridable bicycles URB 1-4. The first bicycle URB 1 had an additional front wheel, the second bicycle URB 2 had zero trail and a thin tyre, the third bicycle URB 3 had positive trail and was the most stable, and the fourth bicycle URB 4 had negative trail. The last bicycle URB 4 was the hardest to ride and came close to being unridable.

References

[1] Jobst Brandt. *The Bicycle Wheel*. Avocet Inc., Third edition, 1993.

[2] Sapim Spokes. Butted spokes, 2018. URL https://bit.ly/2y9Pw9b.

[3] C. J. Burgoyne and R. Dilmaghanian. Bicycle wheel as prestressed structure. *Journal of Engineering Mechanics*, 1993.

[4] Damon Rinard. Wheel stiffness test, 2001. URL https://bit.ly/2BYTUKV.

[5] Roger Musson. *Professional Guide to Wheel Building*. www.wheelpro.co.uk., 2009.

[6] Patrick Taylor. Building a bicycle wheel, 2018. URL https://bit.ly/2OhwQOT.

[7] WilliamCycling.com. Bicycle wheel spoke lacing, 2018. URL https://bit.ly/2Pw5jpu.

[8] Parktool. Wheel tension measurement, 2018. URL https://bit.ly/2QqMX9w.

[9] Park Tools. Wheel tension app overview. URL https://bit.ly/2CATDhM.

[10] Ian Smith. Bicycle wheel analysis, 2018. URL https://bit.ly/2JhFR4A.

[11] Rhett Allain. We can prove why extra mass on bike wheels is your worst enemy, June 2016. URL https://bit.ly/2JGmeTz.

[12] Roff Smith. Romancing the trail, 2012. URL https://bit.ly/2MY6TDo.

[13] David Gordon Wilson and Jim Papadopoulos. *Bicycling Science*. MIT Press, Third edition, March 2004.

[14] Daniel Eley. The physics of bicycles, 2014. URL https://bit.ly/2OhU43u.

[15] Terry Colon. Bikes don't turn by leaning, 2008. URL https://bit.ly/2MGvuYx.

[16] James Witts. How far can you lean a bike in a corner?, September 2015. URL https://bit.ly/2OxbpGO.

[17] American Association of State Highway and Transportation Officials. *Guide for the Development of Bicycle Facilities*, 1999.

[18] J. D. G. Kooijman, J. P. Meijaard, Jim, Papadopoulos, Andy Ruina, and Arend L. Schwab. A Bicycle can be Self-Stable without Gyroscopic or Caster Effects. *Science*, 332:339–42, 2011.

[19] David E. H. Jones. The stability of the bicycle. *Physics Today*, September 2006.

6 Brakes

Ideally you would be the most efficient, if you could ride without braking. But this is impractical and you need to brake quite often when riding in a city. Brakes use friction to convert the kinetic energy of the bicycle and rider to heat energy. When you press the brake lever, a pair of rubber or composite brake shoes rub against some surface of the wheel. The harder you press the lever the more tightly the brake shoes rub against the braking surface. In a fixed gear bicycle, the friction between the tyre and the ground is used to stop the bicycle, instead of the friction between the brake shoes and the wheel surface.

You may have seen a car driver who accelerates towards a red light and then brakes sharply followed by another acceleration when the traffic light turns green. The cycle of repeated acceleration and braking not only stresses the brake and other components, it also does not save much time. While a car driver may not tire of repeated acceleration and braking, a cyclist cannot sustain this style of riding. A cyclist will tire following many cycles of acceleration and braking. Minimizing the use of brakes by coasting saves energy and extends the life of brake pads.

6.1 Types of Brakes

Assuming that you are no longer pedalling, the frictional force due to the brake shoes rubbing on the wheel surface acts like a drag force that will reduce your velocity to zero, if you come to a complete stop. The two popular types of bicycle brakes are rim and disc brakes. Both act on the same principle of pushing a brake pad against some wheel surface to dissipate the kinetic and potential energies of the bicycle and rider.

Broadly, the two types of brakes on a bicycle are defined by the location of the braking force – the rim or hub. The first two brakes on the left and center of Figure 6.1, the V-brake and the caliper brake, act on the rim of the wheel. The right brake, a disk brake is close to the hub of the wheel. All three types of brakes work by applying a force on either side of a rim or a rotor that is attached to the hub.

6.1.1 Rim Brakes

The three types of common brakes are the V-brake, caliper brake, and disk brake. The caliper brake is often found on road bicycles which typically have narrow wheels and is attached to the frame with a single pivot centered over the wheel. The cable to the brake lever is located on the left side of the brake and another name for this type of brake is a *side pull* brake. When the brake lever on the handle bar is squeezed, the left and right arms of the brake are pulled closer to the wheel rim. The right caliper arm has a pivot off center to the right of the wheel while the left caliper arm has a pivot centered over the wheel. A cam connects the brake arms so that they move by the same distance in opposite directions.

Ideally, you would like to use a brake with relatively high mechanical advantage, i.e. a low force on the brake lever translates to a high force of the brake pad on the rim. However to move the brake pads close to the rim with a high mechanical advantage brake may require a large brake lever that will be bulky and heavy. The brake levers on the handle bars have limitations on the distance that the lever can be squeezed before touching the handle bar and therefore mechanical advantage cannot be made arbitrarily high.

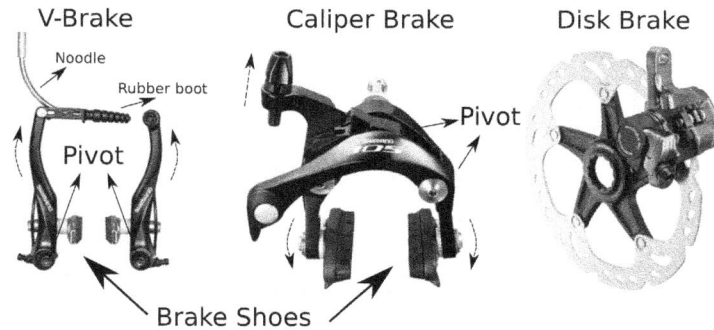

Figure 6.1: Three Types of Common Brakes (courtesy Shimano)

While caliper brakes can be attached with a single bolt to the frame, V-brakes use two bolts to attach the brake to the seat stays (rear brake) or the fork (front brake). Since V-brakes are not centered over the wheel, they can accommodate wider wheels like the ones found on mountain bicycles without an excessively long brake arm. Once V-brakes are installed correctly, they have enough stopping power for most bicycles including heavy touring bicycles. The mechanical advantage of a V-brake is fixed by the manufacturer since the lengths of the brake caliper and lever cannot be altered. The cable and pivot are located at fixed distances above and below the tyre respectively.

V-brakes are quite popular and relatively simple to adjust. The brake shoe can be moved vertically and rotated such that it does not touch the tyre and is positioned correctly on the rim. The two lever arms on either side of the wheel are long enough to generate sufficient braking force. The adjustment screws on each lever arm are used tighten or loosen a spring to center the brake pads. If the adjustment screw is turned clockwise, then the lever arm moves away from the rim and vice versa.

6.1.2 Disk Brakes

Rim brakes have been popular on road bicycles for some time because of their low weight and aerodynamics. Few modifications were necessary to adapt a frame to use a rim brake beyond the braze-ons to install the brakes. However on a heavier mountain bicycle on a downhill, the need for sturdier brakes with a higher braking force led to the development of the disk brake for bicycles. A mountain bicycle rim is also exposed to harsher road conditions and is likely to be muddy and wet leading to poor rim braking performance.

Further mountain bicycles tend to have much wider tyres than road bicycles and a rim brake would require a large wide caliper that arches over the tyre. It also takes a little more force at the lever to activate a large caliper. Since in a disk brake a disk is mounted near the hub, the thickness of the tyre does not factor in the transfer of the lever force to the caliper.

Disk brakes for road bicycles are more common now that the road bicycles are being designed for on-road and off-road surfaces. The width of tyres on road bicycles has increased to make riding more comfortable with fewer vibrations. As a result, most new bicycles of any type are more likely to have disk brakes to appeal to the rider who would like to ride on multiple surfaces with a high braking force.

Speed   One of the advantages of disk brakes is that the rotor of a disk brake rotates at a lower speed compared to rotation of the rim. The diameters of a typical road bicycle wheel and a disk brake rotor are 622 mm. and 203 mm. respectively. At a speed of 20 kmph. (or 5.5 m/sec.), the number of wheel rotations per hour of a tyre of diameter 700 mm. ($R_{wheel} = 350mm.$) is $\frac{20000}{\pi \times 0.7} = 9095$. While the rim of diameter 622 mm. ($R_{rim} = 311mm.$) rotates at a speed of about 5 meters per second., the rotor of diameter 203 mm. ($R_{rotor} = 101.5mm.$) rotates at 1.6 meters per second, almost a third of a speed of the rim. At the double the speed (40 kmph.), the rim and rotor rotate at speeds of 9.9 and 3.2 meters per second respectively. The brake pads for disk and rim brakes are located a distance of R_{rotor} and R_{rim} from the center of the wheel (see Figure 6.2).

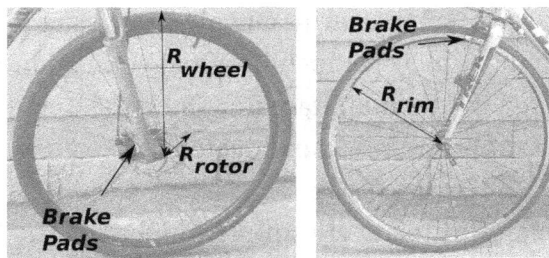

Figure 6.2: Radius of Brake Pads for Disk vs. Rim Brakes

Temperature   The increase in the temperature (flash temperature) of the rotor and the rim due to friction is proportional to the relative speeds of the rotor and rim respectively. Since the rim is moving past the brake pads at a higher speed, the rim will have a higher temperature than the temperature of the rotor. The heat of the rim can lead to punctures in a clincher tyre since hotter conditions increases the air pressure in the tube and may melt the glue used for patches.

Braking Surface Area   The rotor typically is not a solid disk and has holes to allow water to flow out in addition to minimizing the weight of the rotor. A solid rotor with a diameter of 203 mm. has an area of $\frac{\pi}{4} \times 0.203^2 = 0.032m^2$. However, a large part (60%) of the rotor is hollow and is designed in some pattern to let air and water flow through. Therefore, the effective surface area of the rotor is about $0.4 \times 0.032 = 0.013m^2$. The braking surface area of a rim is almost twice as much at about $0.024m^2$ for a 622 mm. diameter rim of depth 12 mm.

Braking Force   With both disk and rim brakes, the total kinetic energy of the bicycle rider must be removed. With a rim brake, a larger fraction of the kinetic energy is dissipated in the form of thermal energy due to the higher temperature of the rim generated from the friction between the pad and the faster moving rim. The remaining energy is transferred to the ground through the brake force F_{bf} (see Figure 6.6) that is applied for some stopping distance. Although less thermal energy is dissipated in a disk brake, the braking force is applied closer to the axis of rotation and must therefore be higher. This braking force is transmitted through the spokes to the rim and finally to the ground.

Consistency   Since the rim is close to the surface of the road, it will be exposed to more water under wet weather conditions than the rotor. The consistency of the disk braking force under different weather and road conditions is one of the reasons to choose a disk brake over a rim brake. Another reason is that the rim of the wheel is not subject to any braking forces and therefore should last longer.

Modulation   The purpose of modulation is to precisely control the magnitude of the braking force. In other words you can accurately control the rate of deceleration with the appropriate force on the brake lever. The rate of deceleration can range from $0.0g$ (coasting) to about $0.5g$ ($4.9\ m/s^2$) at which point the rear wheel will lose traction.

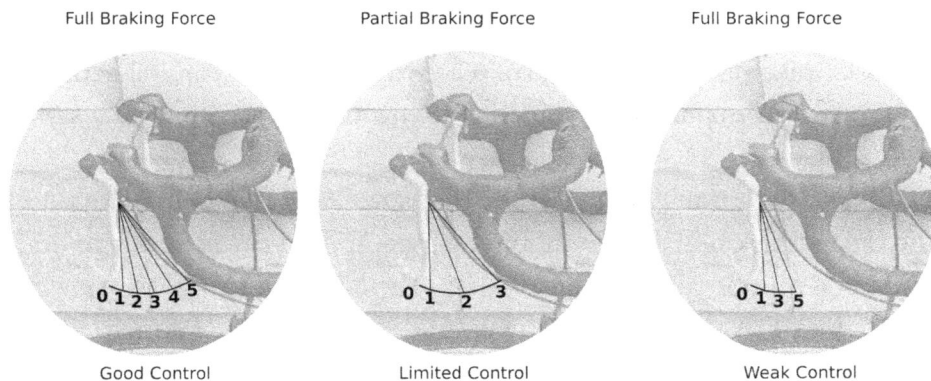

Figure 6.3: Brake Lever Modulation to Control Braking Force

The figure 6.3 shows three cases of how a brake lever controls the braking force. In the leftmost image, the lever moves from position 0 to 5 with increasing braking force in increments of $0.1g$ upto $0.5g$. This is the ideal case where you can use the lever to exert force upto the maximum before losing traction with the braking force increasing steadily as you move the lever till it is very close to the handle bar.

In the middle image, the lever moves the same distance, but the braking force does not increase upto the maximum. This can happen when the brake pads are slightly further away from the rim or the disk, than the recommended distance. In the rightmost image, the brake pads are too close to the rim or disk which limits the distance that the lever can travel. Although you maybe able to apply the maximum braking force, you will not have the fine control of the braking force that you have in the leftmost image.

6.1.3 Disk vs. Rim Brakes

Since a disk brake has a rotor closer to the axis of rotation than the rim of a rim brake, the force applied on the rotor must be higher to generate the same torque as the rim brake. If the ratio of R_{rim} to R_{rotor} is about 3, then the disk brake force will need to be three times higher than the rim brake force for identical braking torque.

How does a disk brake generate a high braking force Since disk brake pads can be placed very close (0.5 mm) to the rotor, the mechanical advantage for a small lever force is higher than the mechanical advantage with a rim brake. In a rim brake, you may need to maintain a higher separation distance (> 0.5mm.) between the rim brake pad and rim. A rim with a minor imperfection or that is not trued will necessarily require a larger distance from the brake pad to avoid rubbing against the rim while riding. Rim brake pads are also toed-in to make greater contact with the rim when brakes are applied. To toe-in a brake pad, it needs to be placed close to the rim on one side (facing the front of the bicycle) and away from the rim on the other side. When a brake pad is contaminated or not toed-in, you may hear squeaks when braking. A toe-in will increase the distance of the rim brake pad from the rim. A rotor is much smaller and thinner than a rim and can be made with fewer flaws. However, a rotor may also become warped over time and need to be replaced. A disk brake pad at a distance of 0.3 mm. from the rotor will generate about three times as much force as a rim pad at a distance of 1.0 mm from the rim for the same lever force.

Small or large disk pads? You would assume that since disk brakes generate more force than rim brakes, the area of disk brake pads should be greater than the area of rim brake pads. Disk brake pads come in a number of sizes and shapes. The area of a single disk pad can range from $500\ mm^2$ to more than $1200\ mm^2$. While a disk brake pad is more squarish, a rim brake pad is wider at about 70 mm. with a depth of 5mm.

Since braking is effectively the removal of kinetic energy, the power that a brake pad must absorb is rate at which this kinetic energy must be dissipated. The kinetic energy (KE) of a bicycle with a total weight of bicycle and rider of 80 kgs. moving at 27 kmph. (or 7.5 m/s) is $0.5 \times 80 \times 7.5^2 = 2250J$. The time ($t$) to stop a bicycle moving at 7.5 m/s with a braking deceleration of 0.5g is $\frac{7.5}{0.5 \times 9.8} = 1.5s$. The power absorption of the brake is $\frac{KE}{t}$ or $\frac{2250}{1.5} = 1500W$. If the area of the brake is $4 \times 70 \times 5 = 1400mm^2$, then the power absorbed per meter squared is $\frac{1500 \times 10^6}{1400}$ or $1.0MW/m^2$.

If you are riding on a downhill of slope 10%, then the potential energy in addition to the kinetic energy will need to be removed. For the same bicycle, speed, and braking deceleration of $0.4g$, the time to stop is about $\frac{7.5}{0.4 \times 9.8} = 1.9s$. The stopping distance is $\frac{v+u}{2} \times t$ where v is the final velocity, u is the initial velocity, and t is the time to stop or $\frac{7.5 \times 1.9}{2.0} = 7.0m$. The potential energy removed is $80 \times 9.8 \times 7.0 \times 0.1 = 548J$ and the total energy dissipated is $2250 + 548 = 2798J$. For the same disk pad area of $1400mm^2$, the power absorbed is roughly the same at $1.0MW/m^2$. The rate of power absorption is about the same on a downhill as a flat road since the time to brake increases in a downhill. The braking deceleration rate of 0.5g is close to the maximum possible without losing traction.

The material used in a disk or rim brake pad can alter the coefficient of kinetic friction (μ_k) making the brakes more or less effective. A rubbery rim brake pad can deform under pressure leading to a variable μ_k. A softer brake pad also may have embedded particles due to the force of the brake pad on the rim which can then scratch the rim.

Feature	Disk Brakes	Rim Brakes
Braking force	High	Low
Effectiveness on a wet road	High	Low
Weight	Heavier	Lighter
Modulation	Better	OK

Table 6.1: Some differences between Disk and Rim Brakes

Strong Fork   The frame of a bicycle with disk brakes must have a fork that can support the higher braking forces. The high disk force is applied on one side of the bicycle unlike rim brakes that apply force on both sides. When the brake lever is pressed, the left side of the fork is pushed forward while the right side remains attached, resulting in a twist that the fork must resist. The braking force is then transmitted through the hub to the spokes and eventually to the rim. A stronger frame and fork to handle these forces adds to the total weight of the bicycle.

Although, the front brake generates most of the braking force without skidding, a rear disk brake must also be built similarly to handle relatively high braking forces. Normally, the seat stays to which the rear disk brakes are attached are narrower than other tubes, but must be strong enough to withstand these forces.

Dished Wheel   If the rotor of a disk brake is placed on the left side of the wheel, then the weight of the wheel on the left side will be slightly higher than the weight of the right side. Although, all the spokes maybe equi-distant from the rim, the tension of the spokes on the left side will have more tension than the spokes on the right side. The number of spokes should also be suffcient to transmit the braking force from the hub to the rim.

A fewer number of spokes can be used in a rim brake based front wheel, since the spokes have to support the weight on the front wheel alone. Further, the spokes can be radially laced to build a symmetric front wheel. Braking forces from the rim are converted to ground forces due to the static friction between the tyre and road.

Modulation   A rim brake pad is installed with a series of concave and convex washers to allow for toe-in and can be moved up or down and rotated. With the ability to move the rim brake pad in three dimensions, there is a likelihood that the brake pad will not stay aligned for a long duration. A disk brake pad cannot be easily moved around in the caliper and stays aligned with the rotor when braking. Without toe-in, the entire area of the disk brake pad makes contact with the rotor at the same time giving slightly more control over the applied braking force.

Wet Roads One of the big problems with rim brakes is the sensitivity to wet weather. A brake pad pressing against a wet rim takes significantly longer to dissipate kinetic energy and therefore the stopping distance correspondingly increases. The rims of both wheels are a few millimeters from the ground and very likely to become wet. To generate a reasonable braking force, the water on the rim has to be first removed to allow the brake pad to make contact with the rim. A rotor is slightly further away from the road surface and also comes in various patterns to let water flow away. Mud and other debris are also less likely to get embedded in a rotor. On the other hand, disk braking performance is very sensitive to contaminants on the surface of the rotor such as chain lube or some lubricant. If disk brake pads are contaminated, then they may have to be replaced.

6.1.4   How do Brakes apply Force?

As you pedal to move the bicycle from left to right in Figure 6.4, the front and rear wheels experience a propelling torque in the clockwise direction. When you brake, the frictional force of the brake pad on the rim, F_{rim}, opposes the clockwise rotation of the tyre and generates a anti-clockwise braking torque. The anti-clockwise torque in turn generates a frictional force, F_{bf}, on the wheel from right to left. The braking force F_{bf} opposes the direction of motion and eventually brings the bicycle to a stop.

If you apply a force (F_{lever}) of 75N for a distance of 3mm. at the brake lever ❶, then the force F_{pad} applied at the rim, where the brake pad is 1mm. from the rim, will be three times greater at 225N ❷. This assumes that all the force is transmitted through the cable from the lever to the brake pad and is based on the idea that work or mechanical energy is conserved. The brake pads are at a much smaller distance of 1.0mm. from the rim than the distance of the lever from the handlebar. If the brake pads are even closer to the rim, the value of F_{pad} can be even higher. A high F_{lever} force is difficult to generate at ❶ since the distance that a lever can move is limited to the range of a hand grip and the distance of the handlebar from the lever.

If F_{pad} is the applied force by the brake pads on the wheel surface, then the frictional force on the rim $F_{rim} = \mu_{rim}F_{pad}$ where μ_{rim} is some fraction that depends on the type of rim surface and brake pad. The value of μ_{rim} depends on the *kinetic* friction between the brake pad and wheel surface, since the surfaces are moving relative to each other. Since a high F_{rim} is desirable to generate a high braking torque, manufacturers have experimented with many types of materials for brake pads and rims. Finding the right combination of materials that works well under both wet and dry conditions has been challenging.

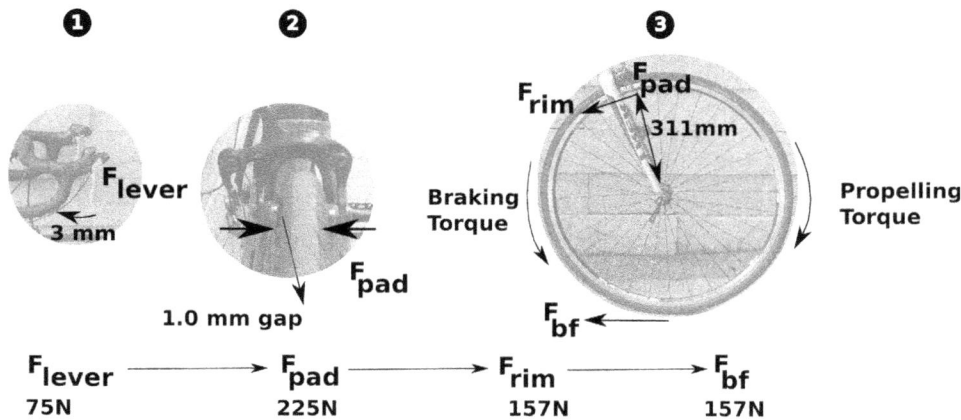

Figure 6.4: Transfer of Brake Force from Front Brake Lever to Front Tyre

When the wheel surface moves in relation to the brake shoe, the value of μ_{rim} increases in proportion to the speed of the relative movement if the pad contact area is also increased [1]. We would like a higher value of μ_{rim} to generate a higher F_{rim} such that the total drag force increases and the stopping distance is minimized. A polished or wet surface will lower μ_{rim}. Steel rims on a wheel that were found to have a relatively low μ_{rim}, have been replaced with aluminium or alloy rims.

Increasing F_{pad} is not always possible because it is difficult to move the brake pads arbitrarily close to the rim, since a slight deformation in the shape of the rim will rub against the pads when pedalling. It is also not necessary to move the brake pads very close to the rim, since the value of F_{pad} is often sufficient to stop within a reasonable distance (see Section 6.3). The actual force transmitted from the lever to the brake pads depends on the condition of the cable and the distance from the lever to the pads. A rear brake pad will have less force transmitted since the cable from the lever to the pad is longer and will have several turns which add to the loss of force transmitted due to friction. Fortunately, the front brake which provides most of the braking force is located close to the lever and the cable to the pad can be installed with few turns. Hydraulic brakes do reduce the frictional losses compared to mechanical brakes and a hydraulic rear brake will be as responsive a front brake.

At ❸ the force on the brake pad, F_{pad} is converted to a frictional force F_{rim} based on the kinetic friction between the rim and the pad. If the value of the coefficient of kinetic friction, μ_k, is 0.7, then $F_{rim} = 0.7 \times 225 = 157.5N$. This force (combined force of both pads) acts at a distance of 0.311m. from the center of the wheel and produces the braking force F_{bf}. The braking torque is $157.5 \times 0.311 = 49N-m$. Since the braking force acts at the same distance of 0.311m. from the center of the wheel, the value of F_{bf} is 157.5N. If a similar force is applied at the rear wheel, then the total braking force is 315N or about $0.39g$ for a combined bicycle and rider weight of 80 kgs.

Since F_{bf} is a reactionary frictional force that is based on the coefficient of static friction between the road and the tyre and the normal force on the tyre, there is a limit to the value of the braking force. This is very apparent in the rear wheel, where the rear braking force must be less than the static frictional force or $F_{br} <= \mu_s \times F_{nr}$. Unfortunately with higher braking forces, the normal force on the rear wheel, F_{nr}, decreases because more weight is transferred to the front of the bicycle. Therefore, the limit for F_{br} also decreases at higher brake forces. When F_{br} exceeds $\mu_s \times F_{nr}$, the wheel is no longer rotating and is being dragged on the road which would make the bicycle hard to control. The value of μ_s is also lower when the road surface is wet. This means that there is a higher chance of skidding on a wet road if you use the rear brake to generate a large fraction of the braking force (see Section 6.2.2 for the limits on rear braking force without skidding).

6.2 Braking Forces

Most riders are not perfectly centered longitudnally over a bicycle (see Figure 6.5). The mass of a typical upright rider is closer to the rear of the bicycle. The center of mass (COM) is defined as a point location from where in every direction the same amount of mass (bicycle and rider) exists. The COM can move depending on the rider's position on the bicycle. If a rider is an aggressive position with more mass towards the front of the bicycle, then the COM will move forward. The values of w and r are the distances between the centers of the two wheels (or wheelbase) and roughly the distance from the center of the rear cog to the center of the crank respectively. In most cases $r < (w - r)$ and the COM is closer to the rear wheel than the front wheel.

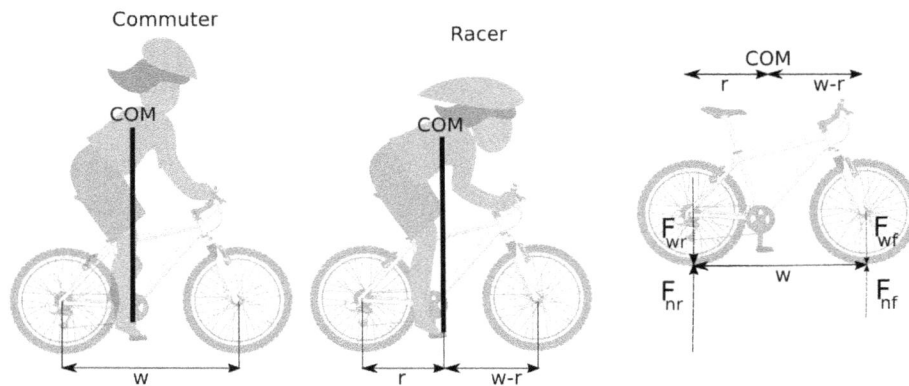

Figure 6.5: Weight Forces on a Bicycle and Center of Mass

On a stationary bicycle, roughly 60% of the weight is supported by the rear wheel since the ratio $\frac{w-r}{w} \approx 0.6$. If the COM acts like a fulcrum, then $F_{wr} \times r = F_{wf} \times (w - r)$ or $\frac{F_{wr}}{F_{wf}} = \frac{w-r}{r}$ where F_{wr} and F_{wf} are the rear and front fractional weights respectively of the combined weight (mg) of the bicycle and rider. The downward weight forces, F_{wr} and F_{wf}, are balanced by equal upward rear and front normal forces F_{nr} and F_{nf}. As r decreases, the ratio $\frac{w-r}{r}$ becomes larger and correspondingly F_{wr} increases as well. When r becomes 0 or the COM is directly over the rear wheel, the entire weight mg is supported by the normal force F_{nr}.

While a rider can change positions to move the COM forwards or backwards, the movement is limited to a small range. However, braking forces on the rear or front wheel can move the COM by a significantly larger distance. A braking force on the front wheel is equivalent to increasing F_{wf} without adding an extra mass in the front. It is convenient to express the deceleration due to a braking force in terms of the acceleration due to gravity g, because the required braking force can then be described as the product of the mass and the deceleration.

6.2.1 Calculating Normal Forces

We measure the braking force in terms of the force pushing the wheel into the ground. The higher the downward force on the wheel, the greater the braking force on the bicycle. Instead of measuring the downward force on the wheel, it is convenient to measure the normal force which is a reactionary force and equal to the downward force. When the normal force becomes zero, then the wheel has lost traction and the bicycle will skid.

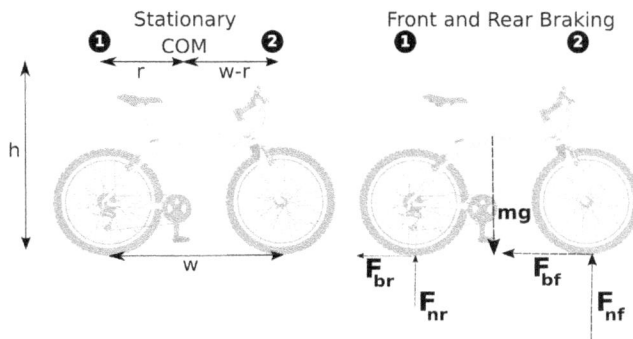

Figure 6.6: Normal forces on a Stationary Bicycle and with Front / Rear Braking

Normal Forces when Stationary In the Figure 6.6, the two downward weight forces F_{wr} and F_{wf} have been replaced with a single force mg in line with the COM. The normal forces, F_{nr} and F_{nf}, can be calculated from the rotational forces (moments) around axes passing through ❶ and ❷. The rotational forces are either in the clockwise or anti-clockwise directions. If the bicycle is not rotating vertically, then the clockwise and anti-clockwise torques must be the same. One of the two unknown normal forces (F_{nr} or F_{nf}) is removed from the equation when calculating moments around ❶ and ❷.

Moment	Clockwise	Anti-clockwise	Normal Force
❶	$mg \times r$	$F_{nf} \times w$	$F_{nf} = \frac{mg \times r}{w}$
❷	$F_{nr} \times w$	$mg \times (w - r)$	$F_{nr} = \frac{mg \times (w - r)}{w}$

Table 6.2: Normal forces on the Front and Rear Wheels for a Stationary Bicycle

Table 6.2 shows the calculation of the normal forces F_{nf} and F_{nr} when the bicycle is stationary. Since r and $(w - r)$ are roughly in the ratio of 4:6, the front wheel and rear wheel support about 40% and 60% of the weight (mg) respectively.

197

Normal Forces when Braking On the right of Figure 6.6, two brake forces, F_{bf} and F_{br} have been added to the front and rear wheels respectively. These forces represent the brake forces when the front and rear brake levers have been applied. The normal forces, F_{nf} and F_{nr} can be calculated by taking moments around around ❶ and ❷ respectively.

By equating the clockwise and anti-clockwise torques as before in Table 6.3, the value of F_{nf} has increased by $\frac{(F_{br}+F_{bf}) \times h}{w}$ compared to the same force when the bicycle was stationary. If $F_{br} + F_{br} = F_b$, where F_b is the combined braking force, then the additional normal force on the front wheel is directly proportional to the total braking force and the ratio of the height of the COM (h) to the wheelbase (w). Since the braking force is measured in terms of gs, it is equal to $x \times mg$ where x is some fraction between zero and 0.6 for a bicycle.

Moment	Clockwise	Anti-clockwise	Normal Force
❶	$(F_{br} + F_{bf}) \times h + mg \times r$	$F_{nf} \times w$	$F_{nf} = \frac{(F_{br}+F_{bf}) \times h + mg \times r}{w}$
❷	$F_{nr} \times w + (F_{br} + F_{bf}) \times h$	$mg \times (w - r)$	$F_{nr} = \frac{-(F_{br}+F_{bf}) \times h + mg \times (w-r)}{w}$

Table 6.3: Normal Forces on the Front and Rear Wheels when using Front and Rear Brakes

While F_{nf} increases, the value of F_{nr}, the normal force on the rear wheel, decreases by the same amount - $(F_{br} + F_{br}) \times h$. Effectively, some weight has been transferred from the rear wheel to the front wheel. When the amount of weight transferred equals $mg \times (w - r)$, the rear wheel has a zero normal force and will skid. As the values of the braking forces F_{br} and F_{bf} increase, F_{nr} will decrease and when the total braking force reaches a maximum, F_{nr} becomes zero.

Normal Forces with Front or Rear Brakes Alone When the front brake alone is used, the value of F_{br} is zero and we can calculate moments around ❶ and ❷ as before. The normal forces F_{nf} and F_{nr} are identical to the expressions shown in Table 6.3. Similarly, when the rear brake alone is used, the value of F_{bf} is zero and the same expressions can be derived for F_{nf} and F_{nr}. Therefore regardless of which brake is used, the normal force on the front wheel will increase and the normal force on the rear wheel will decrease by the same amount.

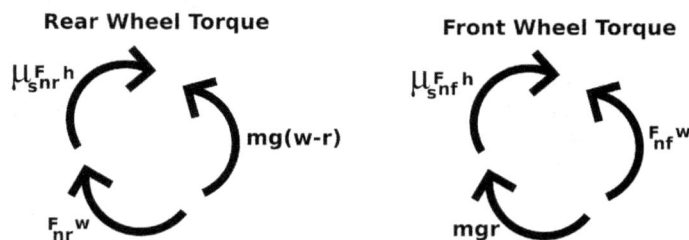

Figure 6.7: Torques on Front and Rear Wheels

The rear wheel and front wheel torques in Figure 6.7 show why the normal force on the rear wheel F_{nr} will be less than the normal force on the front wheel F_{nf}. The brake forces F_{br} and F_{bf} have been replaced with the product of μ_s and the respective normal forces. In the rear wheel, the brake force torque $\mu_s F_{nr} h$ acts in the same clockwise direction as $F_{nr} w$ to oppose the anti-clockwise torque due to $mg \times (w - r)$. The torque due to mgr and the brake force torque $\mu_s F_{nf} h$ in the front wheel together oppose the anti-clockwise torque $F_{nf} w$ in the front wheel. Since F_{nf} contributes to a torque in both directions, it can be made significantly higher than F_{nr} which acts in the clockwise direction alone. When the torques are in equilibrium, the front and rear normal forces F_{nf} and F_{nr} are inversely proportional to $(w - \mu_s h)$ and $(w + \mu_s h)$ respectively. The value of $(w - \mu_s h)$ is significantly smaller than $(w + \mu_s h)$ and therefore the front normal force is larger than the rear normal force.

6.2.2   What is the Maximum Braking Force?

There are three ways of applying brakes, the rear brake alone, the front brake alone, and both front and rear brakes. The maximum braking force is the highest braking force that you can apply before you lose control of the bicycle. You would naturally expect that when both brakes are used, the maximum braking force would be the highest compared to the cases where the front and rear brakes are used alone. However, there are occasions where the rear brake cannot be used with the front brake, without skidding.

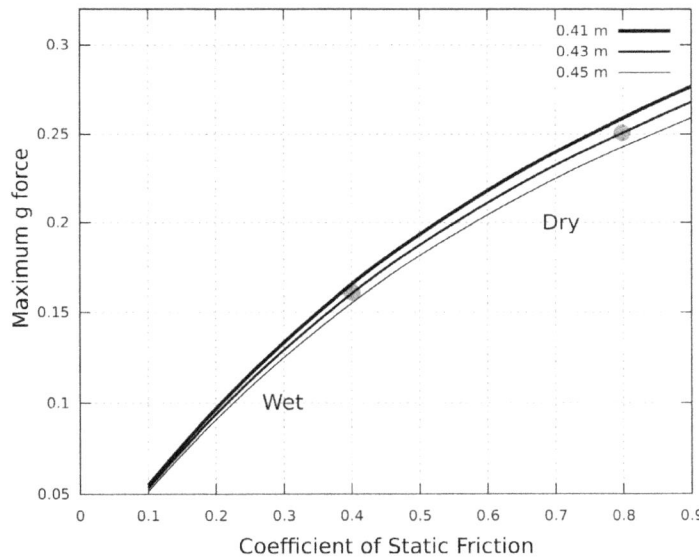

Figure 6.8: Maximum g force from Rear Brakes Alone vs. Coefficient of Static Friction for 3 Values of r

The maximum braking force occurs when F_{nr} becomes zero or from Table 6.3 when $(F_{br} + F_{bf}) \times h = mg \times (w - r)$. If the maximum braking force $F_{bmax} = F_{br} + F_{bf}$, and F_{bmax} is measured in terms of mg, then $F_{bmax} = \frac{w-r}{h}$. If w, r, and h are 1.067m, 0.432m, and 1.143m respectively, then $F_{bmax} = 0.56g$. Therefore, the maximum braking force can be increased by lowering the height h of the COM or increasing the difference between w and r.

Rear Brake Alone   When the rear brake alone is used, we can use the moments around ❷ from Table 6.3 to calculate the maximum value of F_{nr} as a fraction of the weight mg. The maximum rear brake force is $F_{br} = \mu_s F_{nr}$ where μ_s is the coefficient of static friction and F_{nr} is the normal force.

Setting $F_{br} = \mu_s \times F_{nr}$, $F_{bf} = 0$ and taking moments around ❷ in Table 6.3, we can solve for $F_{nr} = \frac{mg \times (w-r)}{\mu_s \times h + w}$. Figure 6.8 shows how the maximum g force that you can apply with rear brakes alone changes with distance of the crank to the rear cog (r) and the coefficient of static friction (μ_s). The calculations are made assuming a wheelbase (w) of 1.05 meters, a COM height (h) of 1.15 meters, and three values of r: 0.41m, 0.43m, and 0.45m. On a wet road with a lower coefficient of static friction, the maximum rear braking force ($\mu_s F_{nr}$) is lower.

When μ_s is 0.8 on a dry road, the maximum braking force is about $0.25g$ which is reduced to $0.16g$ on a possibly wet road with μ_s of 0.4. Smaller values of r for a crank which is closer to the rear cog will have a marginally higher braking force that is larger on dry roads than on wet roads. You can also physically move to the rear of the bicycle when braking, to position the COM closer to the rear cog which in turn increases F_{nr}.

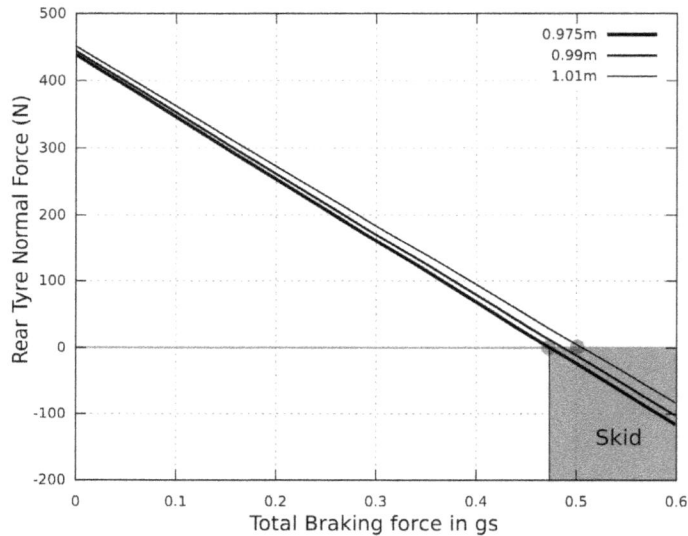

Figure 6.9: Decreasing Rear Tyre Normal Force (N) vs. Total Braking force in gs for three Wheelbases

Front Brake Alone   In theory, you can apply a high braking force on the front tyre alone without skidding. However, the rear tyre will lose traction when the braking force is excessive (typically $> 0.5g$). If we take moments around ❶ as before with $F_{br} = 0$ and $F_{bf} = \mu_s \times F_{nf}$, then the maximum F_{bf} occurs when $F_{nr} = 0$ or $\mu_s F_{nf} \times h = mg \times (w - r)$.

If $F_{nr} = 0$, then the entire weight force is supported by the front tyre alone or $F_{nf} = mg$ or $F_{bf} = \mu_s mg$ and therefore $F_{bf} = \frac{mg \times (w-r)}{h}$. For the same w,r, and h values of 1.067m, 0.432m, and 1.143m respectively as before $F_{bf} = 0.56mg$. Therefore, the maximum front braking force alone is the same as the limit for both front and rear brakes.

Front and Rear Brakes In Table 6.3, the values of F_{nf} and F_{nr} increase and decrease respectively as the total braking force increases. Starting from $0.0g$, the braking force F_b can be steadily increased till F_{nr} becomes zero.

Figure 6.9 shows how the normal force on the rear wheel steadily decreases with increasing braking force. If a bicycle with a total weight of 80 kgs. has dimensions of 0.975m, 1.15m, and 0.43m for the w (wheelbase), h (COM height), and r (distance of rear cog from crank) respectively, then the front and rear tyre normal forces can be calculated using the formulas in Table 6.3 for a number of brake forces.

With higher braking g forces, more weight is shifted to the front tyre till the rear tyre has zero normal force. For a wheelbase of length 0.975 m (49 cms. frame), the maximum braking force is $0.47g$. With a longer wheelbase of 1.01m (61 cms. frame), a slightly higher maximum braking force of $0.5g$ can be applied before the rear tyre loses traction.

Since observing the g force while braking is not easy, you can get an idea of your braking limits by repeatedly applying the front brakes harder till the rear wheel begins to skid. If you do continue to apply the same braking force, the rear wheel will begin to skid sideways to the left or right, making it very hard to control the bicycle. One approach to effective braking suggests a roughly 3:1 braking force on the front and rear brakes respectively.

In general, a bicycle is less stable than other motorized vehicles. The wheelbase of a motor cycle or scooter is larger and the center of mass is lower than on a bicycle leading to more stability in a motorized two wheeler. On a car, the wheelbase is even longer and the center of mass is still lower making it very difficult to flip a car by simply braking. The maximum braking force on a car or motor cycle is significantly larger than $0.5g$.

6.2.3   Distributing the Rear and Front Brake Force

In addition to a limit on the total braking force F_b, there are limits on how much of the total braking force can be handled by the rear brake. If the rear brake force is excessive, then the rear wheel may skid even though the total brake force is within limits. The rear brake can be safely used exclusively for low brake forces ($< 0.1g$) without any risk (see Figure 6.10). The values below are for a specific speed of 25 kmph. with no wind and on a flat road.

On a dry road, once the total braking force exceeds $0.24g$, the front brake must be used along with the rear brake. The fraction of the braking force that the rear brake can handle decreases with higher braking forces till the total braking force reaches $0.5g$. The rear braking force depends on the product of the coefficient of static friction, μ_s, between the road and the tyre and the rear normal force, F_{nr}.

When the road is wet, the rear brake becomes even less effective since μ_s will be lower and therefore for the same rear normal force, the maximum rear braking force is lower. On a wet road,

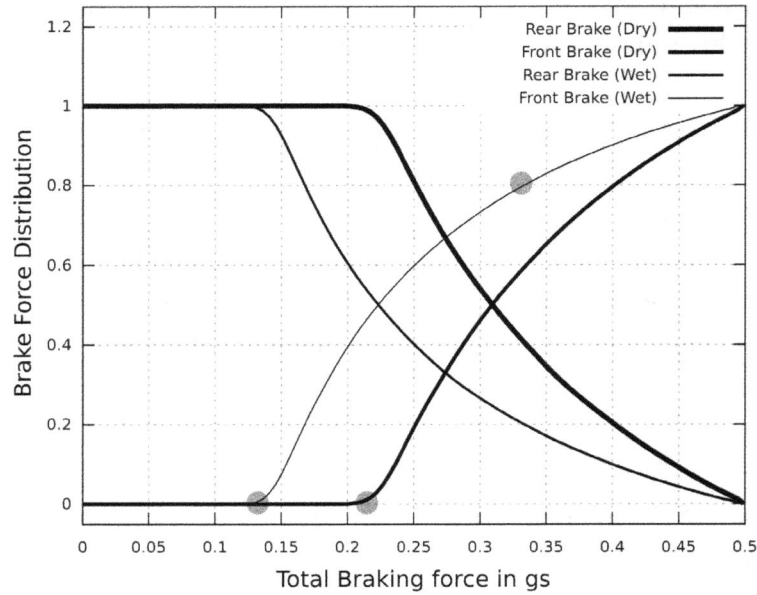

Figure 6.10: Fraction of Rear and Front Brake Forces for Wet and Dry conditions vs. Total Braking Force

the rear brake cannot be used alone, once the total braking force exceeds $0.14g$. If the braking force exceeds $0.33g$, then the front brake must handle over 80% of the braking force. When the road is wet, the value of μ_s may change substantially depending on the type of tyre compound. The rear brake will be more effective at higher gs with a slick tyre that has a higher μ_s on a dry road.

Table 6.4 shows the minimum percentage of brake force that needs to be generated from the front brake alone under dry and wet conditions for a given total braking force measured in gs. As the total braking force increases, a larger fraction of the force must be generated from the front brake. In a *panic stop* where the braking force is the maximum (just before losing traction on the rear wheel), the front brake provides all of the braking force. The rear brake is useful when relatively minimal braking forces are required. Keeping track of which brake to use when stopping is not something that comes naturally.

g Force	Braking Effects	Front % (Dry)	Front % (Wet)
0.1	Slowing down	0	0
0.2	Gradual stop	0	40
0.3	Quick stop	45	75
0.4	Hard stop	80	90
0.5	Panic stop	100	100

Table 6.4: Minimum Percentage of Brake Force from the Front Brake alone in Dry and Wet conditions

In general, you would use the front brake and not the rear brake when you need to brake more forcefully. Olsson [2] calculated braking performance of the rear and front brakes using simulations with somewhat similar results, but a higher maximum braking force of close to $0.7g$. Wilson [1] computed a value of $0.56g$ for the maximum braking force by equating clockwise and anti-clockwise torques when the rear normal force was zero. This value is dependent on the dimensions of the bicycle such as the length of the wheelbase and the height of the COM. However, the dimensions cannot be changed substantially to make a large change in the maximum braking force.

6.2.4 Head-over-Handlebar

One of the hazards of braking too hard is the risk of falling over the handlebar. It is best to avoid this painful experience and the accompanying risks of a collision with another vehicle or even the road. In section 6.2.1, the weight that was transferred to the front of the bicycle was proportional to the braking force. Since the COM is a point from which the mass in all directions is the same, the COM also moves forward (see Figure 6.11) when you brake. The value of F_{wr} reduces by x and that of F_{wf} increases by the same amount x.

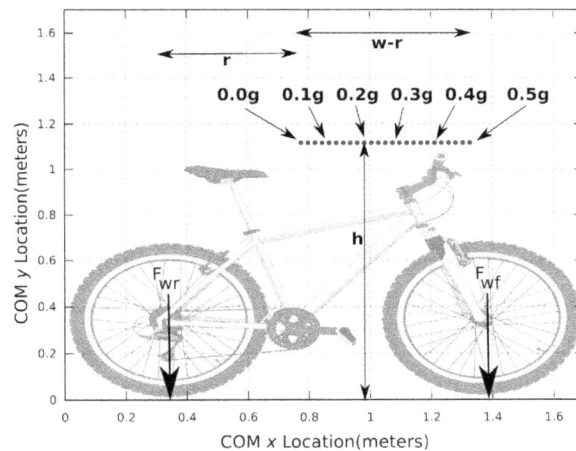

Figure 6.11: Movement of Center of Mass with Increasing Brake Force (in gs)

The value of x is computed as $\frac{F_b \times h}{w}$ where F_b is the total braking force, h is the height of the COM, and w is the length of the wheelbase. If the following values of 0.975 m., 1.15 m., and 0.43 m. are assumed for the length of the wheel base, COM height, distance of the crank center to the rear cog, then at $0g$ (no braking force), $F_{wr} = 438N$. When the braking force is $0.2g$ and $0.4g$, F_{wr} decreases to $253N$ and $68N$ respectively.

The value of r is calculated from the equal clockwise and anti-clockwise torques, $F_{wf} \times (w - r) = F_{wr} \times r$, taken about the COM. The value of r gives the x location of the COM. In Figure 6.11, the y location of the COM is constant at h. This assumption will not be true if the rider changes from a sitting to a standing position.

The problem with the COM moving forward is that it increases the chances of falling over the handlebar on a hard stop. Since a rider is not strapped to the bicycle, when the bicycle comes to a stop, the rider will continue to move forward (due to the inertial force) and with a COM over the front wheel it is difficult to stop the fall. Some recommendations include riding low, moving back on the saddle to keep the COM closer to the rear wheel, and using your arms to brace yourself before the bicycle comes to a stop.

When you ride on a downhill, there is a higher risk of falling over. A slope force of $mg \times sin(\theta)$ (see Figure 6.12) acts in the same direction of motion. The slope force adds to the inertial force which acts in the same direction when braking, making it harder to stay on the saddle.

To Skid or Not   Common wisdom recommends that the use of skidding to stop is risky. One of the main reasons to avoid skidding is that it is hard to control and if the bicycle is tilted too far to the left or right, then you are likely to fall over. Fixed gear bicycles are appealing because of their simplicity with no cables or gears. While the use of gears is not necessary for a bicycle, all types of practical bicycles must have some type of braking mechanism.

In a fixed gear (fixie) bicycle without front brakes, the deceleration of the bicycle is provided from the kinetic friction between the road and tyre unlike other bicycles which use the kinetic friction between brake pads and a rotor or the rim. The rear wheel does not have a freewheel which means that you cannot coast on a fixie and you have to be either pedalling forwards or backwards. When you pedal backwards, you are actually trying to reverse the direction of motion of the bicycle and if you are moving forward, then the rear wheel will skid. All the braking force comes from the kinetic friction of the rear wheel being dragged on the road. This of course increases the wear and tear of the rear tyre.

One of the issues to consider when riding is whether braking through a skid is sufficient to stop within a reasonable distance without falling over. The stopping distance for a fixie is calculated in Section 6.3. Since there is no brake lever in a fixie, you have to control the braking force by the force on the pedals.

6.2.5   Braking on a Slope

On a downwhill, not only will you need a higher braking force to stop, the normal force on the rear wheel is less than on a flat road, increasing the risk of falling over the bicycle. Gravity acts as a negative drag force on a downwhill and therefore the braking force must increase to compensate for the lower drag force. Conversely on an uphill, the positive drag force due to gravity will reduce the braking force necessary to stop.

On a downhill of slope with angle θ the normal forces (F_{nr} and F_{nf}) and the weight force (mg) are altered. The sum of the normal forces matches the reduced weight force of $mgcos(\theta)$, instead of mg. The other component of the weight force, $mgsin(\theta)$, is a negative drag force and reduces the total drag force. The braking forces, F_{bf} and F_{br}, act as before to oppose the direction of movement.

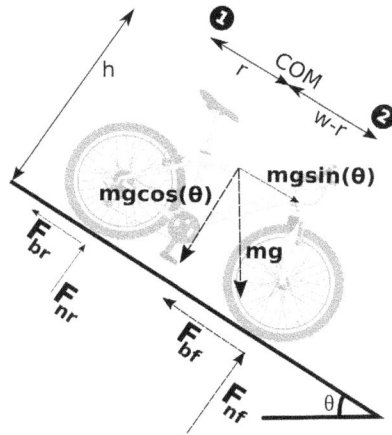

Figure 6.12: Braking and Normal Forces on a Downhill

Since F_{nr}, the rear wheel normal force on a slope, is less than the normal force on a flat road, the maximum braking force before the rear wheel loses traction will also be less. If we take moments around ❷ as before (see Figure 6.12) and equate the clockwise and anti-clockwise torques, we get

$$F_{nr} \times w + (F_{br} + F_{bf}) \times h = mgcos(\theta) \times (w - r) \text{ or } F_{nr} = \frac{mgcos(\theta) \times (w - r) - (F_{br} + F_{bf}) \times h}{w}$$

The value of F_{nr} is identical to the value in Table 6.3 on a flat surface with the replacement of mg by $mgcos(\theta)$. The maximum braking force ($F_{bmax} = F_{br} + F_{bf}$) occurs when F_{nr} becomes zero or when $F_{bmax} = \frac{mgcos(\theta) \times (w-r)}{h}$. In Figure 6.13, the maximum braking force decreases with steeper slopes since the normal force, $mgcos(\theta)$, decreases with higher values of θ (slope percent). The F_{bmax} for three wheelbases of 1.01m, 0.99m, and 0.975m at a -10° slope is $0.496g$, $0.48g$, and $0.466g$ respectively for a fixed h of 1.15m and r of 0.43m. Since F_{bmax} is directly proportional to the wheelbase w, a longer wheelbase corresponds to a larger F_{bmax}.

From the maximum braking force, we can calculate the minimum stopping distance on a given slope. Figure 6.13 shows the minimum stopping distance in meters for a range of slopes from a downhill of 10% to an uphill of 10%. On a wet road ($\mu_s = 0.35$), the stopping distance on a downhill of 5% is 8.7 meters compared to a lower 6.5 meters on a uphill of 5% for a speed of 25 kmph. and wheelbase of 0.975m. The corresponding maximum braking g force for the downhill and uphill grades is 0.31 and 0.30 respectively and is computed assuming that the rear brake generates 25% of the total braking force. When the rear brake force is reduced, you can generate a higher total braking force without skidding.

On a downhill the braking force over the stopping distance, must not only be sufficient to remove the sum of the kinetic and potential energies, but also overcome the negative slope resistance. By contrast on an uphill, you gain potential energy and the positive slope resistance acts in the same direction of the braking force increasing the total force to stop the bicycle.

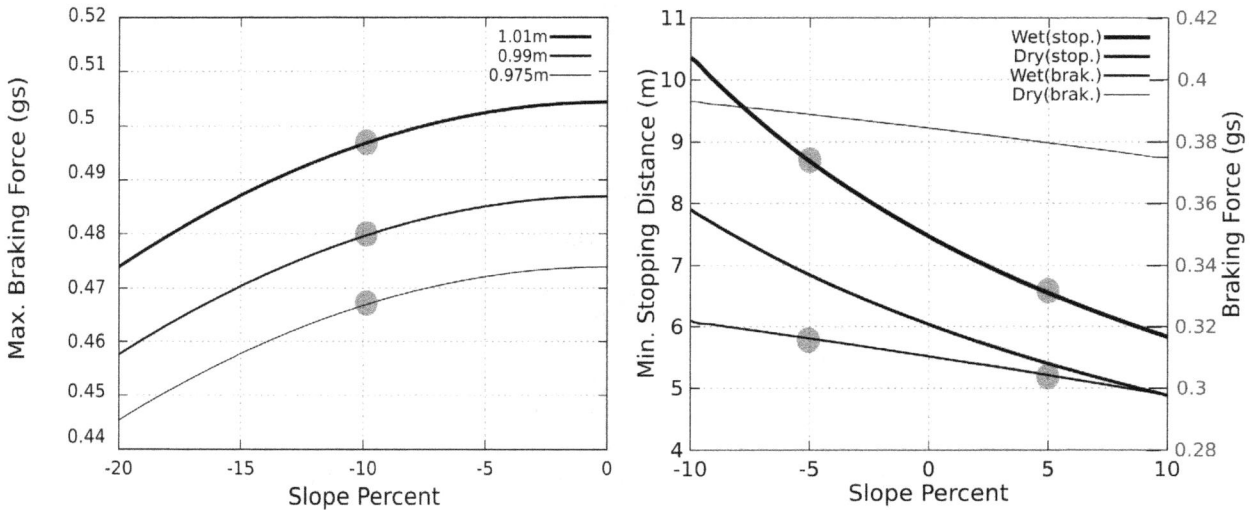

Figure 6.13: Maximum Braking Force F_{bmax} in gs for Three Wheelbases and Minimum Stopping Distance in meters vs. Slope Percent at a Speed of 25 kmph.

The coefficient of static friction and speed are major factors that contribute to the magnitude of the stopping distance. Under wet conditions where $\mu_s = 0.35$, the stopping distance of 10.3 meters on a downhill of 10% at 25 kmph. increases to over 25 meters at 40 kmph. At 25 kmph., on a 10% downhill, the stopping distance for a dry road increases from 7.9 meters to 10.3 meters for a wet road. Neither a disk brake nor a rim brake can exert a braking force larger than F_{bmax} without losing traction at the rear wheel.

Feathering and Wheel Locking On a long downhill, besides the wheel rim and rotor becoming excessively hot, you also need to control the bicycle with a sustained grip on the brake handlebar. Here hydraulic brakes can help by transmitting more force to the pads and will be less of a strain on your grip of the brake lever. However, hydraulic brakes have other issues such as the maintaining the temperature of the hydraulic fluid and ensuring that there is no leak.

Under normal circumstances a bicycle wheel rotates without slipping due to the static friction between the tyre and road. However, when a high braking force that exceeds the static frictional force is applied to the wheel, the tyre stops rotating and begins sliding (wheel locking) on the surface. This leads to two problems - a sliding wheel is difficult to control and it may take longer for the bicycle to slow down since $\mu_s > \mu_k$.

Feathering the brake is a process of alternately pressing and releasing the brake lever. This may not be possible in a panic stop, but on a long descent, you can retain better control over the bicycle compared to constant pressure on the brake lever. Feathering may actually increase the stopping distance since brake force is not being uniformly applied. The purpose of feathering is to avoid skidding and not necessarily reduce the stopping distance.

6.2.6   Braking in a Turn

Approaching a turn, it is instinctive to slow down. The main reason that it is difficult to take a turn at high speed is that some of part of the frictional force that was earlier entirely used to move forward must now be used to generate the centripetal force to make the turn. The friction between the tyre and the road is limited based on the normal force and the coefficient of static friction. This frictional force is part of the braking force as well as the force to turn the bicycle. If the static frictional force is not sufficient to both brake and turn, then the tyre will skid.

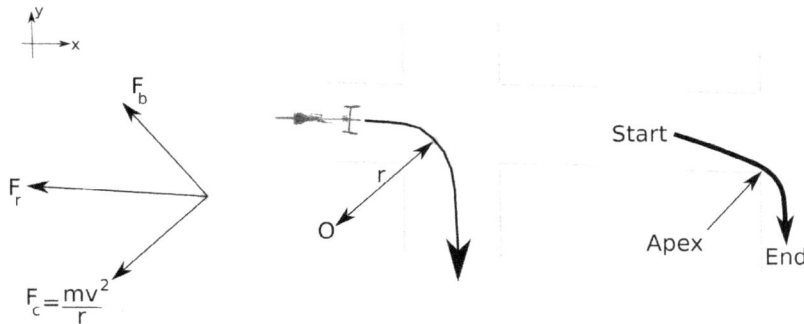

Figure 6.14: Braking and Centripetal Forces on a Turn

Flat Road Turn   Suggested ways of avoiding skidding on a turn include braking *before* a turn and releasing the brakes while turning. If you attempt to make a turn too fast, the tyres start to skid since there is insufficient static friction to generate enough centripetal force to stay on a controlled path in the turn. Both the coefficient of static friction and the normal force need to be large enough for a safe turn.

On a turn without any pedalling force, the two horizontal forces acting on the bicycle and rider are the centripetal force F_c and the braking force F_b (see Figure 6.14). The radius of the turn is assumed to be r and the corresponding centripetal force is $\frac{mv^2}{r}$ where v and m are the velocity and mass respectively. The resultant force is $F_r = \sqrt{F_c^2 + F_b^2}$. As long as the bicycle is not skidding the source of F_r is from the static frictional force, $\mu_s N$, between the tyre and the road. The centripetal force F_c increases with the square of the velocity v and therefore at a high velocity the centripetal force will be a large fraction of F_r.

To safely make a turn, you would apply the braking force F_b before the turn to ensure that the centripetal force F_c can be generated from static friction alone. On a dry road with a μ_s of 0.7 and N of 800N, the maximum frictional force is 560N. If the bicycle and rider weigh 80 kgs. on a turning radius of 6 meters, the maximum velocity before skidding without braking is 6.5 m/sec. or about 23 kmph. When the velocity is higher than 23 kmph., the centripetal force on the rider will exceed the maximum frictional force and therefore the rider will skid away from the apex.

In Figure 6.15 the maximum safe speed in kmph. on a turn when braking, reduces at higher brake forces. The dimensions of a typical bicycle is assumed, where the wheelbase is one meter,

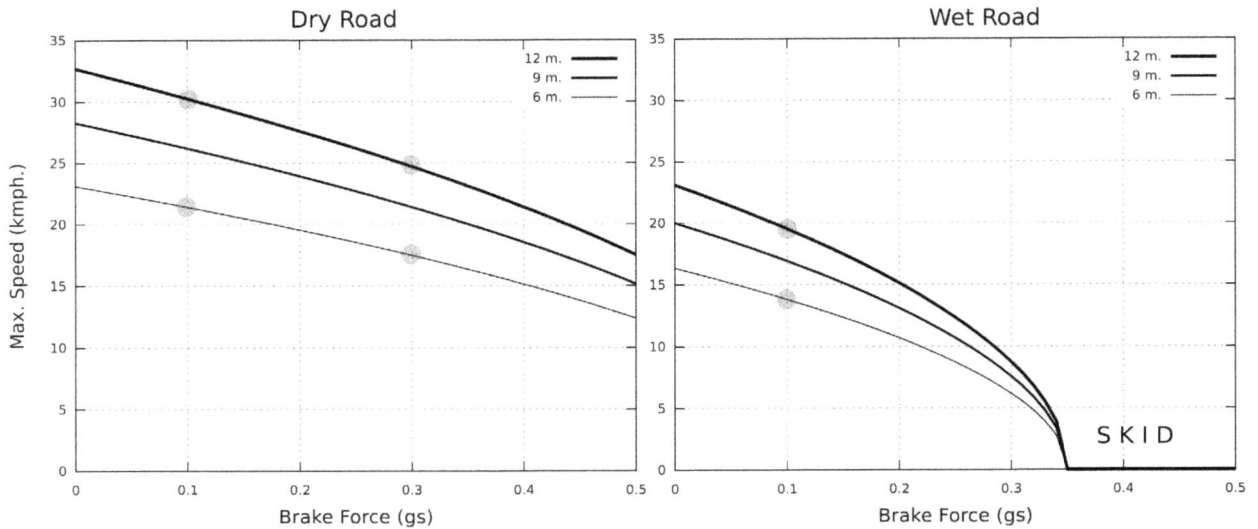

Figure 6.15: Maximum Safe Speed (in kmph.) before Skidding vs. Brake Force (in gs) for a Dry and Wet Road on turns of radii 6 m., 9 m., and 12 m.

the rear crank is 0.4 meters from the rear cog, and the COM height is 1.1 meters. On a turn of radius 6 meters, the maximum safe speed reduces from about 21 kmph. to 17 kmph. when the braking force increases from $01.g$ to $0.3g$. With a much wider turn of radius 12 meters, the maximum safe speed reduces from about 30 kmph. to 25 kmph. for the same brake force range from $0.1g - 0.3g$. When the radius of the turn is doubled, the maximum safe speed increases by over 40%.

On a wet road the lower coefficient of static friction (0.35 vs. 0.7 on a dry road), reduces the maximum safe speed further. Without any braking force on a 6 meter radius turn, the maximum safe speed is about 16 kmph. On a wet road, the maximum safe speeds at $0.1g$ on turns of radii 6 meters and 12 meters is about 14 kmph. and 19 kmph. respectively. It is also practically impossible to apply a brake force greater than $0.35g$ without skidding on turns of radius less than 12 meters.

Banked Road Turn The normal force ($\frac{mg}{cos\theta - tan\theta' sin\theta}$) on a banked road turn is slightly higher than the normal force (mg) on a flat road turn, where θ is the bank angle and θ' is the lean angle. A bank angle of 10° accompanied with a lean angle 10° will increase the normal force by about 5%. Since frictional force is dependent on the normal force and braking force is dependent on frictional force, you will be able to generate a moderately higher braking force on a banked road. The difference between the maximum speeds for different brake forces on a banked road of angle 10° with a lean angle of 10° and a flat road is about 2-3 kmph. When the bank angle is much steeper at 36° and the lean angle is 28°, the normal force will be more than double the normal force on a flat road.

Recommendations for safely making a turn include starting the turn on the outside edge of the road, crossing the apex on the inside edge and then moving to the outer edge at the end of the turn. You may not always be able to follow this style of turning depending on traffic conditions and the best option is to simply slow down before a turn. If you do need to brake in a turn, then doing it as gently as possible and riding straight may prevent a skid.

6.3   Calculating Stopping Distance

The effectiveness of a brake is usually measured by the stopping distance from a given velocity. This assumes that you are using the full power of the brake to stop. Some states have laws that require a bicycle to stop within some minimum distance, given an initial velocity. Massachusetts law requires bicycles to stop within 9.1 meters from an initial speed of 6.7 meters / second (24 kmph.) on a dry, clean, level, and hard surface or a deceleration of $2.46\ m/s^2$ or $0.25g$. In all the methods to calculate the stopping distance below, the weight of the bicycle does not affect the stopping distance. Although a moving bicycle that is heavier will have more kinetic energy than a lighter bicycle moving at the same speed, the braking force applied in the heavier bicycle depends on the normal force which in turn depends on the mass. So a heavier bicycle moving at the same speed as a lighter bicycle will *not* stop in a shorter distance than a lighter bicycle. The main parameters that determine stopping distance are the coefficient of static friction, speed of the bicycle, the gradient, and the direction of the wind. There are at least four different ways to calculate stopping distance. Each method uses a subset of these parameters to calculate the stopping distance.

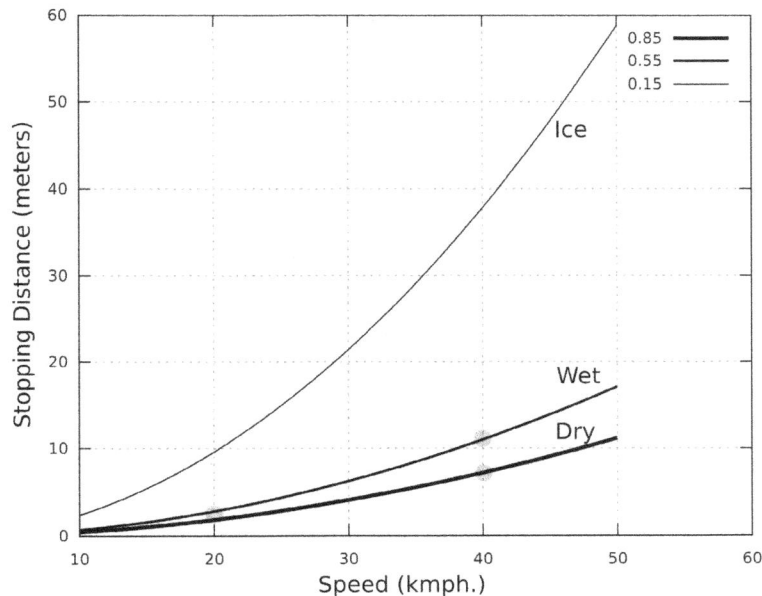

Figure 6.16: Stopping Distance in meters vs. Speed in kmph. for Three Different Surfaces with Three Coefficients of Adhesion

Calculating Stopping Distance 1   The formula used by Wilson [1] uses the velocity of the bicycle, coefficient of adhesion (C_a), and coefficient of rolling resistance (C_r). A relatively high C_r of 0.014 and braking force of $\approx 0.5g$ is assumed. The value of C_a is the maximum value of μ_s, the coefficient of static friction between the two surfaces just before the tyre skids. The formula to calculate stopping distance is

$$S = \frac{v^2}{20 \times (C_a + C_r)}$$

The Figure 6.16 shows the stopping distance for speeds from 10 kmph. to 50 kmph. for three surfaces - ice, dry asphalt, and wet asphalt. The coefficients of adhesion for ice, wet asphalt, and dry asphalt were assumed to be 0.15, 0.55, and 0.85 respectively. As you would expect, when the coefficient of adhesion is low, the stopping distance increases substantially with higher speeds.

At 40 kmph., the stopping distances for dry and wet asphalt are 7.15 and 10.95 meters respectively, corresponding to an increase of 53% for wet over dry. When the speed is halved to 20 kmph., the stopping distances drop by a large margin to 1.78 and 2.74 meters respectively for dry and wet asphalt. In reality, the actual stopping distances maybe higher than the values calculated with the above formula.

Calculating Stopping Distance 2   A book [3] titled "A Policy on Geometric Design of Highways and Streets" published by the American Association of State Highway and Transportation Officials, describes a formula commonly used in road design for establishing the minimum stopping sight distance required on a given road. The stopping sight distance takes into account the perception / reaction time t before a rider decides to brake as well as the actual time to brake. The value of t can vary from one second or less for a young person to more than two seconds for an elderly person. The formula for the stopping distance is

$$s = (0.278 \times t \times v) + \frac{v^2}{(254 \times (f + G))}$$

where v is the velocity in kmph., G is the grade of the slope expressed as a fraction, and f is the coefficient of friction between the tyres and the road. An uphill gradient of 5% and a downhill gradient of 5% would correspond to a G of $+0.05$ and -0.05 respectively. The estimated value of f is 0.7 for a dry road and 0.35 for a wet road. The second expression in the formula is the braking distance computed in the Figure 6.17 and excludes the variable reaction time ($t = 0$).

The figure 6.17 shows the increase in stopping distance for speeds ranging from 10 kmph. to 50 kmph. on five gradients on a dry and wet road. In both dry and wet roads, the stopping distance increases rapidly at higher speeds. On a dry flat (0% gradient) road at 20 kmph., the stopping distance is 2.25 meters, while at 40 kmph. the stopping distance increases by 300% to 9 meters. In contrast, a car takes needs over $2\frac{1}{2}$ times the stopping distance of 2.25 meters for a cycle or 6 meters on a dry road at a speed of 20 kmph [4]. However, this distance includes 4 meters to account for the reaction time of a motorist. When the road is wet, the stopping distances for the same speeds of 20 kmph. and 40 kmph. is 4.5 meters and 18 meters respectively.

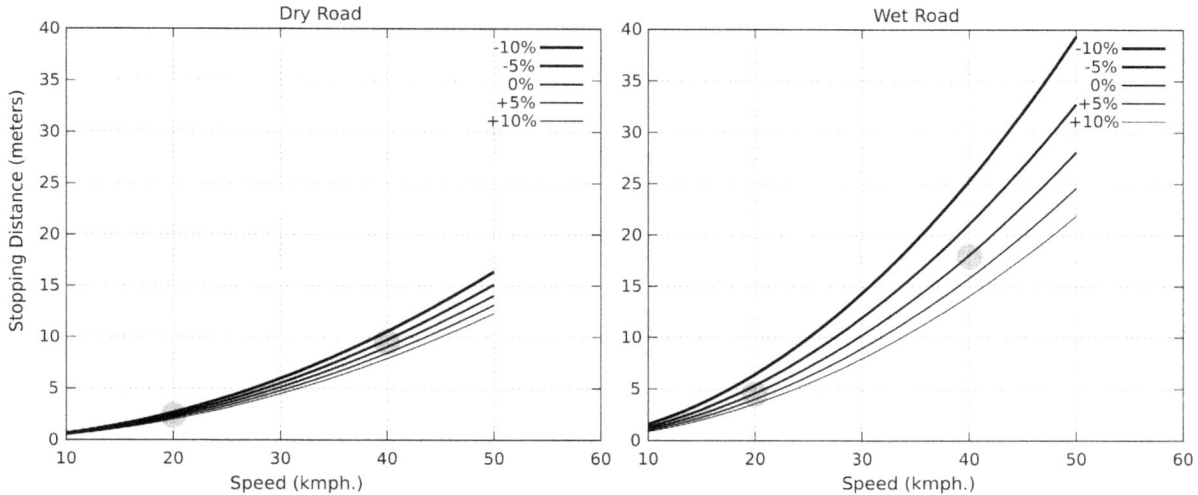

Figure 6.17: Stopping Distance in meters vs. Speed in kmph. for Dry and Wet Roads on Five Gradients using the AASHTO Formula

The rate of increase in the stopping distance is the same (300%) when the speed doubles from 20 kmph. to 40 kmph. and the stopping distance on a wet road is roughly double the distance on a dry road.

These values are the maximum stopping distances based on the speed limit for a given road. The time t can increase the stopping sight distance substantially. At 40 kmph. and a perception / reaction time of 1.5 seconds, the total stopping distance increases by 16.6 meters over the braking distance alone.

Calculating Stopping Distance 3 The third method of calculating stopping distance assumes that the brake generates a drag force to slow down the bicycle in addition to the existing drag forces due to wind, gradient, and rolling resistance. The total drag force that a cyclist will experience when braking, is the sum of all four drag forces - F_{aero}, F_{roll}, F_{slope}, and F_{brake}, where F_{aero} is the air drag, F_{roll} is the rolling resistance, F_{slope} is the slope or gradient resistance, and F_{brake} is the resistance due to the brake force on the wheels.

While the propelling force is generated by the rear wheel alone, the brake force can be generated from the front and rear wheels (provided you have brakes on both wheels). The brake force, F_{brake} can be calculated in terms of g and the coefficient of static friction μ_s between the road and the tyre. If the combined weight of a rider and bicycle is 100 kgs., then on a dry road where μ_s is 0.7, and a deceleration rate of $0.5g$, $F_{brake} = 0.5 \times 100 \times g \times \mu_s = 343N$. This is close to the maximum braking force that you can apply without skidding on a dry road.

Calculating stopping distance on a wet road cannot be accurately computed since the value of μ_s between the road and tyre will depend on the weather conditions and also the value of μ_k between the brake pads and the rotor or rim. If the rim is wet, then there will be a short period when the brake pad removes water from the rim, before the braking force is applied to the road. This period may depend on the type of brake pads and rim material. At a high speed of 40 kmph., if the time to dry the rim is just two seconds, the stopping distance will increase by about 22 meters. At the same time, the lower value of μ_s on a wet road means that the brake force cannot exceed $\mu_s F_n$. Wilson [1] suggests that stopping distance on a wet road could increase by as much as ten times compared to the stopping distance on a dry road.

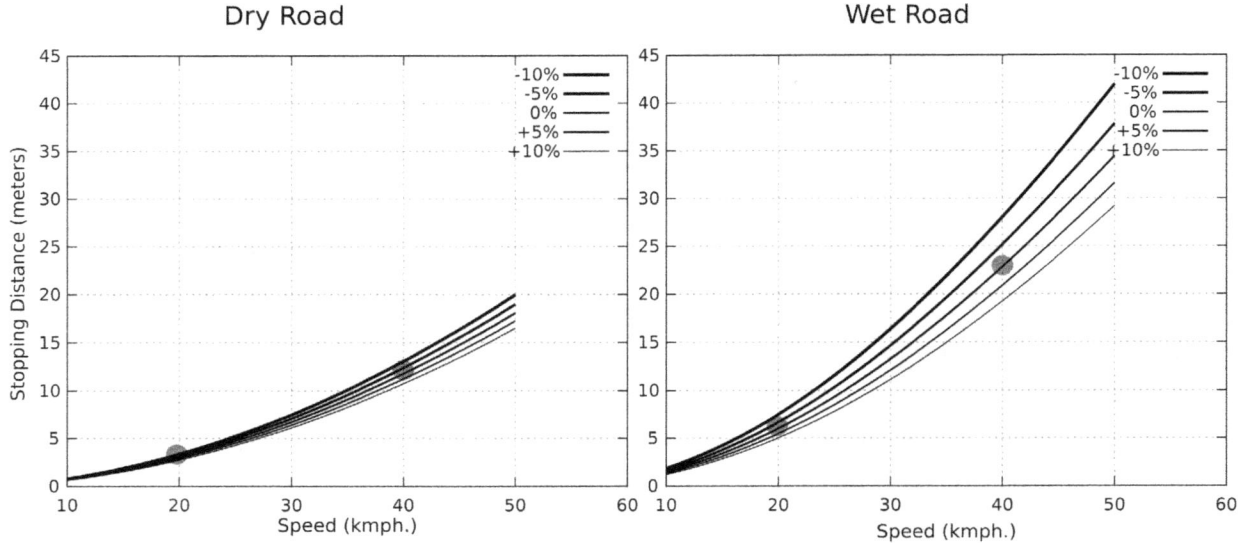

Figure 6.18: Stopping Distance in meters vs. Speed in kmph. for Dry ($0.50g$ braking force) and Wet ($0.25g$ braking force) Roads on Five Gradients using the Sum of Current Drag Forces

The braking force is typically the dominant force on a flat road with no wind. If $F_{total} = F_{brake} + F_{aero} + F_{roll} + F_{slope}$, then the change in velocity ΔV is $\frac{F_{total} \times \Delta t}{m}$ over a small period Δt (0.05 seconds), where m is the combined mass of the bicycle and rider. In each time step i, a new velocity V_i is calculated as $V_{i-1} - \Delta V$. The value of F_{total} changes with time, since F_{aero} depends on the current velocity. Eventually, V_i becomes zero and the stopping distance is the sum of the distances in each time step: $S = \sum_i V_i \times \Delta t$.

Similar to the results from the AASHTO formula, the stopping distance in both dry and wet roads increases rapidly at higher speeds. On a dry flat (0% gradient) road at 20 kmph., the stopping distance is about 3 meters, while at 40 kmph. the stopping distance increases by about 300% to 11.75 meters (see Figure 6.18). When the road is wet, the stopping distances for the same speeds of 20 kmph. and 40 kmph. are 5.9 meters and 22.75 meters respectively. A simplistic assumption of half the braking force is made to account for wet road conditions.

Without a reaction time, the AASHTO formula generates a shorter stopping distance. The braking forces in motorized vehicles can exceed $0.5g$, while a bicycle has a maximum limit of $0.56g$ [1] before a rider will be thrown forward. Therefore, the stopping distances for motorized vehicles are more likely to be lower than the stopping distances for a bicycle.

In addition to the braking drag force, both the wind and gradient are drag forces that can either aid or oppose the braking force. On an uphill road (positive gradient), the slope resistance

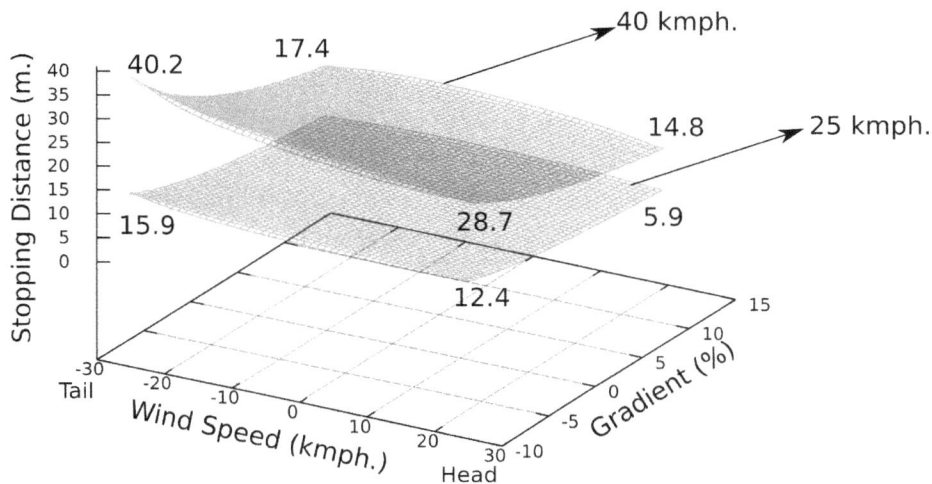

Figure 6.19: Stopping Distance (meters) for Wind Speed (kmph.) and Gradient (percent) for two Speeds of 25 kmph. and 40 kmph. on a dry surface

aids the braking force and stopping distance will be less than on a downhill road. Similarly, a head wind (positive wind speed) will aid the braking drag force.

At a relatively low speed of 25 kmph. on a downhill of gradient -10%, the stopping distances with a tail wind of -25 kmph. and head wind of +25 kmph. are 15.9 meters and 12.4 meters respectively (see Figure 6.19). The difference between the stopping distances is a few meters, since a moderate braking force of $0.25g$ is assumed. The time to come to a stop under these conditions is about four seconds and therefore air drag and the slope resistance do not significantly alter the stopping distance.

However at a higher speed of 40 kmph. and braking force of $0.25g$, air drag due to a tail wind of -25 kmph. increases the stopping distance to 40.2 meters from 28.7 meters with a head wind of +25 kmph. and the corresponding stopping time increases to 7.25 from 5.5 seconds. The longer stopping times at higher speeds gives air drag enough time to lower stopping distances.

On the other hand, the gradient force is a steady drag force and does not depend on current velocity or wind speed. At speeds of both 25 kmph. and 40 kmph., the stopping distance increases by about 67% when the gradient changes from 0% to -10%. (see Table 6.5). The center of each surface in Figure 6.19 where the gradient is 0% and wind speed is 0 kmph. is a standard stopping distance against which other stopping distances can be compared.

Bicycle Speed	25 kmph		40 kmph	
Grade \ Wind	Tail (-25 kmph.)	Head (+25 kmph.)	Tail (-25 kmph.)	Head (+25 kmph.)
Uphill (+10%)	-28%	-27%	-27%	-24%
Downhill (-10%)	+67%	+52%	+66%	+47%

Table 6.5: Stopping Distance Percent Change from Zero Wind and Gradient Conditions for Two Bicycle Speeds of 25 kmph. and 40 kmph.

At both speeds of 25 kmph. and 40 kmph., an uphill gradient of 10% reduces the stopping distance by an average of 26% compared to the stopping distance under zero wind and zero gradient conditions. Similarly, a downhill gradient increases the stopping distance with the largest increase of about 67% under a tail wind of -25 kmph. The effects of changes in gradient are more apparent in stopping distance calculations than the effects from changes in wind speeds.

If the braking force is increased to $0.5g$ at a velocity of 40 kmph. on a -10% gradient, then the stopping distance with a head wind of 25 kmph. decreases from 28 meters to 14 meters and the stopping time decreases from 5.5 seconds to 2.5 seconds. Similarly, with a higher braking force of $0.5g$ (compared to $0.25g$) and a tail wind of -25 kmph., the stopping distance decreases from 40 meters to 15 meters and the stopping time decreases from 7.25 seconds to 2.45 seconds. To reduce stopping distances, not only do you need brakes that are capable of applying a force of $0.5g$, but you also need to apply the braking force in time.

Calculating Stopping Distance 4   The fourth method of calculating stopping distance is based on computing the work required to dissipate the energy of the moving bicycle [5]. This method is based on the idea that the work required to convert the sum of the kinetic and potential energies to heat energy and bring the bicycle to a stop is equal to the product of the sum of all drag forces and the stopping distance.

The kinetic energy of a bicycle is the sum of the translational kinetic energy of the bicycle and the rotational kinetic energy of the wheel. The translational kinetic energy of the bicycle is the product $\frac{1}{2}m_b v^2$ where m_b is the total weight of the bicycle including the wheels and rider and v is the velocity of the bicycle. The rotational kinetic energy of a wheel is $\frac{1}{2}m_w r^2 \omega^2$ where m_w is the mass of a wheel, r is the radius of a wheel, and ω is the angular velocity of the wheel measured in radians per second. The rotational kinetic energy for two wheels is $m_w v^2$ where $v = r\omega$. The total kinetic energy of the bicycle is defined as $v^2(\frac{m_b}{2} + m_w)$.

When braking on a flat road, the loss or gain of potential energy is zero. On an uphill or a downwhill gradient of G, where G is a positive or negative fraction, potential energy is defined as $m_b g h$ where g is the acceleration due to gravity and h is the height lost or gained when stopping. The height h can be expressed as $s \times sin(\theta)$ where s is the stopping distance and θ is the angle of the gradient. For small values ($< 10°$) of θ, the value of $sin(\theta)$ can be approximated as $tan(\theta)$ or G. The potential energy of the bicycle is defined as $m_b gs \times G$ where G is positive and negative for downhill and uphill gradients respectively. The work done by all the drag forces is equal to the sum of the kinetic and potential energies.

$$\sum_i F_i \times s = v^2(\frac{m_b}{2} + m_w) + m_b gs \times G$$

From this expression, we can calculate s the stopping distance from the sum of drag forces F_i that includes a braking force, velocity v, mass of the bicycle m_b, and gradient G. The value of s can be estimated numerically by accumulating a series of smaller distances Δs, where $\sum_i F_i \times \Delta s =$

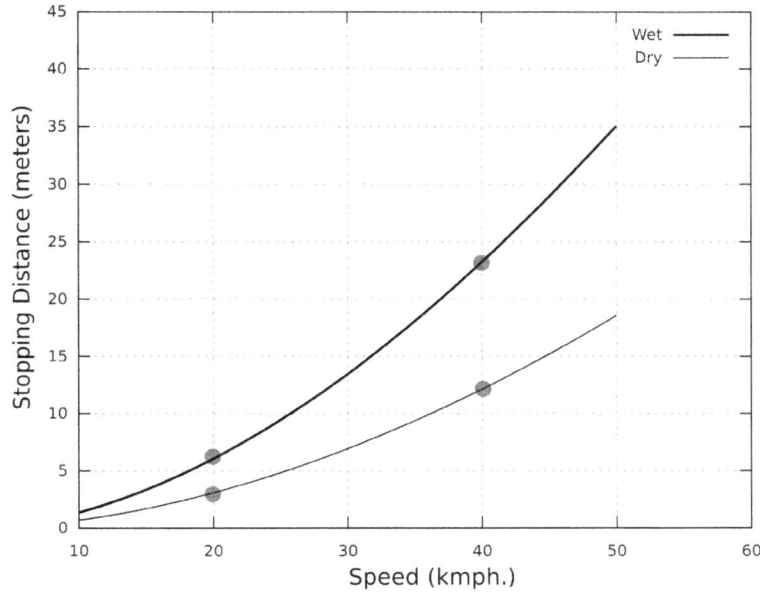

Figure 6.20: Stopping Distance in meters vs. Speed in kmph. for Dry ($0.50g$ braking force) and Wet ($0.25g$ braking force) Roads using a Work-Energy formula

$-\Delta KE + \Delta PE$. The change in kinetic energy $\Delta KE = (v_f^2 - v_i^2) \times (\frac{m_b}{2} + m_w)$, where v_f and v_i are the final and initial velocities respectively, is negative since $v_f < v_i$ when braking. The change in potential energy $\Delta PE = m_b \times g \times \Delta s \times G$ is positive and negative when riding uphill and downhill respectively. The final velocity $v_f = v_i + \Delta V$ where $\Delta V = -0.05 m/s$. At each iteration, the velocity is reduced by $0.05 m/s$ till the final velocity is less than 1 m/s. However the terms ΔPE and Δs are inter-dependent and therefore this method will generate higher stopping distances on downhill gradients.

Figure 6.20 shows the stopping distance for several speeds on a wet and dry road with zero gradient. The braking force of $0.5g$ has been reduced to $0.25g$ on a wet road to account for lower coefficients of friction. Similar to the results from previous methods, the stopping distance increases rapidly at higher speeds. At 20 kmph. the stopping distances for dry and wet conditions are 3 and 12 meters respectively (or an increase of 300%). The increase in stopping distance at 40 kmph. for wet over dry conditions is similarly 290% (23.0 and 5.9 meters respectively).

Calculating Stopping Distance on a Fixie One of the criticisms of a fixie is that it is unsafe, since the braking force generated from skidding the rear wheel is not sufficient to stop within a reasonable distance at a high speed. On a flat road a braking force F_{br} must act over some stopping distance s to dissipate all of the kinetic energy. In a skid the kinetic friction between the tyre and road is the braking force, $F_{br} = \mu_k \times F_{nr}$ where μ_k is the coefficient of kinetic friction and F_{nr} is the normal force on the rear tyre.

A first attempt at calculating the stopping distance s would be to assume the $\mu_k \times mg \times s = \frac{1}{2} \times mv^2$ where $F_{nr} = mg$ and therefore $s = \frac{v^2}{2g \times \mu_k}$. This method can also be used to estimate the velocity v of a vehicle from the length s of the skid marks and a given μ_k. However, the stopping distance from this method was found to be much less than empirical stopping distances for a vehicle [6].

The value of F_{nr} is less than mg, since a fraction of the total weight mg is transferred forward while braking and the rear wheel also supports only part of the total weight. Even though the braking occurs in the rear wheel, there is some weight transfer. Taking moments around ❷ in Figure 6.6, the maximum braking force from the rear wheel alone occurs when $F_{nr} = \frac{-(F_{br}+F_{bf}) \times h + mg \times (w-r)}{w}$ and $F_{bf} = 0$. If $F_{br} = \mu_k F_{nr}$, then $F_{nr} = \frac{mg \times (w-r)}{w + \mu_k h}$ (the braking force here uses the coefficient of kinetic friction μ_k). Now s can be calculated as

$$s \times \mu_k F_{nr} = \frac{1}{2}mv^2 \ \ or \ \ s = \frac{v^2}{2g} \times \frac{w + \mu_k h}{\mu_k(w-r)} \ \ or \ \ s = Kv^2 \ where \ K = \frac{1}{2g} \times \frac{w + \mu_k h}{\mu_k(w-r)}$$

Surface	μ_k	K
Pavement	0.56	0.245
Gravel	0.65	0.224
Grass	0.71	0.213

Table 6.6: K values for a Mercier bicycle based on Coefficient of Kinetic Friction

Empirical estimates of K for different surfaces such as grass, gravel, and pavement were computed using a Mercier Kilo TT bicycle [6]. In the experiment the wheelbase w was 1.02m and the corresponding approximate values for h and r were 1.15m and 0.4m respectively. From the dimensions of the bicycle and estimates of μ_k for different surfaces, the value of K can be computed (see Table 6.6).

As the coefficient of kinetic friction increases, the value of K decreases implying shorter stopping distance for rougher surfaces. From the equation $s = Kv^2$, the stopping distance for different surfaces and velocities can be estimated.

The stopping distance is inversely proportional to the coefficient of kinetic friction and proportional to the square of the velocity. In Figure 6.21, the stopping distance at 15 kmph. is about 4 meters on all three surfaces. At a higher speed of 24 kmph., the stopping distance on a pavement increases to 10.9 meters which maybe less or more depending on the roughness of the pavement. A smooth pavement or lower μ_k means that it will take longer to stop and therefore the stopping distance maybe even higher than 11 meters. MassBike (https://www.massbike. org/laws) recommends that the brakes on any type of bicycle should be good enough to stop within 30 feet (9.1 meters) at a speed of 15 mph. (24 kmph.) on a dry, clean, hard, and level surface. From the model in Figure 6.21, you cannot stop a fixie within the recommended stopping distance. The stopping distance on a fixie is about 20% longer than the distance recommended by MassBike. The model depends heavily on the value of μ_k which can be altered with a rough

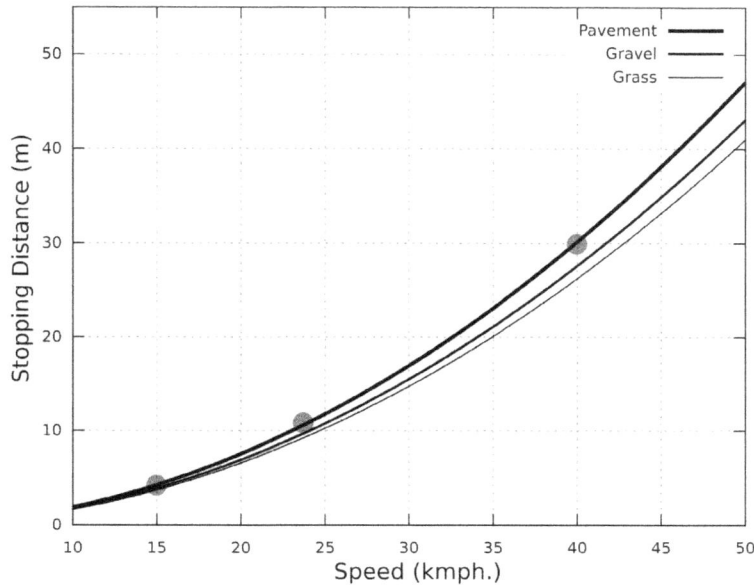

Figure 6.21: Stopping Distance in meters vs. Speed in kmph. for three different surfaces

tyre surface. At 40 kmph., the stopping distance increases to 30 meters compared to about 17 meters with brake pads (see Figure 6.18).

Summary There is more than one way to calculate the stopping distance and each of the methods above generate slightly different results. These methods assume that you are using the maximum braking force without skidding to stop in the shortest distance possible. Table 6.7 shows the stopping distances using the five methods for two speeds, 20 kmph. and 40 kmph., on a dry pavement. The average stopping distance at 20 kmph. and 40 kmph. is about 2.5 m and 10 m respectively with a range of about $\pm 25\%$ of the average in each method. The stopping distance for a fixie is roughly three times as large as the stopping distance for a bicycle with front brakes.

Speed	Aashto	Bicycling Science	Drag Force	Work-Energy	Fixie
20 kmph.	2.2m	1.8m	2.9m	3.0m	7.5m
40 kmph.	9.0m	7.1m	11.7m	12.1m	30.1m

Table 6.7: Stopping Distances in meters for Two Speeds on a Dry Pavement

6.3.1 Brake Force and Stopping Distance

Although a bicycle will come to a stop without brakes, the stopping distance is far in excess of even a modest brake force $(0.1g)$. If you stop pedalling on a bicycle with a freewheel without braking, then you can coast till the drag force eventually reduces your velocity to close to zero.

217

The drag force on a flat road when coasting is just the sum of the air and rolling resistances. The rolling resistance depends on the tyre and surface. It is similar to a frictional force and calculated as the product of the coefficient of rolling resistance (C_r) and the weight force (mg).

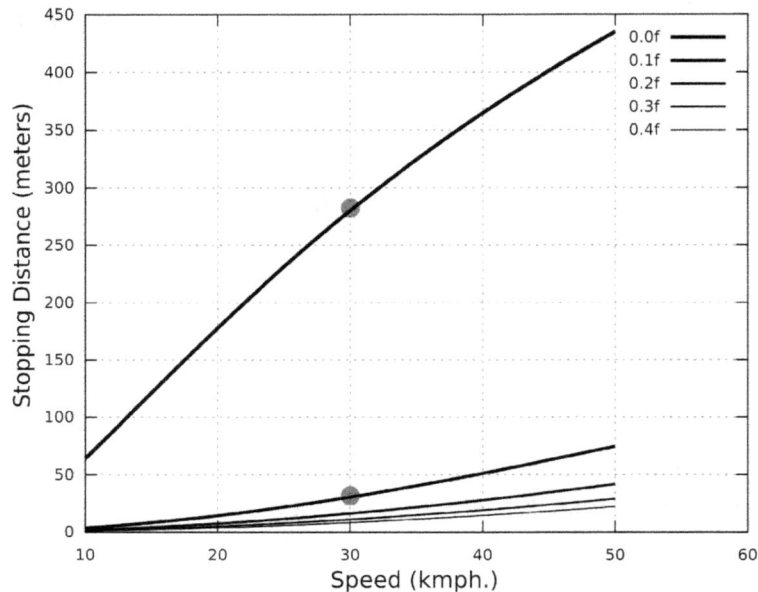

Figure 6.22: Stopping Distance in meters vs. Speed in kmph. for 5 brake forces on a flat road

The value of C_r can range from 0.004 for a tubular tyre on an asphalt road to 0.1 on a mud road. In Figure 6.22, the stopping distances for speeds from 10 kmph. to 50 kmph. are calculated with C_r of 0.005. The stopping distance with a braking force of $0.0g$ or coasting is much higher than the stopping distance with a braking force of $0.1g$ on a dry road (coefficient of friction = 0.7). At 30 kmph. the stopping distance is 281 meters when coasting and just 31 meters with a brake force of $0.1g$. A higher brake force of $0.4g$ can lower the stopping distance further to 8.3 meters.

Another way of looking at stopping distances is to consider the braking g force that you would need to stop within a given distance and speed. The maximum braking force that you can apply on a bicycle without losing traction on the rear wheel is about $0.5g$ (see Figure 6.23). On a dry surface at a speed of 40 kmph., you can stop within 15 and 20 meters with a braking g force of $0.39g$ and $0.29g$ respectively.

However to stop within 10 meters, you would need to a generate braking force $> 0.5g$ which would be difficult to accomplish without losing traction in the rear wheel. On a wet road, the maximum braking force that you could generate would be half as much or less compared to the generated braking force on a dry road.

On a dry surface, if you double your speed from 20 kmph. to 40 kmph., it takes more than three times the braking force to stop within the same distance (see Table 6.8). Stopping distances also increase sharply with speed for the same braking force. When speed is doubled from 20 kmph. to 40 kmph., the stopping distance increases by more than three times from 5.9 meters

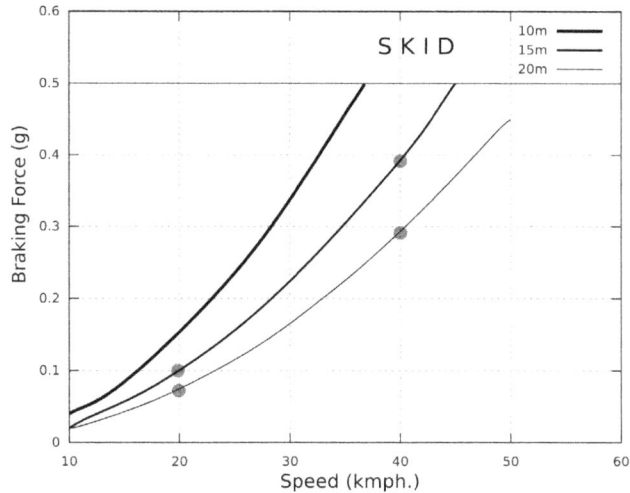

Figure 6.23: Braking Force in gs vs. Speed in kmph. for three Stopping Distances

to 22.7 meters. In reality, the actual braking forces and stopping distances will depend on the condition of the road, tyres, brake pads, and rims. On a wet road, the lower coefficient of static friction between the tyre and the road lowers the maximum possible braking force from the rear wheel before skidding. Even with just front wheel braking alone at a speed of 25 kmph., the time to react and apply the brakes makes it hard to brake within a few meters on a wet road.

Speed (kmph.)	Braking Force to Stop in 15 m	Stopping Distance with 0.25g Braking Force
20	0.10g	5.9m
40	0.39g	22.7m

Table 6.8: Braking Force to Stop in 15m and Stopping Distance with 0.25g Braking Force for two Speeds

6.4 Wheelies

In section 6.2.1 when a brake force was applied, a fraction of the total weight was transferred to the front wheel. The amount of weight transferred was proportional to the brake force. A *stoppie* occurs when the rear wheel no longer makes contact with the ground and you can pivot over the front wheel alone.

When the brake force was high enough ($> 0.5g$), the normal force on the rear wheel became zero and the total weight was on the front wheel. Similarly, when a propelling force is applied, some weight can be transferred to the rear wheel. As before, the amount of propelling force needed to reduce the normal force on the front wheel to zero can be calculated using the moments around a located ❶ above the rear wheel (see Figure 6.24).

219

The dimensions of the bicycle for the wheelbase w, COM height h, and the distance of the rear cog from the crank r are assumed as 1.0 meters, 1.1 meters, and 0.43 meters respectively. The normal forces F_{nr} and F_{nf} act on the rear and front wheels respectively. The propelling force F_p is generated on the rear wheel and the opposing static frictional force $\mu_s F_{nf}$ is due to the front wheel. On a flat road, the entire weight mg of the bicycle is directed from the COM while on an uphill road a part of the weight $mg\sin\theta$ acts along the surface of the road opposing F_p and the other part $mg\cos\theta$ generates a clockwise torque around ❶.

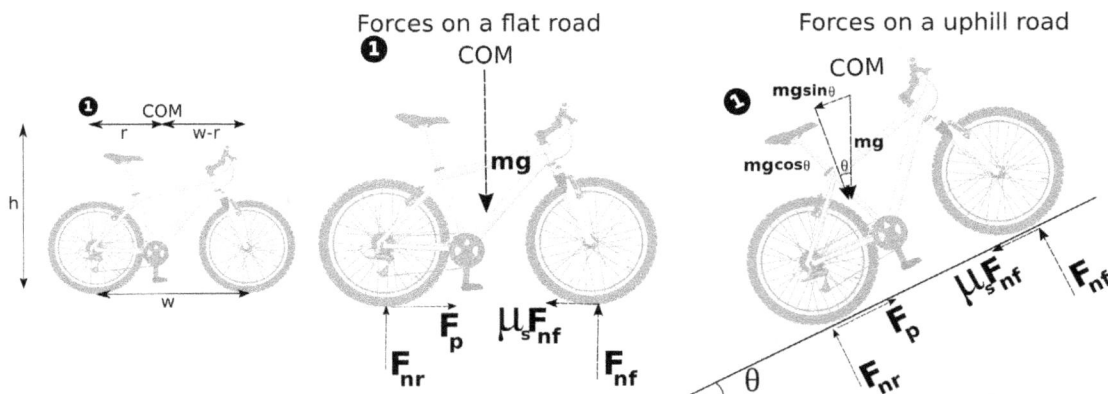

Figure 6.24: Forces on a flat and uphill road while riding

In Table 6.9, the clockwise and anti-clockwise torques on a flat road should be equal before a wheelie. As the propelling force F_p increases the normal force on the front wheel F_{nf} reduces. When the torques are equal, $mg \times r = (F_p - \mu_s F_{nf}) \times h + F_{nf} \times w$ and $mg \times r - F_p h = F_{nf}(w - \mu_s h)$. Therefore F_{nf} becomes zero when $mg \times r = F_p h$ or $F_p = \frac{mg \times r}{h}$. The required propelling force F_p can be lowered if r becomes smaller (or the COM moves closer to the rear wheel) and if h becomes larger (or the COM is higher). Since F_p is proportional to the mass m, a heavier person and bicycle will require a higher F_p. When the total mass of the rider and bicycle is 60 kgs. and 80 kgs., the required F_p is 230N and 306N respectively. Each additional kilo added to the mass adds roughly 4N to the required F_p. The values of F_p will vary depending on the dimensions of the bicycle.

On an uphill road, one part of the weight force $mg\sin\theta$ opposes the propelling force F_p. This is the same drag force (or slope resistance) that makes riding uphill more difficult than on a flat road. Using the same method of equating clockwise and anti-clockwise torques, the required propelling force F_p such that the front wheel normal force F_{nf} becomes zero can be found. The value of F_{nf} can be computed from $mg\cos\theta \times r + mg\sin\theta \times h - F_p h = F_{nf} \times (w - \mu_s h)$. When the left hand side of the equation is zero, $F_p = \frac{mg(\cos\theta \times r + \sin\theta \times h)}{h}$.

For every additional degree of uphill slope, the propelling force increases by about 11-13 N. For a mass of 80kgs. at 5° and 10° slopes, the required F_p increases from 306N (0° slope) to 374N and 438N respectively. This may appear counter intuitive since doing a wheelie on an uphill does seem to be easier on a flat road. One reason it may appear to be easier is that the COM can be moved closer to the rear wheel which does reduce the required F_p. Moving the COM back by 15 cms. from 0.43 meters to 0.28 meters changes the required F_p to 240N and 267N on 3° and 5° slopes respectively. The value of F_p can be reduced by about 6-8N for each centimeter that the COM is closer to the rear wheel, depending on the total weight of the rider and bicycle.

Road	Clockwise	Anti-clockwise
Flat	$mg \times r$	$(F_p - \mu_s F_{nf}) \times h + F_{nf} \times w$
Uphill	$mg\cos\theta \times r$	$(F_p - \mu_s F_{nf} - mg\sin\theta) \times h + F_{nf} \times w$

Table 6.9: Calculating Moments around ❶ on a Flat and Uphill Road

On a downhill, the slope resistance $mg\sin\theta$ acts in the same direction of the propelling force F_p and acts as a negative drag force. With same methods to equate torque, the value of F_p can be calculated from the degree θ of the slope. However on a downhill, a high F_p may increase your speed past a safe limit. A simpler way is to reduce the value of r. On a 5° downhill with $r = 0.2m$, the required F_p is about 75N and when r is further reduced to $0.1m$, F_p becomes almost zero.

6.5 Rim Temperature Risks

Although the stopping distance on a heavier bicycle will not be less than on a lighter bicycle, the braking force to stop a heavier bicycle is higher. Both the kinetic and potential energies depend on the combined mass of the bicycle and rider and will be higher on a heavier bicycle. To stop a bicycle, brakes convert kinetic and potential energies to heat energy and therefore the heat generated when braking will depend on the mass of the bicycle. The challenge for any braking system is to ensure that heat is dissipated as efficiently as possible without excessively heating the brake pads, the disk, or rim. A brake pad will become ineffective when the temperature of the pad is high and will feel "mushy" instead of "responsive". The rated temperatures for a brake pad on a metal or carbon rim can range from 180°C to 320°C. Once the temperature of the rim exceeds the rated temperature of the brake pad, it will begin to disintegrate and leave material on the rim.

Heat is more evenly distributed on a disk brake than on a rim brake [7] and not as dangerous. On a downhill ride, the constant braking force can not only raise the temperature of the rim, but also the temperature of air in an inner tube. Since hot air expands, the pressure of air in your tyre may increase by 10-15 psi. While braking constantly, the temperature of a rim may be high enough to melt the cement of a patch on a tube.

Wilson [1] has used a thermal model to estimate the increase in the temperature of the rim depending on the mass m of the bicycle and rider, the coefficient of rolling resistance C_r, the slope degree, the aerodynamic drag factor K_a, the cooling area of the rim, and the speed v of the bicycle. On a downhill slope of degree θ the downward force is $mg\sin\theta - mgC_r\cos\theta - K_a v^2$ where $mg\sin\theta$ is negative drag force due to the slope, $mgC_r\cos\theta$ is drag force due to rolling resistance,

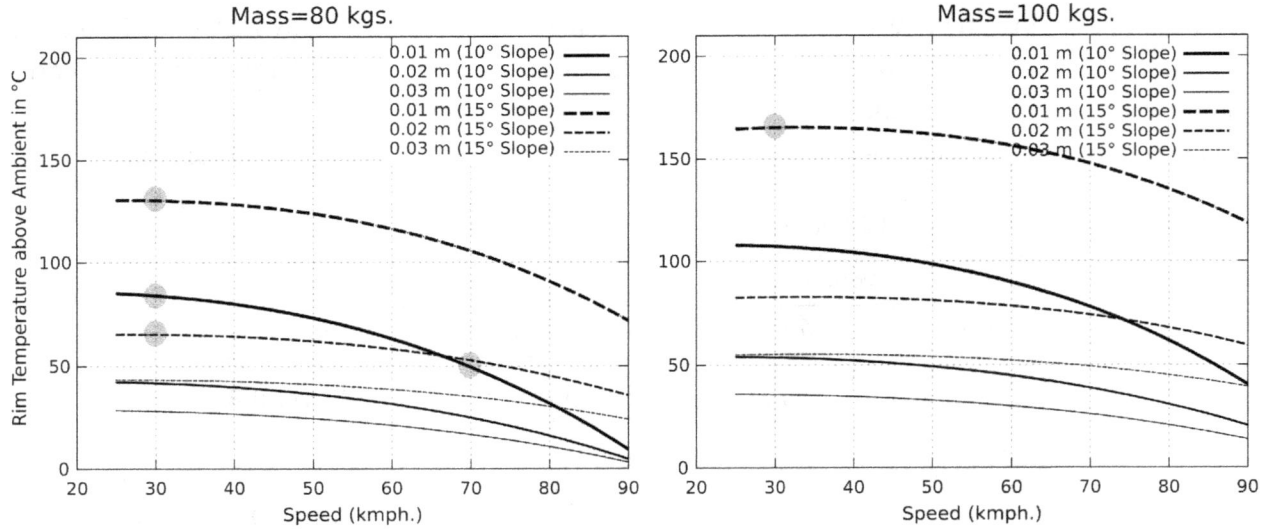

Figure 6.25: Change in Rim Temperature in °C vs. Speed in kmph. for Two masses, Three rim areas, and Two slopes

and $K_a v^2$ is the drag force due to air resistance. The change in temperature using the thermal model is

$$\Delta T = \frac{mg\sin\theta - mgC_r\cos\theta - K_a v^2}{100 \times area \times (0.5 + 1.125 \times (1 - 0.0632rv))}$$

where $area$ is the cooling area of the rim or rotor and r is the radius of the rim. The left figure 6.25 shows the increase in rim temperature for a total mass of 80 kgs for two slopes and three rim areas. A steeper slope of 15° will have a higher downward force than the downward force due to a slope of 10° and therefore the value of ΔT is higher for the same rim area. At 30 kmph., the value of ΔT is 83°C and 130°C for 10° and 15° slopes respectively. For a 15° slope, doubling the $area$ from $0.01m^2$ to $0.02m^2$ halves the value of ΔT from 130°C to 65°C.

The other parameters that can be changed include the aerodynamic drag factor K_a and the coefficient of rolling resistance C_r. When K_a is increased, the downward force will decrease and therefore ΔT will be lower. Similarly, a lower C_r will lower the drag force and also lower ΔT,however changing K_a has a larger influence on ΔT than changing C_r.

At higher speeds, the cooling effect due to air flow reduces the rim temperature substantially. At 70 kmph. on a 10° slope with a cooling area of $0.01m^2$,the value of ΔT is reduced to 50°C from 83°C at 30 kmph. However while riding at high speeds does cool the rim, there is a risk that the heat energy generated while braking will be excessive.

A heavier mass of 100 kgs. raises the ΔT for all slopes and rim areas. At 30 kmph. the ΔT for a 100 kgs. mass increases to 107°C from 83°C for a 80 kg. mass. The mass here includes the weight of the rider and bicycle. A higher weight will increase the downward force on a slope and therefore also increase ΔT.

References

[1] David Gordon Wilson and Jim Papadopoulos. *Bicycling Science.* The MIT Press, 2004.

[2] Gunnar Olsson. Brake performance and stability for bicycles, 2013. URL http://bit.ly/2GuDTwV.

[3] American Association of State Highway and Transportation Officials. *A Policy on Geometric Design of Highways and Streets.* 2018.

[4] Random Science Tools.com. Car stopping distance calculator, 2020. URL http://bit.ly/33iZORJ.

[5] 123HelpMe.com Website. Physics of stopping a bike, 2019. URL https://bit.ly/2EGaUFl.

[6] Cali Warner. Stopping distance of a fixed-gear bicycle skid-stop, 2014. URL http://bit.ly/2L4Ivzt.

[7] Todd Downs. *The Bicycling Guide to Complete Bicycle Maintenance and Repair: For Road and Mountain Bikes.* Rodale Books, sixth edition, 2010.

7 Navigation and Tracking

On a bicycle, it is not very difficult for an average rider to cover distances of 50-100 kilometers or more in a day. A well prepared rider will have a route in mind to reach a destination. Some of the things to keep in mind when following a route is awareness of the current location, distance covered, and speed. On a route consisting of multiple junctions and roads, you also need to know when to make a turn to stay on track. One method is to maintain a list of *waypoints* along the route. Each waypoint is a landmark or location that can be identified on the route. Identifying a few key waypoints on a route where you may need to change direction will make it less likely that you will lose your way and also divides the entire route into segments that can be completed in sequence.

7.1 Global Positioning System

Earlier, riders memorized or saved routes on paper. With the help of technology, we have come to rely heavily on devices to keep track of routes. However, all these devices with the exception of stand alone cyclocomputers depend on fetching current location information from Global Positioning System (GPS) satellites. A stand alone cyclocomputer can only track parameters such as distance travelled, average speed, current time, duration of ride, maximum speed, and current speed. While this information is useful in itself, you would also need to know your location to infer if you had strayed off route.

7.1.1 History

We take for granted the availability of GPS on a handheld device. But it has taken many years of research to reach this stage. In the late 70s, the GPS receiver was a large bulky backpack with an antenna, together weighing close to 11 kgs. From the initial ideas in the 60s and 70s to the full scale development in the 80s leading to a fully operation system in 1995, a number of researchers worked to build the world's first fully global navigational system [1]. The first commercial hand held GPS receiver in 1989 cost about 1000 USD. By 1998, GPS devices for navigation were implemented in cars and by 2005 GPS enabled smart phones like the HP ipaq became popular. Once the potential of GPS was recognized, chip manufacturers shrank the GPS circuitry to a tiny 2mm. × 2mm. chip that cost about one USD. The use of GPS became so popular that the number of GPS receivers grew to over a billion by 2010.

7.1.2 How does GPS work?

The problem of identifying your location is deceptively simple, yet it is a much harder problem than one would imagine. A GPS receiver must periodically convert GPS signals into a latitude and longitude that represents the current location. GPS uses a signal that travels at close to the speed of light (c) to calculate the distance (d) to a satellite from the receiver. It takes about 70 milliseconds (t) for a signal to travel from the satellite to the receiver, where $d = c \times t$. However, without an atomic clock in the receiver that is synchronized with the atomic clock on the satellite, the value of t will not have the high level of precision required to accurately calculate the distance d. An error difference between the satellite clock and receiver clock of just 10 milliseconds would mean that error in the distance calculated would be close to 3000 kms.

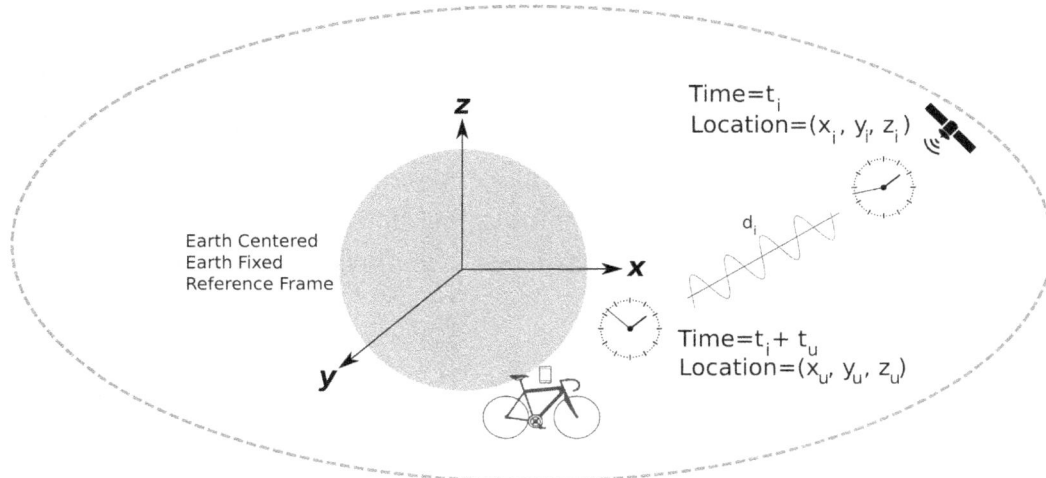

Figure 7.1: Signal Transmission from a GPS Satellite to a GPS receiver

Figure 7.1 shows the use of GPS in the simplest form. A satellite i sends a signal at a given time t_i from a location (x_i, y_i, z_i) to a receiver that observes the signal at some time $t_i + t_u$. The distance d_i (or range) of the satellite from the receiver at time t_i is then $c \times t_u$. Each distance d_i can be considered as an arc from satellite i and the intersection of multiple arcs representing the distances from other satellites will represent the location of the receiver (see Figure 7.2). The elliptical shape of the orbit of the satellite is exaggerated in the figure and the actual orbit is almost circular.

You could consider the x, y and z values analogous to the latitude, longitude, and altitude respectively. While the latitude and longitude values are computed to the sixth decimal place (about 0.1 meter accuracy), the altitude values are not as accurate. Part of the accuracy of the measurements of the latitude and longitude depends on the distance measurements from a set of satellites spread across the sky. While measuring altitude, the spread of satellites is limited, since signals from satellites beyond the horizon cannot be detected.

Figure 7.1 also shows one of the key problems that had to be solved to make GPS affordable and popular. The atomic clock on a GPS is not only extremely accurate, but also expensive as well. Requiring a synchronized atomic clock on millions of GPS receivers would increase the cost of the receiver far beyond the current cost of roughly one USD.

The less expensive (crystal) clock on a GPS receiver must be synchronized to the GPS atomic clock with close to nanosecond accuracy. [11] Correcting the receiver clock to match the satellite

---

[11] Light travels about 0.3 meters in one nanosecond.

clocks is fortunately identifying a single unknown, since all satellite clocks are assumed to be synchronized to a nanosecond. In total, there are four unknowns - x, y, z,and b where b is the receiver clock bias or time difference between the receiver and satellite clocks. Therefore, at least four satellites are needed to get any estimate of the unknowns.

The accuracy of the last unknown, the bias b, depends on having redundant distance values from 7-8 satellites whose distance arcs can be iteratively adjusted by altering the bias b till a convergence of the arcs at a single location is found. Since the atomic clocks on the satellites are moving at about 14,000 kmph., satellite time slows down relative to earth's clocks. GPS clocks lose about $7\mu s$ per day due to the high orbit speed and gain $45\mu s$ due to the reduction in gravity at the orbit height of about 20,000 kms. The time correction of $38\mu s$ is implemented with the satellite atomic clocks losing $45\mu s$ to account for the reduction in gravity and the GPS receiver gaining about $7\mu s$ based on the speed of a particular satellite [2].

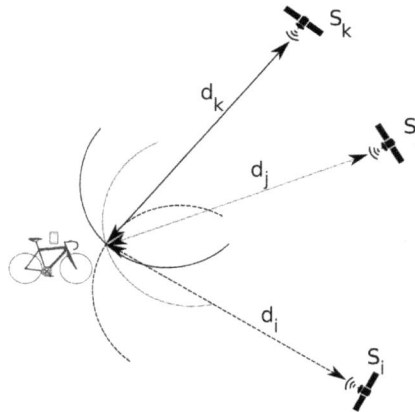

Figure 7.2: Distance arcs from three satellites intersecting at a location

In three dimensions, at least three distances d_i, d_j, and d_k from three different satellites i, j, and k are used to compute the latitude and longitude(see Figure 7.2). A fourth distance is also needed to account for the time difference between the receiver clock and satellite atomic clocks. The accuracy of the readings will improve with data from more than four satellites since the bias b will be more accurately computed when more than 4 arcs intersect at a location.

Other problems included maintaining the orbits of the fast moving GPS satellites to an accuracy of within a meter and synchronizing the atomic clocks of all GPS satellites to a billionth of a second. During the roughly 10 year life of a GPS satellite, the orbit must stay close to the predicted values to maintain reasonable accuracy. The predicted accurate locations and times of GPS satellites are maintained in the GPS receiver. The satellites broadcast a signal at about 1.5 Ghz. where the interference of clouds in the transmission of the signal is minimal.

Typically, the GPS receiver uses a common antenna for WiFi and GPS embedded in a circuit board to receive signal. It is quite remarkable that a tiny GPS antenna can detect a weak signal (-125 dbm or 3.16E-16 watts) from a satellite at a distance of about 20,000 kms. Every GPS satellite has an unique id that is embedded in the signal. Satellites from other countries use a different range of ids that is detected by the receiver. From the unique id and accompanying timestamp, the position of the satellite is estimated based on the expected locations of the satellite. The distance of a GPS satellite can range from 20,000 kms. when it is at the zenith to about 25,560 kms. when it is on the horizon and the respective time delays range from 66 msec. to 86 msec.

Roughly 30 GPS satellites from US, 30 from Europe and Russia and 15 or more from China, India, and Japan are in orbit. In total 80 or more satellites are used for navigation with a majority of the satellites used for civilian purposes. Some GPS receivers will process signals from Glonass (Russia), Galileo (Europe), Beidou (China), QZSS (Japan), and IRNSS (India). The specifications of a phone will usually include the list of GPS signals that the receiver can use.

The satellites of other nations and Europe transmit GPS signals to a specific region of the globe and a cell phone with a GPS receiver in almost any part of the world will receive some signal. Of the 30 US GPS satellites, 24 are used with six reserve satellites in medium earth orbit. Each satellite can cover roughly $\frac{1}{3}$ the earth's surface and makes two revolutions of the Earth per day. You would expect to receive signals from eight or more satellites at any given time. However, the number of satellites from which you can receive signals may fluctuate from 5 to 25.

Since the time to initially acquire a signal can take a minute or more, the initial signal information is received from a closer cell phone tower. Ground based augmentation of can improve the accuracy of GPS to 1 meter from 10 meters. The location of a cell phone tower is known very accurately since it can be precisely computed once. The location of your receiver will be computed based on the strength of the signal from your phone.

7.1.3 GPS Errors

The Android OS includes a *LocationRequest* that can be set to periodically request GPS readings. The settings include a minimum distance and minimum time separation between two consecutive readings. You can request a new GPS reading if the previous GPS reading is at least five meters away or was received five seconds ago. There is no guarantee that you will receive GPS readings precisely every five seconds or when the distance between two readings is exactly five meters. The actual time and distance between two GPS readings can vary substantially. Every GPS reading includes a reported accuracy in meters.

Figure 7.3 shows the reported accuracies from about 1000 GPS readings of a bicycle ride in good weather. A majority (almost $\frac{2}{3}$) of the reported accuracies are within five meters and the remaining one third of accuracies are distributed over a wide range upto 30 meters or more. A small number (4 out of 1000) of reported accuracies of 96 meters are excluded in the figure. The right side of Figure 7.3 shows the same reported 996 accuracies distributed in two dimensions. The black dot in the center at $(0,0)$ represents the correct location and the remaining 900+ gray dots show readings based on the reported accuracy. The figure shows the location of a hypothetical set 1000 GPS readings for a given location and the given accuracy frequencies. With the same GPS receiver and weather conditions, we would expect the same distribution of GPS read-

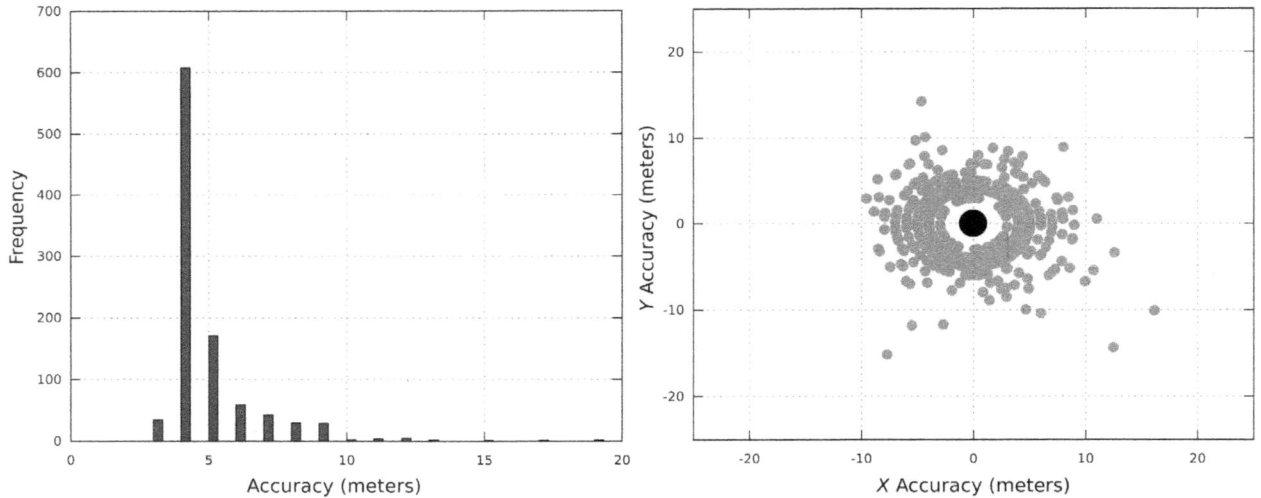

Figure 7.3: Frequency of Reported GPS Reading Accuracy for 1000 Readings

ings for another location. An uniform distribution in both the x and y directions based on the accuracy frequencies is also assumed, which may not always be true.

Distance Measurement Errors Measurements of speed, gradient, and power generated are strongly dependent on the accuracy of the distance measured. The total distance covered in a trip is the sum of the individual distances between two consecutive GPS readings. A ride or trip is represented by a series of positions $p_1, p_2, ..., p_n$ where each p_i represents a location (x_i, y_i). The location is a cartesian transformation of the World Geodetic System coordinates used by GPS. Since consecutive positions are relatively close to each other, distortions due to the projection in cartesian coordinates are minimal.

Every position p_i has a corresponding timestamp t_i and $t_1 < t_2... < t_n$. You can also assume that the route between two consecutive positions p_i and p_j is a straight line. This assumption is a source of error, if GPS readings are not frequent enough on a route with many sharp turns (see Section 7.2.1 for a comparison the accuracy of GPS and Cyclocomputer distance readings).

One of the complaints of GPS based applications is that the distance computed is frequently over estimated. Since distance is used to compute speed, one of the most frequent calculations, an error in distance computation will propagate to speed calculations as well. If two locations A at $(0, 0)$ and B $(50, 0)$ are known to be exactly 50 (d) meters apart (see Figure 7.4) and located with the same accuracy distribution shown in Figure 7.3 using GPS, then d_i is one possible distance between a random pair of readings selected from the set of readings for A and B. If $d_i > d$, then the distance between A and B has been over estimated.

From a random set of about 10,000 pairs of readings for A and B, the fraction of distance readings (d_i) that were over estimated compared to the actual distance (d) can be computed. Figure 7.4 shows that the fraction of over estimated distance calculations is high (>75%) when

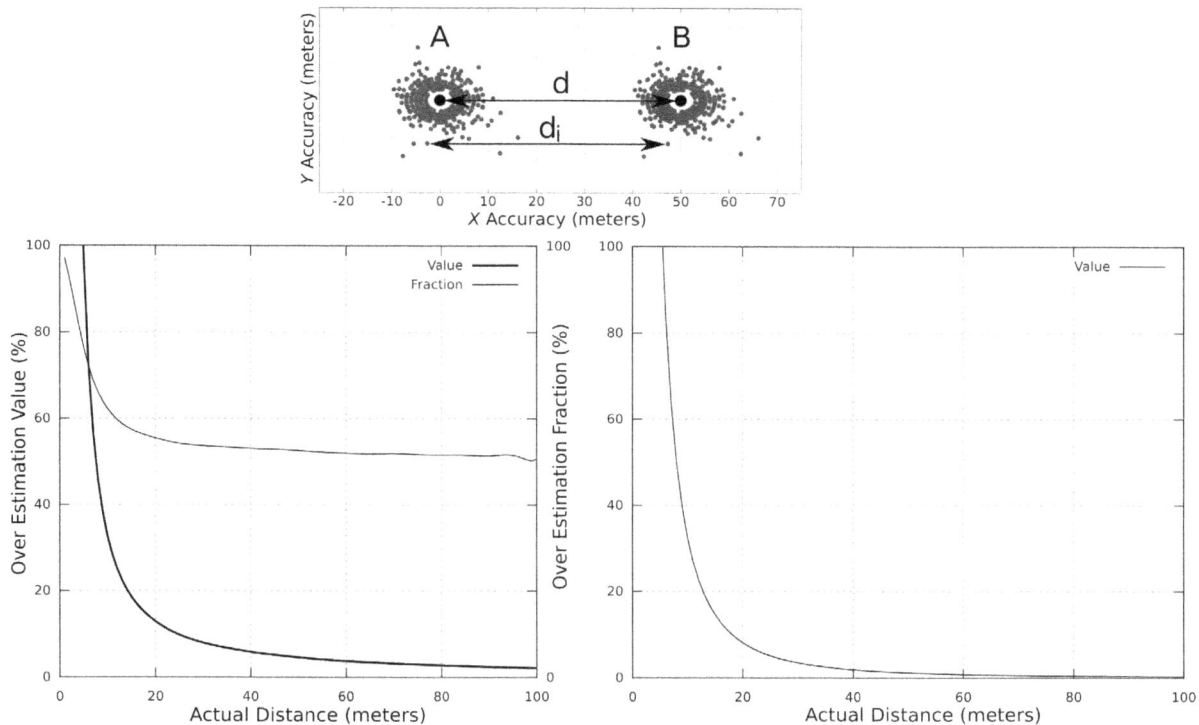

Figure 7.4: Over estimation of distance percentage and fraction for actual distances (d) from 2-100 meters (Empirical and Ranacher)

the actual distance is less than five meters. The value by which the distance d_i is over estimated is also high for actual distances of 2-3 meters. This does seem logical since the range (4-35 meters) of accuracies is greater than the actual distance.

When the actual distance between two locations is 20 meters or more, over estimation of the distance stabilizes at close to 50%. You would expect an over estimation and under estimation of the distance between A and B roughly half the time, when accuracy is small enough relative to the actual distance. Beyond 20 meters between A and B the degree of overestimation drops from 12% to about 2% of the actual distance. The above evaluation is based on empirical accuracy data reported by a GPS receiver and also assumes that the accuracy distribution is consistent for each location.

Ranacher [3] used a theoretical approach to show that GPS distances are over estimated. Their formula computes the over estimation of distance (oed) as $\sqrt{(d^2 + V - C)} - d$ where $V = 2\sigma_x^2 + 2\sigma_y^2$. The variance σ_x^2 is the mean square difference between a GPS reading's x coordinate and the actual x coordinate. The variance σ_y^2 is similarly calculated in the y axis.

C is defined as $2 \times (Cov(X_1, X_2) + Cov(Y_1, Y_2))$ where X_1 and Y_1 are random variables representing the x and y coordinate accuracies respectively for location A. The x and y coordinates for another location B are similarly represented by the random variables X_2 and Y_2. If the covariance $Cov(X_1, X_2)$ is positive and if X_1 is some negative number, then X_2 is also a similar negative number. In other words, covariance is positive, if x accuracy values of two locations are both positive or both negative. The values of C ranged from -7 to +5 with about 40% positive values.

The values for σ_x^2 and σ_y^2 using the given accuracy distribution were 14.6 m. and 14.3 m. respectively and the corresponding value of V was 57.8. Since the value of V is much higher than the value of C, the over estimation of distance is always positive. The right side of Figure 7.4 shows that the percentage of over estimation of the distance decreases sharply with increase in the actual distance between A and B. However when C is zero, there will always be an over estimation in the distance calculation. If $C > V$, then the distance calculated will be an under estimate. This can happen when the correlation C between accuracies of locations is high relative to the variance of accuracies V.

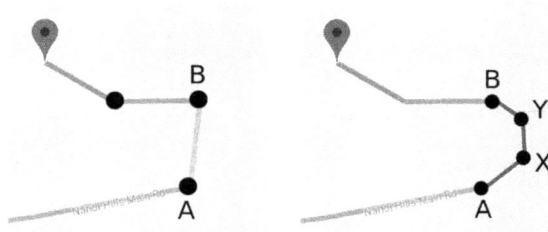

Figure 7.5: GPS Distance Estimates without and with Sufficient Readings on a Curve

Distance Errors on a Climb or Descent   On a climb, a road may have multiple turns (hair pin bends) as shown in Figure 7.5. In the left side of the figure, GPS measurements are not frequent enough to estimate the actual distance of the turn. The straight line distance between A and B in the left hand side of the figure will under estimate the actual distance.

In the right hand side of the figure, two additional measurements at X and Y have been added to more accurately estimate the actual distance. This error can occur on narrow winding roads where a turn is sharp and may take just a few seconds to navigate on a bicycle. The interval between GPS readings is not short enough to accurately estimate the distance between points A and B on either side of the curve. The total distance computed on such a road will under estimate the actual distance.

Correcting GPS Distance Errors   Google and Microsoft provide *Snap to Road* APIs that can be used to fix GPS reading errors when the route is on a known road. If you are riding on a narrow road that is 2-3 meters wide, then it is very likely that some of the GPS readings will indicate a location that is off-road, given the five meter accuracy limitation of GPS.

The purpose of the "snap to road" APIs is two fold. On a route that was entirely on a known road, all off road GPS readings will be corrected to on road. This not only reduces error, but also appears more accurate on a map, since the shape of the route will follow the road very closely and not have any zig zags that are obvious in raw GPS readings (see Figure 7.6). Two, the API also includes optional interpolation location data to add points such as X and Y in Figure 7.5. This will significantly improve accuracy of distance calculations on a route with many sharp turns.

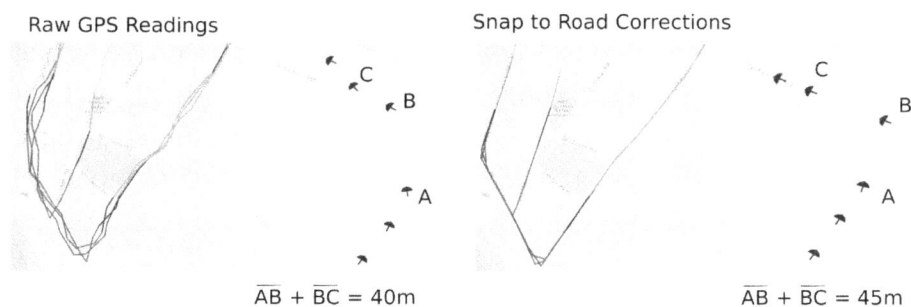

Figure 7.6: GPS Readings before and after Snap to Road Corrections

The documentation in the Google Snap to Road API [12] describes a URL that will accept upto 100 locations with the associated latitude and longitude, which then returns a JSON document with a corrected set of latitudes and longitudes. The left side of Figure 7.6 shows the map view of GPS readings before the corrections.

The right hand side of Figure 7.6 shows the locations of a set of six readings before and after corrections. The set of raw GPS readings do not follow the shape of the road. Following the snap to road corrections, the set of three readings A,B,and C are more closely aligned with the shape of the road. The movement of the readings also changes the calculations of the distances. Following the correction, the cumulative distance from A to C via B is longer by about five meters. The distances calculated following the corrections should be more accurate since the route more closely follows the contour of the road.

An optional *interpolation* parameter in the API will add new readings in addition to making corrections to the set of existing readings. The new readings are mainly added on turns in the road where it is likely that readings will be missing due to the time period between successive readings. The number of added readings will depend on the route. On a ride of about 26 kms. with a few turns, 14% new interpolated readings were added to an existing set of 468 readings for a total of 533 readings. The addition of the new readings increased the total ride distance from 25.9 kms. to 26.11 kms.

Although you would expect fewer corrections when the reported GPS accuracy is higher, there is no obvious relationship between reported accuracies and the magnitude of the corrections. Since the reported accuracy is not with 100% confidence, corrected distances maybe larger than the accuracy distance. If the reported accuracy for a location x is five meters, then about 50% of all readings for x will be within five meters of x and 90% of all readings will be within ten meters of x. Receivers that are more expensive than the ones found on normal cell phones are more likely to have higher (less than five meters) accuracies.

---

[12]This service requires an API key which must be provided for every request. A limited number of locations can be corrected per day using the free subscription.

Is Signal Strength better in Rural Areas? The accuracy of a GPS reading depends on the number of satellites and the strength of the signals received. In rural areas you would expect to receive signals from more satellites, since there are fewer obstructions such as buildings to block signals. The strength of a signal from a satellite that is directly above near the zenith is expected to be higher than the strength of a signal from a satellite on the horizon. In a rural area, you are more likely to receive signals from the entire sky. A sample test of 1000 rural and urban readings shows that indeed on average, a GPS receiver in a rural area can detect signals from more satellites with slightly higher strength and higher reported accuracy (see Table 7.1) per reading.

Area	Avg. No. of Satellites	Avg. Signal Strength (db)	Avg. Accuracy (m)
Rural	13.7	28.6	4.6
Urban	10.2	25.8	9.3

Table 7.1: Average number of satellites, signal strength, and accuracy from 1000 readings

7.2 Cyclocomputers

A cyclocomputer or cycle computer is a device that is typically installed on your handle bar with a display that shows statistics such as distance covered, time elapsed since the start of the ride, the average speed, the maximum speed, and other parameters from sensor data. The basic cyclocomputer will show distance, time, and speed for a ride. The distance shown will include the distance for a ride as well as the distance covered since the cyclocomputer was initialized. The average, maximum, and current speed are also included with the current speed often shown in a larger font. Finally, the time elasped since the start of the ride and the current time are also shown.

While you can capture the same information from a GPS device, a calibrated cyclocomputer will capture the distance covered and speed more accurately than a GPS device. The distance covered is captured from the number of rotations of the wheel (see Figure 7.7). The three parts of a cyclocomputer include the device mounted on the handlebar, the sensor, and a magnet attached to a spoke.

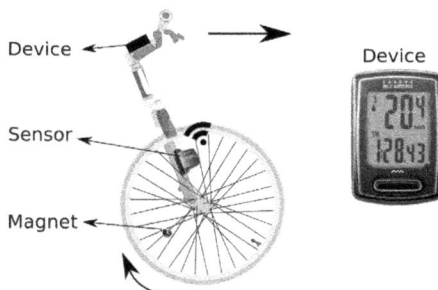

Figure 7.7: Parts of a Cyclocomputer (courtesy CatEye)

The magnet attached to the spoke moves in a clockwise direction and crosses in front of the sensor. The sensor transmits the number of times the magnet crosses in front of the sensor. Since the magnet is small, the distance between the sensor and magnet has to be sufficiently small for the sensor to detect the movement of the magnet. A sensor can also transmit the same information over a wire connected to the device. A wired sensor tends to be simpler and does cost less than a wireless sensor. The benefits of a wireless sensor are easy installation and it is not very hard to transfer the whole cyclocomputer to another bicycle besides a cleaner look without zip ties for the wiring from the sensor to the handlebar. Both wired and wireless cyclocomputers are accurate and not very difficult to install, however wired cyclocomputers are not susceptible to errors due to dropped signals or interference from other signals.

7.2.1 Distance Calculation Errors

While cyclocomputers calculate distance and speed more accurately than a GPS device, the level of accuracy depends on calibration of the cyclocomputer. Every cyclocomputer manufacturer provides a table of tyre sizes and the associated circumference. If you are riding on a $700 \times 25c$ tyre, then the recommended circumference maybe 2.11 meters, while for a $700 \times 32c$ tyre the recommended circumference maybe 2.17 meters.

This difference of 0.06 meters in the circumference measurement may not seem significant, but does change the distance calculation by a fair amount (see Table 7.2). If the correct circumference is 2.15 meters and you ride for an hour at an average speed of 15 kmph., then the tyre should make 6976 rotations and 11,627 rotations at an average speed of 25 kmph. When the circumference is under estimated by 0.04 meters at 2.11 meters, the distances measured at speeds of 15 kmph. and 25 kmph. are under estimated by 281 meters and 468 meters respectively. Similarly if the circumference is over estimated by 0.02 meters at 2.17 meters, then the distances measured at 15 kmph. and 25 kmph. are over estimated by 138 meters and 230 meters respectively.

Average Speed (kmph.)	2.11 m (wrong)	2.15 m (correct)	2.17 m (wrong)
15	14,719 m	15,000 m	15,138 m
25	24,532 m	25,000 m	25,230 m

Table 7.2: Distances reported for 2 speeds after 1 hour with 3 circumference settings

Calibration Even small differences in the circumference setting of the cyclocomputer can make a large difference in the distance calculated, if you ride for a reasonably long period (> 1 hour). The easiest way to get the correct circumference is to ride several kilometers on a route of known distance and check if you are over estimating or under estimating the actual distance. Then, you can iteratively adjust the circumference till the distance calculated is close to the actual distance. On some cyclocomputers you maybe able to set the circumference with millimeter precision. Another method is to mark a location on the tyre, move the bicycle in a straight line to make a single rotation of the tyre using the marked location, and measure the distance between the start and end points.

7.3 Sensors

The speed or distance sensor is just one of the sensors from which you can collect data to analyze the performance of your ride. Sensor data can reveal whether the intensity of a workout is sufficient to improve performance or may lead to injury. Current or average speed alone is not enough to indicate training intensity. A distance sensor is relatively simple and transmits the number of wheel rotations per second. Other sensors include a cadence sensor, a heart rate monitor, power sensor, altimeter, temperature and wind sensors (see Figure 7.8). These sensors periodically transmit data to the device which captures the data and display the results to the rider. One of the issues with data acquired from multiple sources captured at different time intervals is integrating the data to a common time scale.

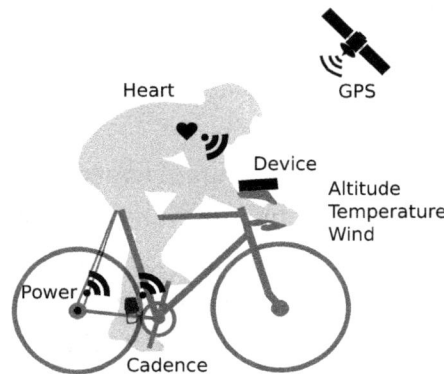

Figure 7.8: Data from Multiple Sensors captured on a Device

7.3.1 Cadence

The cadence sensor measures the number of times the crank rotates in a minute and is mounted on the chain stay. A magnet on the pedal arm like the magnet attached to a spoke periodically crosses the cadence sensor based on the number of crank rotations. The cadence data is reported in terms of rotations per minute (rpm).

One reason to monitor your cadence is to check whether you are riding efficiently. There is no single rpm that is ideal for all riders, but most recommendations for efficient riding are between 70-100 rpm. Sustained cadence in this range implies that you are not over working your muscles and will be less likely to fatigue on a long ride. A slower cadence may not necessarily mean that you are applying a force that you cannot sustain for long. Some rides are slow and a high cadence is not necessary.

When measuring speed, some devices differentiate between moving and stopping times. The total time for a trip is the sum of the moving and stopping times. The average speed maybe computed using the moving time or the total time and the result will be higher with the former. Similarly, there will be times when your cadence is zero (coasting) and the average cadence typically excludes these samples. The average cadence is computed using data captured when you are actively pedaling.

7.3.2 Altitude

One sure way to increase your heart rate is to ride uphill. Keeping track of the number of meters climbed is not just necessary to calculate the power that you generated, but is also a good measure of the difficulty of the ride. Altitudes are stated in the number of meters above mean sea level (MSL)[13].

Why is GPS Altitude not accurate?   A GPS receiver measures the elevation of a location relative to an ellipsoid surface that covers the earth. The ellipsoid is a smooth surface that does not exactly represent the MSL (see Figure 7.9). It was assumed that the ellipsoid, a surface with the equal gravity, would be identical with MSL. Unfortunately, MSL across the globe is not at an uniform distance from the center of earth. Some areas of earth are more dense than others and the MSL maybe higher at these locations.

The geoid (dotted line) is another surface covering earth if it was entirely made up of water. Unlike the ellipsoid, the geoid is not a smooth surface and the irregularities are due to the non-homogenous composition below the surface of earth. The altitude of any location X is the vertical distance H of the location to the geoid. The GPS reported distance of h from the ellipsoid must be corrected to the distance H. A sophisticated GPS receiver may maintain the location of the geoid at a high resolution to make an accurate correction. Other receivers may approximate the value of the correction $(H - h)$.

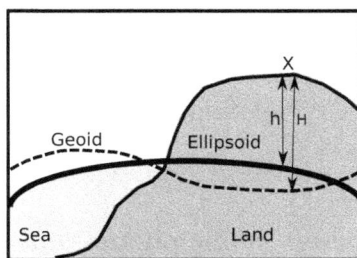

Figure 7.9: Geoid and Ellipsoid Heights for a Location X

Correcting GPS Altitude Errors   Microsoft and Google provide APIs to return an altitude for a given location. A digital elevation model of the earth's surface has a pre-computed altitude at a given location or cell (the altitude of a cell is an average altitude for the cell area). The number

---

[13]Mean sea level is the average hourly level of the sea to cancel out the differences in sea level due to the tides, measured over several years.

of locations with associated altitudes will depend on the resolution of the model. A model with a 10 meter resolution (or $10 \times 10\ m^2$ cell area) will need many more elevation samples than a model with a 50 meter resolution. Given a latitude and longitude for a location x, the API will find the cell y where x is located and return the elevation of y. A high resolution model will be more accurate, since the distance of a given location x will be closer to the model's cell center than the distance to the center of a larger cell in a low resolution model.

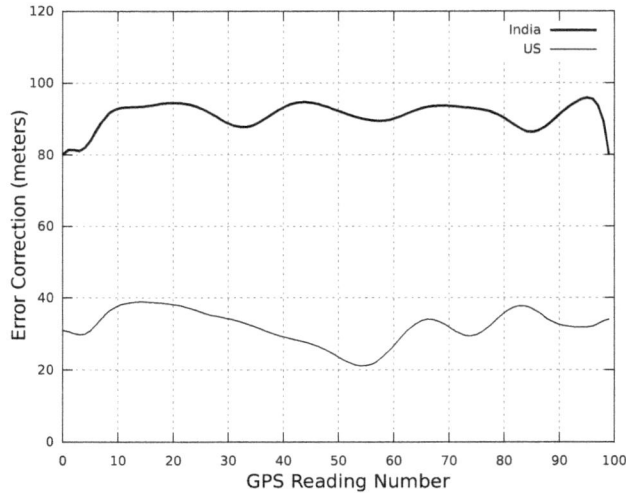

Figure 7.10: Error Correction for GPS Altitudes at Two Locations for 100 Readings

The magnitude of the GPS altitude error depends on the location on earth. The geoid falls below the ellipsoid by over 100 meters in some parts of the Indian ocean while it is above the ellipsoid in some parts of Europe. Figure 7.10 shows the correction to the reported GPS altitude in two locations - India and the US. In both locations, the geoid is below the ellipsoid and therefore the error correction is positive (altitudes are measured in reference to the geoid). However, the magnitude of the correction is higher in India, since the ellipsoid is higher than the geoid in India. The reference altitudes for the GPS readings in Figure 7.10 are based on the values returned by Google Maps API.

Measuring Altitude with an Altimeter   Some phones and GPS devices include an altimeter, which is like a barometer, to measure altitude. The altimeter uses the atmospheric pressure at a location to compute an estimated altitude. The reference atmospheric pressure at MSL is 1013.25 hectopascals or millibars. At higher altitudes the atmospheric pressure will be less. From sea level (0 meters) to 1000 meters, atmospheric pressure decreases by about 0.1 millibar per meter to about 904 millibars. From 1000 meters to 2000 meters, the average decrease in atmospheric pressure is about 0.09 millibar. Therefore given atmospheric pressure, the corresponding altitude can be computed (see Appendix).

Unfortunately atmospheric pressure depends on weather conditions and pressure at the same altitude can be different. At 5°C and 25°C, the computed pressure at 1000 meters is 897 mbars and 904 mbars respectively. Therefore in addition to an altimeter, you would need a temperature sensor to accurately convert the detected pressure and ambient temperature to an altitude. The accuracy of a GPS reading using an altimeter is estimated at ±15 meters about 95% of the time with a rule of thumb that vertical accuracy will be roughly three times horizontal accuracy [4].

7.3.3 Heart rate

During a ride, your real time heart rate is a good indicator of the intensity of your effort. Heart rates are broadly divided into four (see Chapter 9) or five zones [5]. The heart rate ranges for the five zones are based on the type of activity and age/sex of the person. Cycling may have a slightly higher range than swimming but a slightly lower range than running.

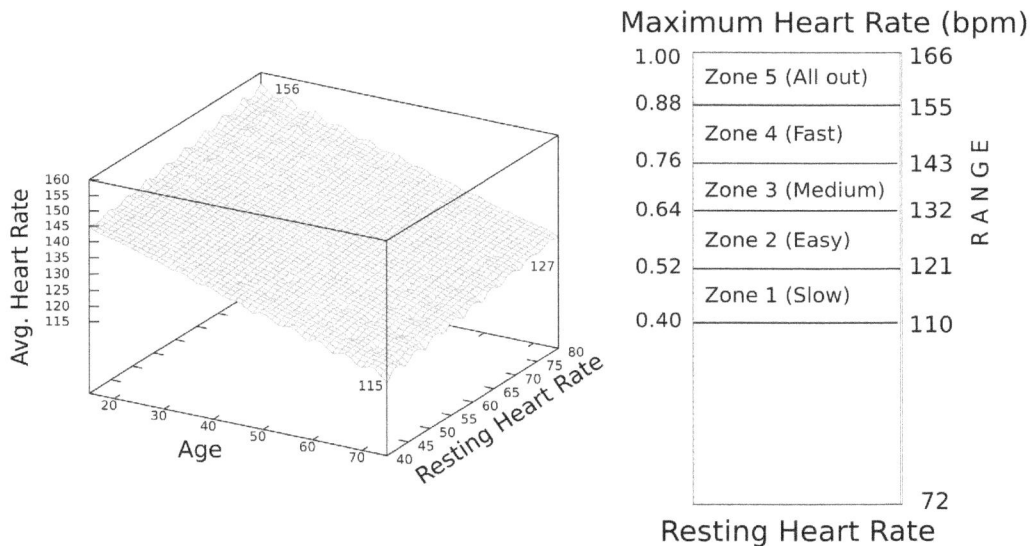

Figure 7.11: Average female heart rate (bpm) in zone 3 (medium intensity) by age and resting heart rate

The maximum heart rate or HR_{max} is computed using the formulas $206 - age \times 0.8$ and $201 - age \times 0.7$ for a male and female respectively. The resting heart rate of HR_{rest} is the number of beats per minute (bpm) when you are at rest. For a 50 year old female (HR_{max} of 166 bpm) with a HR_{rest} of 72 bpm, the range from 72 to 166 bpm (94 bpm) is divided into five zones (see Figure 7.11). The first zone is from $72 + 94 \times 0.40$ to $72 + 94 \times 0.52$ and the second zone is from $72 + 94 \times 0.52$ to $72 + 94 \times 0.64$ and so on.

Each of the five zones represent a level of cycling intensity. Zones 1 and 2 represent intensity levels that a cyclist could sustain for a few hours. In zone 3, a rider would find it difficult after an hour or so while in zone 4 it may take minutes before exhaustion and finally efforts in zone 5 maybe sustainable for seconds.

The range of heart rates from the resting heart rate to the maximum heart rate depends on the age, sex, and resting heart rate. In Figure 7.11, the average heart rate in Zone 3 for females from ages 15 to 75 and with resting heart rates from 40 to 80 is the least (115 bpm) for the highest age (75) and lowest resting heart rate (40). A 15 year old female with a resting heart rate of 80 bpm has the highest heart rate of 156 bpm in Zone 3.

While it is not necessary to constantly monitor your heart rate in real time, if you do know your current heart rate and zone, you will know how long you can sustain a ride at the same effort level. Upto the age of 50, males have a higher heart rate range for the same resting heart rate with the largest difference in the youngest (Figure 7.12). After the age of 50, females have a higher heart range that increases with age.

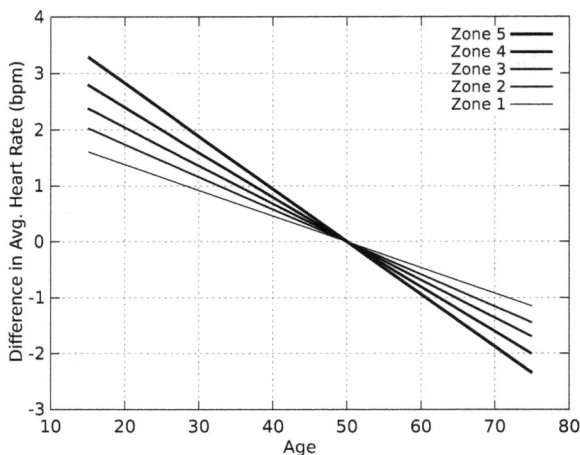

Figure 7.12: Difference between average male and female heart rate for five zones vs. age

7.3.4 Power

Even though heart rate does indicate the intensity of a workout, there are many other reasons that can be attributed to a higher heart rate such as illness, dehydration, altitude, and lack of sleep. Power output is a more accurate indicator of the level of training intensity. One formula for power is $f \times v$ where f is the force applied and v is the velocity. In a bicycle, force is applied to rotate the crank and therefore the circular definition of power $\tau \times \omega$ is used where τ is the torque applied and ω is the angular velocity. While angular velocity (cadence) is not very difficult to compute, measuring torque is harder. There are several locations where a power sensor can be installed that include - pedals, crank arm, rear hub and chain ring. The sensor must detect force as well as angular velocity.

One of the issues with a sensor on a chain ring or pedal is that the power on one side alone is measured. In many cases, a cyclist will generate roughly the same power in each leg. Therefore simply doubling the power generated on one side will give the total power with some accuracy. Some manufacturers claim that a sensor installed on one pedal or chain ring can measure power on both sides. Other manufacturers use separate sensors on each side to measure power and can reveal if there is an imbalance in the power generated by each leg. A hub based sensor detects force applied at the rear hub which will be very close to the force applied at the pedal on a bicycle with a well maintained drive train. Since force is detected at the hub, there is no way to differentiate between the force applied by each leg.

While speed is a convenient measure to compare performance, your speed can vary substantially depending on weather conditions and grade of the road given the same effort. On the other hand power is a measure that is independent of external factors and represents your effort on a ride. A power sensor will transmit the instantaneous applied pedal force to a device that will convert the data to power generated based on the velocity.

Measuring Power with a Sensor   While cadence can be accurately measured with a magnet and a simple sensor, force is harder to measure as accurately. Most power sensors use a strain gauge to measure force. A strain gauge converts an applied force into an equivalent change in resistance. When you apply a force to the pedal, thin foils containing wires in the strain gauge will deform (become longer and thinner) and therefore resistance will increase. The change in resistance must be accurately calibrated with the associated force for a precise measurement of power.

Unfortunately, power sensors remain one of the most expensive sensors compared to an altimeter, GPS, cadence, or heart rate sensor. Power sensors are not being manufactured on the same scale as a GPS receiver or heart rate monitor and therefore will remain expensive relative to the cost of an average bicycle. However, you can estimate your power output from GPS readings, changes in altitude, and wind speed. While you can compute power without a sensor, these estimates will not be as accurate as the results from a power sensor.

An alternate method to measure power is to estimate air drag with a wind sensor. From the air drag and the remaining two drag forces (rolling resistance and slope resistance), power can be computed as the product of all forces including drag and acceleration with current velocity. Devices using former method of calculating power are often called direct force power meters while the latter method is used in opposing force power meters.

Measuring Power without a Sensor   Without a strain gauge to measure applied force, the sum of drag forces and acceleration can be computed using data from other sensors. The drag force has three components - the aerodynamic drag F_{air}, rolling resistance $F_{rolling}$, and slope resistance F_{slope}. Of the three drag forces, the slope resistance can be calculated very accurately. Rolling resistance is less accurate but not a large component of the drag forces at velocities above 20 kmph. The aerodynamic drag force is the least accurate to calculate since it depends on knowing the instantaneous speed and direction of wind relative to the motion of the bicycle. Unfortunately aerodynamic drag is also the largest of the drag forces at higher velocities.

Aerodynamic Drag Factor K_a The aerodynamic drag factor K_a is defined as $\frac{1}{2} \times C_d \times \rho \times A$ where C_d is the coefficient of drag, ρ is the density of air, and A is the frontal area. Air drag is the air resistance in the direction of the travel (V_{cycle}) and is measured as $V_{app}^2 \times cos(\theta) \times K_a$ where V_{app} is the apparent direction of wind and θ is the angle of V_{app} with V_{cycle} (see Figure 7.13).

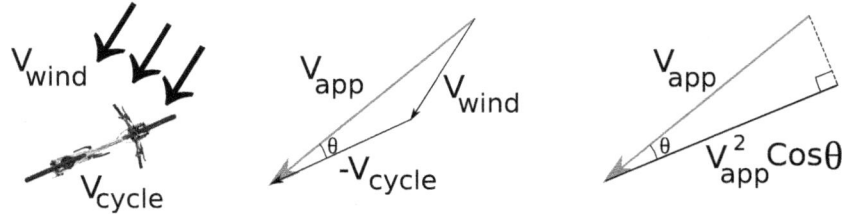

Figure 7.13: Calculating Wind Factor W_a for Air Drag from Wind and Bicycle Velocities

Changing the front area A can significantly change the air drag which in turn affects the calculation of power. Moving from a commuter to a racer position changes the coefficient of drag as well as the front area and will reduce the power you need to generate to ride at a given speed. Since front area is essentially considered as a 2D surface, both the direction and magnitude of the opposing wind will change air drag. Generally, the side or lateral front area will be more than the fore or longitudnal front area with the side area being about 20% larger than the front area The ratio of the side area to the front area μ is approximately 1.2.

Isvan [6] suggests that K_a is a function of the angle θ between the apparent wind vector V_{app} and the vector V_{cycle} and the actual aerodynamic drag factor K_a' can be defined as -

$$K_a' = (cos^2\theta + \mu sin^2\theta) \times K_a$$

When θ is zero, V_{app} directly opposes V_{cycle} and K_a' is the same as K_a. As θ approaches 90° (a cross wind), K_a' approaches the maximum μK_a and then decreases back to K_a when θ becomes 180° (a tail wind). The purpose of this correction to K_a was to correct the under estimation of power when θ is a cross wind.

Wind Factor W_a The apparent wind V_{app} is the wind that you feel when riding along V_{cycle} and is equal to $V_{wind} - V_{cycle}$. If V_{wind} was zero, then the apparent wind would be just $-V_{cycle}$. Air drag is proportional to the component of the square of the apparent wind projected along the direction of travel (the wind factor $W_a = V_{app}^2 \times cos(\theta)$).

The wind factor can change significantly depending on the magnitude and direction of the head wind. Figure 7.14 shows why you rarely experience negative air drag. If a rider is headed north east at an angle of 45° at a speed of 20 kmph., then you will only experience a negative wind factor when the wind is blowing between 200° and 250° with a magnitude greater than 20 kmph. When the wind blows between 200° and 250°, you would need to generate power to overcome the difference of the sum of the slope and rolling resistances with the air drag force. It is very rare

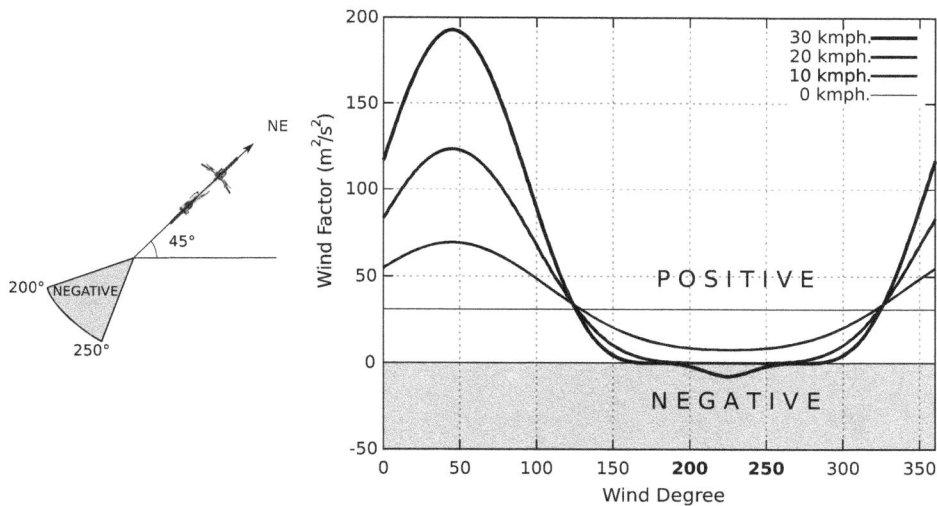

Figure 7.14: Wind factor (m^2/s^2) vs. Wind Degree for Four Wind Speeds

to experience negative air drag since not only must wind magnitude be larger than the bicycle velocity, but it must also blow in the same direction of the travel within $\pm 25°$.

As you would expect, the wind factor peaks at 45° when the wind completely opposes the direction of motion and then drops sharply with change in wind degree. Since the wind factor is based on the square of the apparent wind, the change in wind factor is magnified by the change in head wind speed.

The value of air drag force $F_{air} = K_a \times W_a$. Since both K_a and W_a depend on front area and the wind vector respectively that change in real time and therefore the accuracy of the air drag force will be less than accuracy of F_{slope} and $F_{rolling}$. The values of both F_{slope} and $F_{rolling}$ are proportional to the combined weight of the bicycle and rider (mg) which can be calculated accurately. The calculation of the grade of the road is also accurate since the resolution of the digital elevation model from Google is 10 meters or more (it takes about 5 seconds to cover 27 meters at 20 kmph.). Therefore, elevation readings taken at 5 second intervals should be reasonably accurate. Rolling resistance is proportional to a constant depending on the type of tyre.

Converting Power to METs   One of the issues with using power to evaluate your effort is that it depends on your weight. A heavier person maybe able to generate more power than a lighter person and instead of comparing raw power measurments, a fairer comparison uses the power generated per kg. The estimated values for a fit person, top amateur, and an elite professionals are 3 w/kg., 5 w/kg., and 6 w/kg. respectively. So, if you are reasonably fit and weigh 70 kgs., you should be able to generate about 210 watts for an hour.

An alternate method of evaluating effort is the metabolic equivalent or MET. One MET is the energy you would use if you are resting. Light activity such as walking slowly and light work requires 1-3 METs. It takes 3-6 METs for slightly more intense work such as walking briskly or

241

riding a bicycle leisurely. Beyond 6 METs, the work load is vigorous and very intense workouts can consume 12 METs or more.

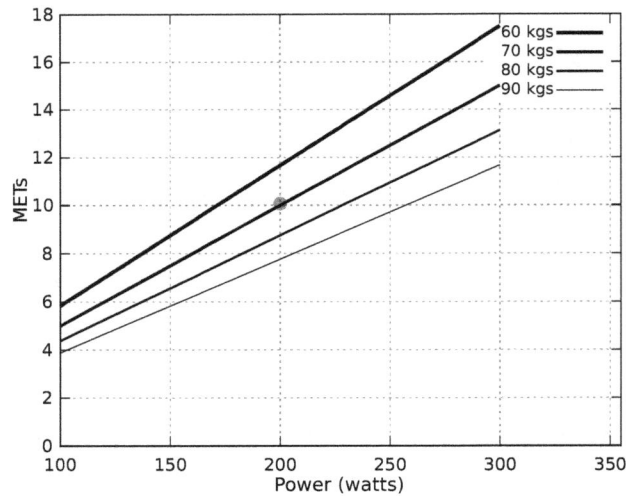

Figure 7.15: Metabolic Equivalent vs. Power for Four Different Weights

In Figure 7.15 the MET value is computed for a given power and weight. The power x watts generated is first converted to $y = x \times 0.84$ kcal / hour [7]. Next the kilocalories per hour is normalized by weight and body efficiency (0.24). If a rider weighing 70 kgs. generates 200 watts, the equivalents METs is $\frac{200 \times 0.84}{70 \times 0.24} = 10.0$. The efficiency of the human body (0.24) is the ratio of the work output to the energy required or oxygen consumed and estimated values for efficiency range from 0.18 to 0.26.

As the efficiency of your body increases, it takes fewer METs to generate the same power and therefore you will be able to sustain a higher power output. Increasing weight with the same power output reduces METs. At 250 watts, increasing body weight by 50% from 60 kgs. to 90 kgs. decreases METs by 33% from 14.6 to 9.7.

Normalized Power   While average power does indicate the average effort level for a ride, it does not tell you of any intense efforts or the duration of such efforts. Instantaneous power can fluctuate significantly over a wide range of the average power and the normalized power measures the difficulty or the intensity of the ride. Normalized power is also an average, but represents the constant power that you could have sustained over the duration of the ride for the same "physiological cost". In a ride with changing intensity over periods of 30 seconds or more, the total work expended in a ride, product of average power and ride duration, will be different from the product of the normalized power and ride duration.

A ride with an average of 200 watts on a flat road maybe steady with instantaneous power between 190-210 watts. On a hilly road, the instantaneous power may fluctuate from 400 or more watts on an uphill to 0 watts on a downhill with the same average of 200 watts. Even though the average power was the same, the ride on a flat road will seem easier compared to the ride on a hilly road.

If a ride had a large variability in generated power, then the normalized power is likely to be higher than the average power. Variance is one of the statistical methods used to find the variability of a data series. It is average of the squared differences between the mean (average power) and individual values. The square root of variance, the standard deviation, is useful to characterize what fraction of data lies within one or more standard deviations, when data in a series is normally distributed.

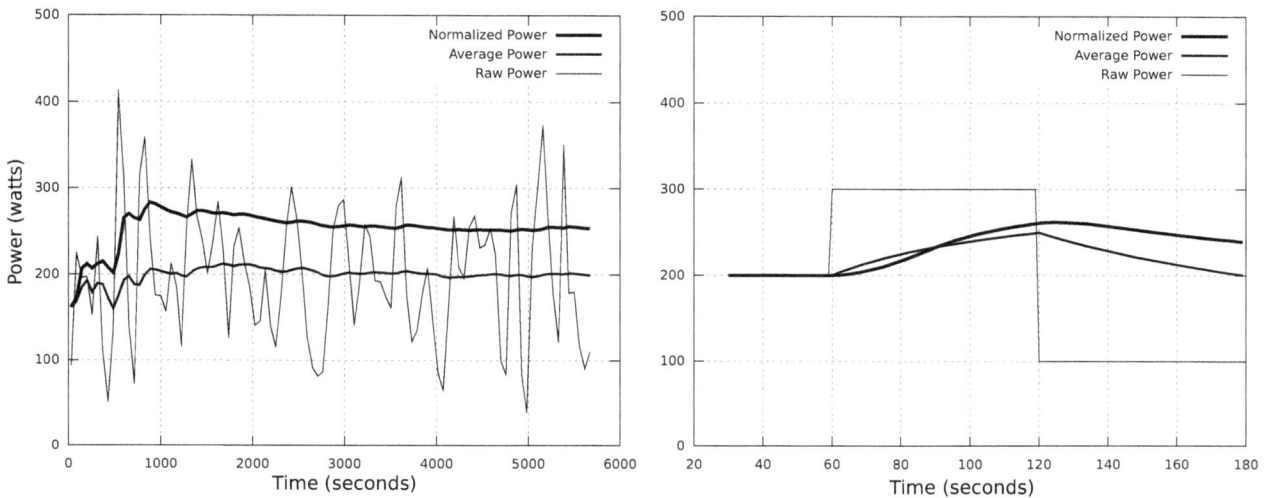

Figure 7.16: Normalized, Average, and Raw Power in watts vs. Time in seconds

However, power data appears to be noisy and is not necessarily normally distributed (see raw power data in left Figure 7.16). The algorithm to calculate normalized power is based on some observations. First the moving average over a period of 30 seconds is computed. [14] An interval of 30 seconds was used to calculate a moving average since the physiological response to changes in intensity for a shorter duration are the same as the average power [8]. The response to a 300 watt effort over 10 seconds followed by 0 watts over 10 seconds was roughly the same as the response to 150 watts over 20 seconds. The physiological response to an increase in intensity is not instantaneous and is detected after a period of about 30 seconds. Therefore, a change in intensity needs to be sustained for longer than 10 seconds to generate a physiological response. The moving average over a 30 second period can be computed if the power data is available in one second intervals. P_i is the moving average power at the ith second where i ranges from 30 seconds to n seconds, the duration of the ride. The normalized power is then defined as

$$P_{normal} = (\frac{\sum_{i=30}^{n} (P_i)^4}{n - 30})^{0.25}$$

---

[14]A moving average is commonly used with time series data to smooth out short-term fluctuations and highlight longer-term trends or cycles.

where P_{normal} is the fourth root of the average of the moving average power raised to the fourth power. In Figure 7.16, the moving average increases the height of valleys and lowers the height of peaks which in turn spread the effects of a peak or recovery over many locations. Experiments [8] with cyclists riding close to their lactate thresholds [15] revealed that the percentage of current lactate compared to the lactate threshold for a cyclist was proportional to the percentage of current power to the power at the lactate threshold, raised to the fourth power. The normalized power then becomes the fourth root of the mean lactate percent relative to the lactate threshold.

The right plot in Figure 7.16 shows the raw, average, and normal power for a hypothetical 3 minute ride with the first minute at 200 watts, second minute at 300 watts, and third minute at 100 watts. The average power is 200 watts and does not reveal the change in intensity in the second and third minutes. The normalized power increases from 200 watts in the second minute and stays above the average power since the moving average distributes the effects of the increase in intensity. At the end of three minutes the normalized power is about 240 watts. If this pattern of a period of high intensity follow by a period of recovery is repeated many times in this hypothetical example, normalized power will be roughly 15% greater than average power. A shorter period of high intensity power compared to a longer period of recovery will reduce both the average and normalized powers. Normalized power will be roughly 25% more than average power if the power output in the intense period is increased to 400 watts.

7.3.5 Temperature

Other sensors that are used less often include sensors to measure temperature and wind. Changes in temperature change air density which in turn changes aerodynamic drag. Since aerodynamic drag is significant at high speeds, a low K_a will improve performance. While frontal area A can be reduced by crouching, air density ρ is dependent on the weather (temperature, relative humidity, atmospheric pressure, and altitude). More than one formula can be used to calculate ρ with similar results. Table 7.3 uses a formula [9] from a popular Web based calculator to calculate air density and another formula is shown in the Appendix.

Parameter	Range	**Density Range**	Conditions
Pressure	800 to 1200 mbar	**0.96-1.45**	500 m, 50% RH, 15°C
Altitude	0 to 1500 meters	**1.22–1.02**	15°C, 50% RH, 1000 mbars
Temperature	-10 to +40°C	**1.32–1.10**	500 m, 50% RH, 1000 mbars
Relative Humidity	1% to 100%	**1.209–1.201**	500 m, 15°C, 1000 mbars

Table 7.3: Air Density Range in $kg./m^3$ for Pressure, Altitude, Temperature, and Relative Humidity Ranges

---

[15]Lactate threshold is the maximum power that you can generate for a short period before the accumulation of lactic acid in the muscles makes it difficult to generate power at the same level.

Each row of Table 7.3 shows the change in density associated with a change in a parameter given certain conditions. Changes in atmospheric pressure generate the largest range of densities from 0.96 to 1.45 $kg./m^3$. Lower pressures are associated with lower densities and vice versa. At higher altitudes, the density of air reduces as you would expect. Colder temperatures have a higher air density than warmer temperatures and relative humidity by itself does not change air density as much. Often when weather conditions change, more than one parameter will change – relative humidity may increase with an increase in temperature.

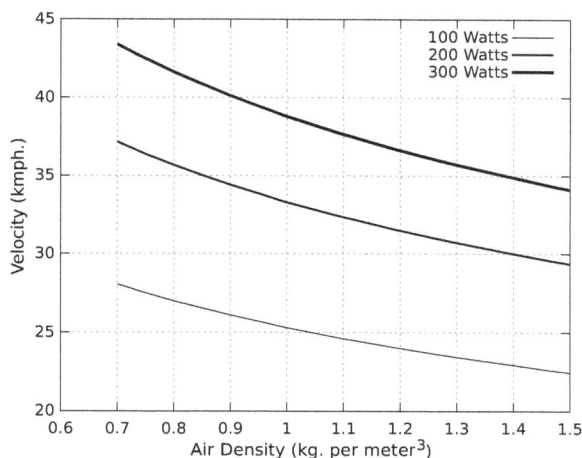

Figure 7.17: Air Density in $kg./m^3$ vs. velocity for three different powers

Since air density is a factor in computing aerodynamic drag, changes in density will also change your riding speed, given generated power. In Figure 7.17, a 70 kg. rider on a 10 kg. bicycle on a flat road with no wind is assumed. As density increases, the air drag also increases and therefore the speed that you can ride at decreases. On a ride, air density is unlikely to change in the range shown in Figure 7.17 and therefore the affects of changes in air density are probably minimal.

Even though an increase in temperature is associated with a lower density and therefore lower air drag, it is unlikely that you will be able to ride faster. At a higher temperature, you will be less efficient and may spend more energy staying cool than generating higher power.

7.3.6 Pairing Sensors

With multiple sensors on a bicycle, ensuring that all the sensors are correctly paired with a smart phone or device is important before beginning a ride. Two of the popular protocols that most sensors use to transmit data are ANT+ and Bluetooth. While ANT+ is possibly easier to use than Bluetooth, it is less secure, since sensor data is broadcast. With Bluetooth, every sensor must be paired with a particular device. Ensuring that a device is receiving data from all sensors is dependent on the manufacturer and there are any number of problems that you may face with a device or sensors.

Since data from multiple sensors is saved, both cyclocomputers and smart phone apps use elaborate screens to display the data. The collected data is mostly numeric and there are a number of ways to display the data on graphs, plots, and so on. Some of the results are intended for display during a ride and the remainder for analysis post ride. It is safer to view most of the collected data with the exception of speed, distance, and location after a ride.

Integrating Data   All sensor data is collected in the form of a time series. A heart rate monitor may transmit the current heart rate every second to a device. Similarly, the cadence may also be tranmitted to a device every second. Concurrently, a GPS receiver may receive signals every five or ten seconds.

Ideally, all sensors should be synchronized with a single clock and transmit data to a receiver in identical periods. Since this is not practical, data from multiple sensors on different time scales must be integrated. The left plot in Figure 7.18 show power data received in roughly five second intervals. The right plot shows the same data interpolated between the time intervals such that the interval is one second between each reading. Linear interpolation is the simplest method to compute values in the interval between two consecutive readings.

Table 7.4 shows raw and interpolated data for a period of four seconds. The power reading at 0 seconds and 4 seconds was 200 watts and 300 watts respectively. The difference $\Delta power$ (positive or negative) between the end and start reading was 100 watts and the increase per second is computed as $\frac{\Delta power}{\Delta time}$ (+25 watts per second) where $\Delta time$ is the difference (4 seconds) between the end and start times. The increase of 25 watts per second is uniform over the interval and may not accurately reflect the rate of increase in power in reality.

Time (secs.)	Raw Power (w)	Interp. Power (w)	Raw Location	Interp. Location
0:00	200	200	39.02°/77.20°	39.02°/77.20°
0:01	-	225	-	39.03°/77.21°
0:02	-	250	-	39.04°/77.22°
0:03	-	275	-	39.05°/77.23°
0:04	300	300	39.06°/77.24°	39.06°/77.24°

Table 7.4: Interpolated and Raw Power and Location Data

This same method of linear interpolation can be used to compute intermediate locations between two consecutive GPS readings. A GPS reading will return a latitude / longitude, say 39.02° / 77.20° at time 0 seconds. Locations for the next 3 seconds are computed using the data for the new GPS reading, 39.06° / 77.24° received after four seconds.

7.4   Cycling Apps

A review of all cycling apps is beyond the scope of this book. A short description of some of them and their features is mentioned below. Some corporations such as Garmin, Cateye, and Wahoo have specialized in building the hardware for devices and sensors and others (Strava,

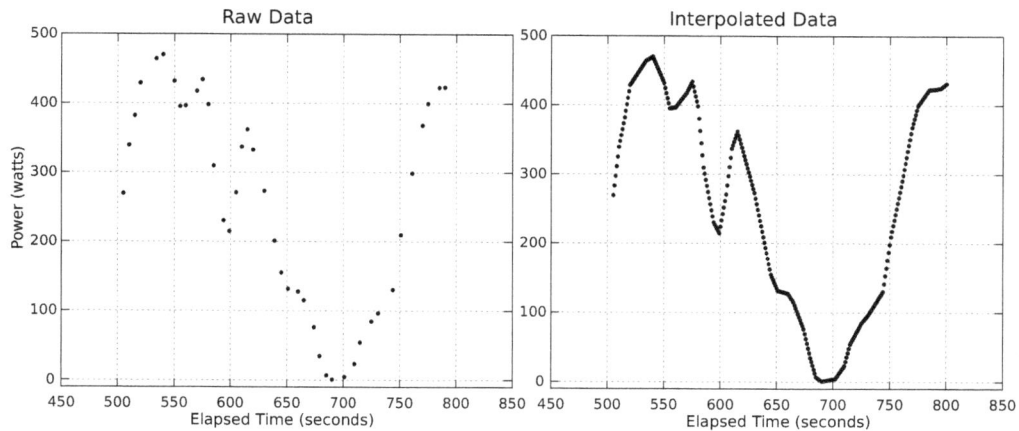

Figure 7.18: Raw and Interpolated Power Data for a Given Period in a Time Series

RidewithGPS, and MapMyRide) providing cloud based services to manage and share collected data. The devices and services offered are usually not exclusively for cycling and often include other physical activities such as running. Many of the services are based on the freemium model, i.e. a subset of the features are available for free and the rest of the features are included with a paid subscription.

Service providers rely on a large number of subscribers with a small fraction paying for the premium services. Hardware manufacturers sell devices and therefore do not need a very large base of users. Many of the services providers include options to import data from a number of devices. With a GPS device and an app, you can monitor, analyze, and save all your rides. Besides apps from cloud based providers, hardware manufactures such as Wahoo and Cateye also include apps along with their devices. The following list includes some of the features for a cyclist that are likely to be useful in a mobile app

- A database of rides that includes the route, distance covered, duration of the ride, start time, end time, number of GPS readings, number of meters climbed, number of meters descended, the times spent on ascent/descent, the lowest and highest location, power, heart rates, cadence, bearing, grade, and average speed.

- GPS reading data including reported accuracy, elevation, period between readings, distance between readings, number of satellites with a detected signal, and strength of the signal.

- Maps and plots of the route with speed, altitude, and grade. On a desktop, both the map and plot can be viewed side by side.

- Split a route into segments or automatically detect segments on a route. A route that overlaps may represent loops on a route that are repeated for convenience to cover some fixed distance longer than the distance of a single circuit.

- Playback a route to evaluate speed, distance covered, time, and location on a ride. Two or more rides can be compared to detect changes in speed and performance.

- Export, import, and backup routes in different file formats that can be viewed on a desktop, mobile app, or device.

- Mark waypoints or points of interest on a route. An audio warning when approaching a waypoint or intersection will help prevent losing track on a route.

- Set GPS parameters such as the minimum distance and time between two readings to control the frequency of readings to preserve battery capacity.

- On a long route that takes many hours, an app should periodically commit a part of the route that has been completed to avoid losing a large number of readings due to low battery or some other malfunction.

- Predict a ride's timings given current weather conditions (wind) using a prior ride's power data.

An organization for a professional cycling team collects large volumes of data that are more detailed and include some of the same sensor data mentioned such as the heart rate, instantaneous power, speed, cadence, and location [10]. However, collected data is analyzed in real time by coaches to make decisions during a race. A coach has to use information, without overloading, to prepare a cyclist mentally and physically for a race.

While an amateur rider may collect data every five seconds or longer, the sensors for a professional cyclist will transmit data more frequently. A million records or more can be collected for a long rides of many hours for a team of cyclists.

7.4.1 Strava

If you simply want to keep track of your rides, a GPS Tracker app (see Section 7.4.3) is sufficient. The purpose of building a cloud based service went far beyond just monitoring individual rides. Strava has a large subscriber base with many millions of cyclists and runners and encourages social interactions between members which in turn motivates users to add posts. A 2015 study [11] of the activity data of 4500 Strava profiles found that social interactions such as kudos and comments encouraged users to post more activities. Strava mainly attracts cyclists and other fitness enthusiasts and therefore is less likely to face issues such as nasty comments and negative feedback that are often found on a general social media website.

Besides social interaction with other cyclists, Strava also includes other features to analyze a ride, compete with other riders, and find new routes. While group rides are common, many riders cycle solo since it is convenient and the route/distance is an individual choice. Since many riders will cover the same route at different times, you can compare your performance (mostly speed) with the performance of others on the same route. The routes of riders may not necessarily be identical but instead have parts or segments that overlap.

Segments One of the innovations of Strava was to let riders build a *segment* or part of a route. A segment could start and end on a popular bike route or a main road and could also represent a loop in a route. The route of every ride that overlapped the segment would include the time to complete the segment. The leaderboard kept track of every rider who attempted the segment and ranked timings if the same segment was covered on multiple rides. Strava coined terms like "King/Queen of the Mountain" (KOM or QOM), "Personal Record" (PR), and "Course Record" (CR) for individual riders to compare performance.

While a segment has a precise start and end location, your start and end locations may not perfectly match. Strava analyzes the GPS readings from your ride to find matching segment start and end locations within some short distance (several meters). This process can be computationally demanding since there are millions of segments online. The start and end location for your segment is the location on your ride that is closest to the start and end segment location. The distance of the segment is known and the duration for the segment is extracted from the timestamps associated with the start and end locations. Since the period between GPS readings is often 5 seconds or more, interpolation (see Table 7.4) will approximate the locations to within one second intervals for more accuracy.

On Strava, you can follow the rides of other users and in turn allow others to follow your rides. You have the option of making your ride available to the public, just your followers, or private. Strava users are frequently interested in comparing their performance with the performance of peers as well as past rides. While speed is one metric to compare rides, generated power is a better metric, since bad weather or a strong head wind will slow you down. Since routes are plain text files in a common format like GPX (see section 7.5), it is not very difficult to manually alter timestamps or locations to increase speed. The elevation readings can also be changed to indicate a segment with an unrealistic 30% grade or more. Such manipulations of file data will inevitably be detected or ignored by an experienced cyclist.

Routes that are popular become known quickly and it is not very difficult for a new comer to find routes in an area. Besides the route itself, you can post photos, tags, and a detailed description similar to the features you would find on a general social media site. You can search for segments based on the grade and distance in a local area. Strava's services depend on its ability to monetize fitness activity and features such as segments, collaboration, and ranking.

7.4.2 RidewithGPS

RidewithGPS has a few common features with Strava and is based on the same freemium model. Some of the features are available for free and others require a subscription. Like Strava, you can find routes by location, distance, or elevation. You can also share routes, comments, and photos of rides.

One of the most useful features of RidewithGPS is planning a route. On a map, you can draw a route that you have planned and build an associated cue sheet. A cue sheet provides detailed instructions of distances and turns on a route and is useful when planning group rides. Along with the map of the route, the cue sheet becomes a useful reference on a new route. If you save the route in a GPX or TCX file format, you can upload the file (route) to a device or cyclocomputer.

A calendar tracks your rides over weeks or months and you can monitor the distance that you have covered over some period. Similar to Strava, RidewithGPS also includes functions to build segments from parts of a route that can be shared. The leaderboard will show the rides of other cyclists on the same segment.

Both RidewithGPS and Strava have apps for a mobile device as well as the desktop. The desktop is more suited for functions such as plotting a route or detailed analysis which require more screen space. The mobile screen is convenient for functions that summarize a route or are needed during a ride.

7.4.3 GPS Trip Analyzer

GPS trip analyzer [12] is an alternate free Android app to maintain detailed information on rides. A ride or a trip is defined as $T = \{S : d_1, d_2, ...d_n\}$ where S contains a Summary record of the trip and d_i is a detail record for a particular location. A trip is made up of a Summary record S and n detail records for n GPS readings. A Summary record S contains information that is common for the entire trip such as

- Trip ID, start time, end time, distance covered, average latitude, and average longitude

- Minimum, maximum, and average of reading accuracy, time between readings, and distance between readings

- Duration, speed, and distance of each segment of the trip

- Moving average speed and category (walk, jog, cycle, or other) of the trip

Each detail record d_i contains information for a single GPS reading such as -

- Associated trip ID, segment number, latitude, longitude, bearing, speed, and an index number

- Battery status, timestamp, altitude, gradient, and accuracy

- Satellite signal strength and number of satellite signals received

- Optionally local weather conditions at the location such as temperature, humidity, wind speed, wind degree, and generated power

The current location (latitude and longitude) is first detected before starting a trip. The latitude and longitude are returned from a *Location* services API with 6 decimal point precision which is more convenient for calculations than precision in minutes and seconds. The Eiffel tower is located at 48.858093°N 2.294694°E where the directions N and E are usually omitted. All locations to the south of equator have a negative latitude and likewise locations to the west of the prime meridian line are negative. While each degree of latitude is roughly about 110 kms. apart across the globe, the distance between each degree of longitude depends on the location. At the equator, the distance is about 111 kms. and gradually reduces to zero as you approach the poles. In theory, the accuracy of a location is a millionth of 111 kilometers or about 11 centimeters.

Menus and Functions   The main screen of the app has 6 functions (see Appendix) that include the list of trips, a calendar showing the distance covered per day, the satellites whose signals can be detected, settings for various functions, a compass, and a power/speed calculator. The *Trips* function lists trips by date and includes a list of menus for other trip related functions (see Figure 7.19). The *Trip List* shows a list of trips in a table with the associated ID, date, duration, distance covered, average speed, and category of trip. One or more trips must be selected to perform trip related functions such as viewing the map, plot, or analysis of the trip. Selecting a date on the calendar will show the trips for the associated date and the trip category.

There are many other functions that can be applied to a trip such as -

- A side by side playback of two trips showing the distance covered and time to compare performance on the same route

- Playback of a trip on a map showing the time, distance, and speed

- Correcting the location of GPS readings using the *Snap to Road* API (needs a Google Maps API key)

- Correcting the altitude of GPS readings using the *Elevation* API (needs a Google Maps API key)

- Downloading the weather at a location using the *OpenWeatherMap* API (needs an Open-WeatherMap API key)

- Manually moving and deleting locations of GPS readings on a map to correct a route

- Split a trip into two or more sub-trips or combine two or more trips into a single trip

- Add waypoints on a trip that will be followed

- Export and import trips in the GPX, CSV, or XML formats.

- Repair a trip by removing GPS readings with poor accuracy, redundant readings at a single location, and automatically identifying segments and U-turns.

The *Satellites* function shows the location of satellites whose signal can be detected and some brief information on the satellite. Even though, your current location maybe identified using the *Location* API, it is possible that the location was cached and will be invalid once you set off on a ride. If the app receives signals from four or more satellites, then the GPS readings on your trip are more likely to be accurate. It is very rare that fewer than five satellites will be visible at any given time and the most likely cause of error is a problem with the GPS receiver or device.

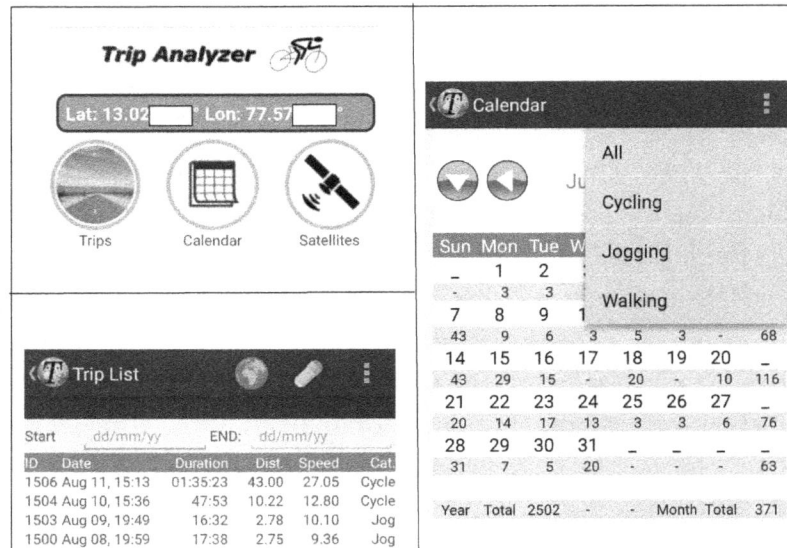

Figure 7.19: Viewing and Managing Trips on GPS Trip Analyzer

From the *Settings* function, you can set the number of trip rows per screen, the maximum number of GPS readings per trip, the maximum duration of a trip, the minimum distance and time interval between GPS readings, the export file format, API keys, average walking speed, and bicycle specific settings such as the weight, rolling resistance, and coefficient of drag.

Detecting Trip Category The average walking speed, average speed, and maximum speed is used to automatically identify the category (Walk, Jog, Cycle, or Other) of a trip. If the average walking speed is 5 kmph., then the ranges of the average speed for a Jog depend on how much larger the maximum speed is than the average speed. When the maximum speed is not significantly larger than the average speed, the range for average speeds for the category is larger. The assumption behind using the maximum speed to determine the category is that in most cases if the maximum speed is twice the average speed, then the range of average speeds for the category is more limited.

Maximum Speed	Walk	Jog	Cycle	Other
$1.25 \times Speed_{avg}$	0.0 - 7.1 kmph.	7.2 - 11.9 kmph.	12.0 - 43.0 kmph.	43.1+ kmph.
$1.50 \times Speed_{avg}$	0.0 - 6.6 kmph.	6.7 - 9.9 kmph.	10.0 - 39.9 kmph.	40.0+ kmph.
$1.75 \times Speed_{avg}$	0.0 - 5.7 kmph.	5.8 - 8.5 kmph.	8.6 - 34.2 kmph.	34.3+ kmph.
$2.00 \times Speed_{avg}$	0.0 - 4.9 kmph.	5.0 - 7.4 kmph.	7.6 - 29.9 kmph.	30.0+ kmph.

Table 7.5: Category Ranges for a Trip with Average Walking Speed of 5 kmph.

The ranges in Table 7.5 may be too low for an athlete. If the walking speed is adjusted to 7 or 8 kmph., then the ranges will also be higher for each category. If the maximum speed is 1.25 times the average speed, then at an average walking speed of 7 kmph., the average speed ranges for jogging and cycling are 9.2 - 16.7 kmph. and 16.8 - 51.0 kmph. respectively.

Maps For every trip, a Google Map track in color shows values in a range of colors from red to blue for a number of parameters. Touching the track at a particular location shows more detailed information from the nearest reading and touching outside the track shows summary information (see Figure 7.20).

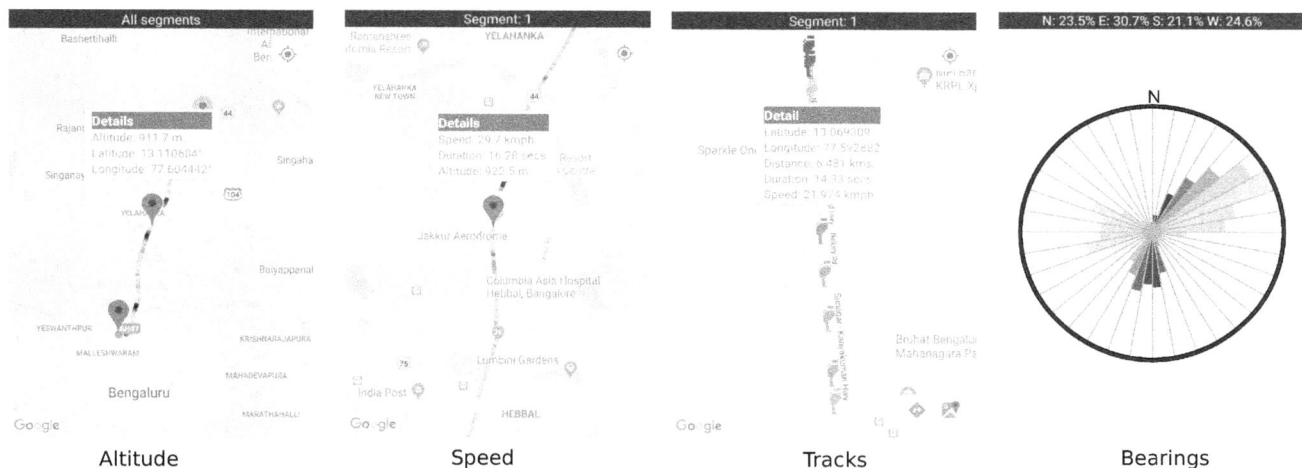

Figure 7.20: Maps of Altitude, Speed, Tracks, and Bearings for a Trip

- **Altitude:** The track is red and blue where the altitude was the highest and lowest respectively. Touching a part of the track shows the altitude, latitude, and longitude at the nearest location.

- **Speed**: The track is red and blue in locations where speed was the highest and lowest respectively. Colors between red and blue show intermediate speeds. The map is shown separately for each segment which is useful when tracks are overlaid. Touching a section of the track shows the speed, duration, and altitude at the nearest reading on the track. Touching outside the track shows summary information for the segment or trip.

- **Tracks:** A series of footprints are shown in the map in the direction of the trip with blue and red colored footprints representing slow and fast speeds. Touching a footprint reveals information such as the location, distance covered, duration, and current speed from the GPS reading.

- **Bearings:** The percentage of time spent in a ride in a particular direction is proportional to the length of the radius of the shaded circle segment. At the top of the bearings map in Figure 7.20, the percentage of the spent riding in each of the four directions is shown. The maximum time (30.7%) was spent riding eastward, between north east (45°) and south east (135°).

- **Other** maps not shown include similar tracks for power, gradients, and satellite information at a particular location.

Plots GPS Trip Analyzer includes 15 types of plots, not all of which would be of interest for any given trip. Most plots include the raw data, the moving average, and a smoothed version of the data. The line plots show the speed, GPS reported accuracy, altitude, battery percentage, gradient, signal strength (in db.), number of satellite signals, trip speed, segment speed, power in watts, vertical climbing speed, head wind speed, time between readings, and distance between readings vs. trip time. Two other plots show a power histogram and the pace (time per kilometer) in a bar chart (see Figure 7.21).

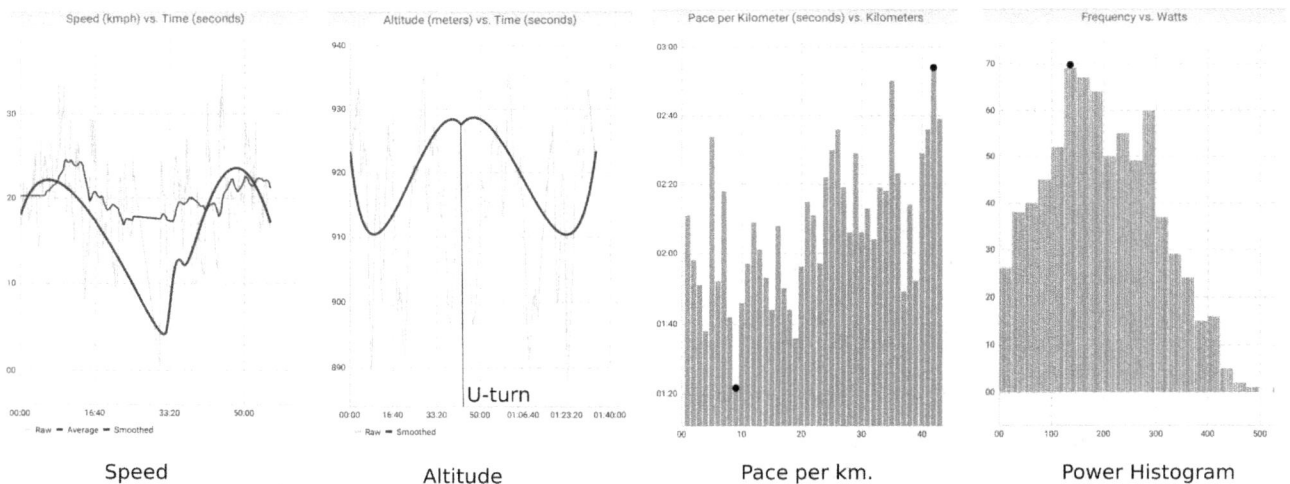

Figure 7.21: Power Histogram and Speed, Altitude, and Pace vs. Time

The plots in Figure 7.21 include -

- **Speed:** A line plot of speed in kmph. vs. time in seconds for the duration of the trip. Raw speed data appears noisy since speed can change substantially from one reading to another. The moving average speed is less noisy and the smoothed speed even more so. Speed by segment plots show the change speed in a separate plot for each segment.

- **Altitude:** The raw data for change in altitude vs. time appears to be noisy as well. However, the smoothed data shows a clearer mirrored pattern that you would expect in a round trip ride with a u-turn between 33 and 50 minutes.

- **Pace:** The bar chart of the pace per kilometer shows how the time to ride a kilometer varies from one minute and 20 seconds (the fastest pace) to more than two minutes and 50 seconds. The average pace per kilometer is displayed on a long screen press.

- **Power Histogram:** Like the plot of speed, the line plot of power is also noisy and no obvious pattern is visible. A power histogram which groups and counts the number of occurrences of power in some range of power is a better visual of power generation than a simple line plot of power. The range for a single bar is 25 watts in Figure 7.21 and the most frequent (70) power output is in the range of 125-150 watts.

- **Other** plots not shown include plots of accuracy, battery usage, gradient, satellite signal strength, number of satellites, raw power, head wind, GPS reading time intervals, GPS reading distances, and vertical climbing rate (VAM).

A bar chart of power output does not show how long a particular power level was sustained. The height of a bar for a power output of 125-150 watts is simply the number of times that a power between 125 and 150 watts was observed, but not whether it occurred consecutively or in a series separated by intervals. The occurrence of power between 125 and 150 watts for a longer single period is an indicator of a more intense ride than one with many occurrences spread over the duration of the ride.

Calculating Power Without a power meter, the calculation of power is likely to be imprecise. Still, a reasonable estimate of power can be computed using drag and acceleration forces. Power generated for a location i is computed using the data from two detail records d_i and d_{i-1}. The detail record d_i includes the altitude, temperature, and relative humidity from which the air density ρ can be computed (see Appendix). Also included in d_i are bicycle speed (v_{bike}), bicycle bearing (d_{bike}), wind speed (v_{wind}), and wind degree (d_{wind}). Isvan [6] describes a correction to the front area A based on the wind and bicycle directions. Following the correction to the front area, the aerodynamic factor $K_a = \frac{1}{2}C_d\rho A$, where the coefficient of drag C_d is set in the bicycle parameters.

Bicycle parameters such as the total rider's and bicycle weight (m), C_d, rolling resistance, and front area are initialized in the app settings. The three drag forces due to air (F_{air}), slope (F_{slope}), and rolling resistance ($F_{rolling}$) are calculated using the formulas $K_a V_{app}^2 cos\theta$, $mg \times gradient$, and $mg \times C_r$ respectively where V_{app} is the apparent wind vector, θ is the angle between the bicycle and wind vectors, and C_r is the coefficient of rolling resistance. The accuracy of power calculations is dependent on the accuracy of the altitude in each detail record. Without an altimeter, the error of the reported altitude of a GPS readings can be in the range of several meters which substantially changes the value of the slope resistance. This in turn changes the value of the total drag force which results in power output that is unrealistic.

While the v_{bike} and d_{bike} can be obtained from the latitude and longitude for the two consecutive detail records d_i and d_{i-1}, the wind speed and degree must be downloaded from another source. The OpenWeatherMap website provides an API to fetch the wind speed and degree at a location specified by the latitude and longitude. However, the resolution of the wind data is not as precise as the altitude from a digital elevation model which can be in meters.

Despite these issues with the accuracy of downloaded wind speed and degree which do affect the calculation of power, a rough estimate of power generated is possible. The accuracy of slope resistance, rolling resistance, and acceleration are weighted higher than air drag to compensate for the lack of accuracy in wind speed and degree. Finally, power is the product of bicycle velocity and the sum of all drag forces plus the acceleration divided by the efficiency (0.95) or $\frac{F_{air} + F_{slope} + F_{rolling} + F_{accel}}{0.95} \times v_{bike}$.

Calculating Velocity   Given power and all other parameters except velocity, the estimated velocity can be calculated using Newton's method (see Appendix). The equation below uses multiple powers of velocity (v_{bike}) and therefore the closest estimate of v_{bike} that satisfies the equation is computed.

$$power = v_{bike} \times (K_a \times (v_{bike} + v_{wind})^2 + F_{slope} + F_{rolling})$$

The best v_{bike} is computed iteratively till a reasonably good estimate is found. The same bicycle specific settings user earlier such as the weight of the bicycle+rider, rolling resistance, and aerodynamic factor are used in the above equation to calculate K_a, F_{slope}, and $F_{rolling}$.

The power/speed calculator in Figure 7.22 computes either the required power for a given speed or the speed for a given power and set of conditions. The wind degree and bike degree are set using 0° for north and 90° for east. The gradient can be positive or negative and is a percentage. The temperature and elevation are used to calculate the air density.

The speed is re-computed for a given power if any of the other parameters such as wind degree, wind speed, or gradient are changed. If speed is changed, then the equivalent power for the given set of values is calculated. In Figure 7.22, the expected speed is 19.8 kmph. for a given power of 200 watts on an uphill gradient of 1.5% with a 15 kmph. wind from 45° on a bicycle heading 30°, temperature of 25°C and elevation of 800 meters. Changing the gradient to 0% and -1.5% increase the speed to 24.5 kmph. and 29.8 kmph. respectively. This calculator is similar to the calculator at `bikecalculator.com` with the addition of wind and bike degree fields.

Trip Analysis   The Trip Analysis function gives a snapshot of the trip showing most of summary and detail features including the number of meters descended / ascended, time for the descent / ascent, the lowest / highest location on the trip, the number of METs for the trip, the normalized power, and the average moving speed (see Figure 7.22). The normalized power is typically higher than the average power in a trip with varying intensity.

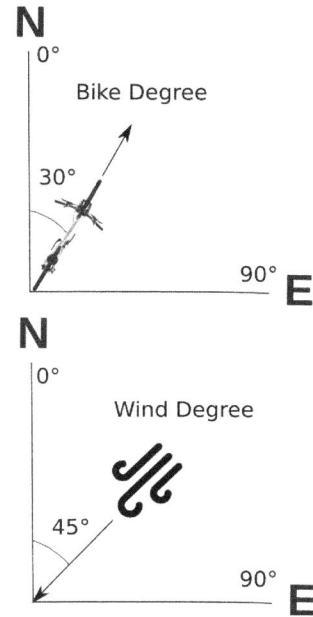

Feature	Value	Feature	Value
Distance	60.6 kms.	Turns	24
Start	14:19:01	End	16:47:37
Readings	1088	Duration	02:28:36
Descent	524.6 m.	Ascent	524.8 m.
Des.Time	01:12:25	Asc.Time	01:16:11
Low Alt	882.0 m.	High Alt	935.0 m.
N.Power	207.4 w	Category	Cycling
Calories	1357	METs	8.0
Used Batt.	6%	Mov.Speed	24.8 kmph.

Feature	Minimum	Maximum	Average
Speed	5.6 kmph.	44.1 kmph.	24.5 kmph.
Accuracy	3.0 m.	45.0 m.	6.7 m.
Time / Fix	0.0 s.	39.9 s.	8.2 s.
Dist / Fix	1.2 m.	164.4 m.	55.7 m.
Gradient	-6.8 %	6.8 %	0.1 %
Satellites	8	28	19.0
Signal	17.8 db	34.4 db	29.3 db
Power	0.0 w	482.6 w	159.8 w
Seg. 1	7.1 kmph.	44.1 kmph.	25.3 kmph.
Seg. 2	5.6 kmph.	20.5 kmph.	22.7 kmph.

Calculator

Power / Speed Calculator

Speed (kmph.)
19.8

Power (Watts)
200

Gradient (%)
1.5

Wind Speed (kmph.)
15

Wind Degree (from North is 0, East is 90, ...)
45

Temperature (C)
25

Elevation (meters)
800

Bike Degree (heading North is 0, East is 90, ...)
30

Figure 7.22: List of Trip Features and Values and Power/Speed Calculator

The moving speed is calculated by dividing the total distance covered in the trip by the total moving time. On some trips, you may have to wait at a traffic light and two consecutive GPS readings at the traffic light maybe separated by a time interval but not in distance. These time intervals where there is no apparent movement, are excluded from the calculation of the moving speed. In all cases, the moving speed will be greater or equal to the average speed.

On the rare occasion that the GPS data is not accurate, the data in Figure 7.22 can reveal the reason. The average accuracy of 6.7 meters for 1088 readings is reasonable and similarly, the average time and distance between a reading is 8.2 seconds and 55.7 meters respectively is also reasonable. The gradient, satellite, signals, and power data can be viewed on track on a map. On a ride where there are two segments of roughly the same distance, the summary will show the segments on which your ride was faster or slower.

Other Trip Functions   While trip analysis does show you the segment with the higher speed, it is worthwhile to playback two trips or segments side by side to see how speed and distance change over time. If you ride the same route on multiple occasions, you maybe curious to know why you were faster or slower on one ride compared to another ride.

Compare Trips   After selecting two trips that you would like to compare from the list of trips, you can playback both rides simultaneously to view the time and distance covered (see Figure 7.23). The playback can be speeded up, slowed down, or played in reverse. The distance difference shows the instantaneous difference between the trip in the upper box and the trip in the lower box at a given time.

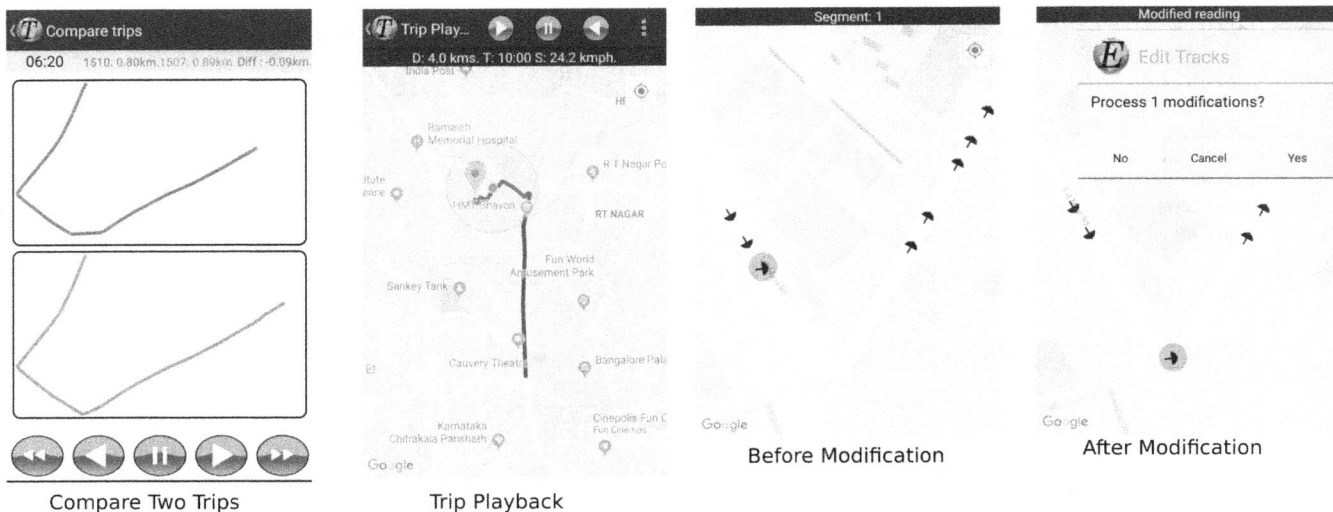

| Compare Two Trips | Trip Playback | Before Modification | After Modification |

Figure 7.23: Compare Two Trips, Playback a Trip, and Modify the tracks of a Trip

Trip Playback   The trip playback function shows the trip track on a Google Map with the real time distance covered and instantaneous speed. The period of time between successive images can be changed from one second to one minute to speed up or slow down the playback. During the playback, you can observe real time speed and identify locations where your speed increased or decreased.

Trip Modification   Every trip is made up of a series of detail records, each of which is associated with the GPS signal received at a specific location. Although, *Snap to Road* does correct the locations to a location on a known road, you can manually make corrections to drag a specific reading such that the map track follows the road precisely.

In Figure 7.23, a reading that was not located at the corner of the road was dragged to the corner. Since a reading can be dragged to any location on the map, a range of 1 kilometer from the original reading is used to limit the distance of the correction. The timestamp of the modified reading is calculated in proportion to the distance to the readings before and after the modified reading. A redundant reading can also be removed the list of readings. If there is more than one modification, then the list of modifications is grouped and processed in sequence. Many of these errors may vanish with devices that have centimeter level GPS accuracy.

Trip Split / Combine   A single trip can be split into multiple trips using automatic or manual segments. U-turns and loops within a trip are automatically detected in a trip and a new segment is created. In a U-turn, the outward and return trips form two separate segments. On a trip of several loops, a new segment is created whenever a location that was seen earlier is detected.

However, you may want to also manually create segments. On a Google map, you can touch a track that splits the trip at that location. The segments of a track are shown in different colors. Once the sub-trips have been created, the original trip is retained along with the smaller trips.

258

Although you could theoretically combine any two trips, it makes more sense to combine two trips whose first and last readings respectively are reasonable close in distance or time. The combined trip for two trips i and j with n and m detail records respectively would become a single trip k with $n + m$ detail records where the nth record of trip i is followed by the first record of trip j.

When two or more trips are combined, they are first sorted by their respective start times. Then a single combined trip is created from the detailed records of the trips sorted in ascending time order. When two trips i and j are combined, the period between the last detail record of trip i and the first detail record of trip j is ignored. The timestamps of all the detail records of trip j are corrected to account for the time difference in the period between the end of trip i and start of trip j.

Waypoints on a Trip   A waypoint is a GPS reading at which the direction of the route changes. Identifying waypoints on an unknown route are useful to stay on track. Clicking on the track near a junction creates a waypoint at the nearest GPS reading. The latitudes and longitudes of the set of waypoints is saved in the Summary of the trip. If you *follow* a trip with waypoints, then you will receive audio warnings when you approach a waypoint. The audio will indicate in which direction to make a turn or follow a route. GPS Trip Analyzer is simplistic and does not re-calculate routes if you ride off route.

Repair a Trip   Following the end of a trip, the Repair trip function calculates the average speed, cumulative distance, other summary properties. The modifications to the trip data include -

- **Clean up:** Often, there maybe redundant detail records that are too close in distance or time. The *Location* API includes settings for the minimum distance and time between successive GPS readings. However, this not guaranteed and GPS readings maybe received more or less frequently than the time/distance setting values. Readings that are too close in time or distance are removed. The *Snap to Road* API does add additional readings that are interpolated to follow a curve and are therefore not removed. The interpolated readings returned by the API can increase the total number of readings depending on the number of turns on a trip.

- **Speed calculations:** Since the time interval between and accuracy of GPS readings is irregular, the calculation of speed at reading $i + 1$ as $\Delta d/\Delta t$ where $\Delta d = dist(l_{i+1} - l_i)$ (the distance between readings $i + 1$ and i) and $\Delta t = t_{i+1} - t_i$ (the time between readings $i + 1$ and i) is not always accurate. At slow speeds, where distances do not change substantially between readings, poor accuracy (high error distance) can result in abnormally high or very low speeds. Corrections include capping speeds to five times the standard deviation of the array of speeds, weighing speeds from GPS readings with high accuracy more than speeds from GPS readings with lower accuracy, and smoothing the speed of reading i using the speeds from three neighbours before and after i. The speeds of neighbours further away have less weight than the speeds of immediate neighbours. As a result speed calculations will differ between GPS Trip Analyzer and other apps which use custom algorithms to compute speed.

- **Detecting Segments:** A ride can cover the same route such as a loop multiple times on a single trip. When such a trip is shown on a map, the tracks will be overlaid and each sub-trip around the loop will not be visible, except for the last. GPS Trip Analyzer creates a separate segment for each loop (with a limit of 16) that can be viewed separately on a map and plot.

- **Other calculations:** The bearing (direction) at a reading i is calculated using the latitude and longitudes of readings i and $i + 1$. The gradient at reading i is calculated from the distance and change in altitude between readings i and $i - 1$. The power at reading i is computed from the drag forces and speed at i.

Detecting Loops A new segment is created when a location l_{i+n} (Y) is close to another location l_i (X) that was passed n readings earlier (see Figure 7.24). Keeping track of distances between all possible pairs of locations can be computationally intensive. The number of unique pair distance calculations for 720 and 1440 readings is over 250 thousand and 1 million respectively.

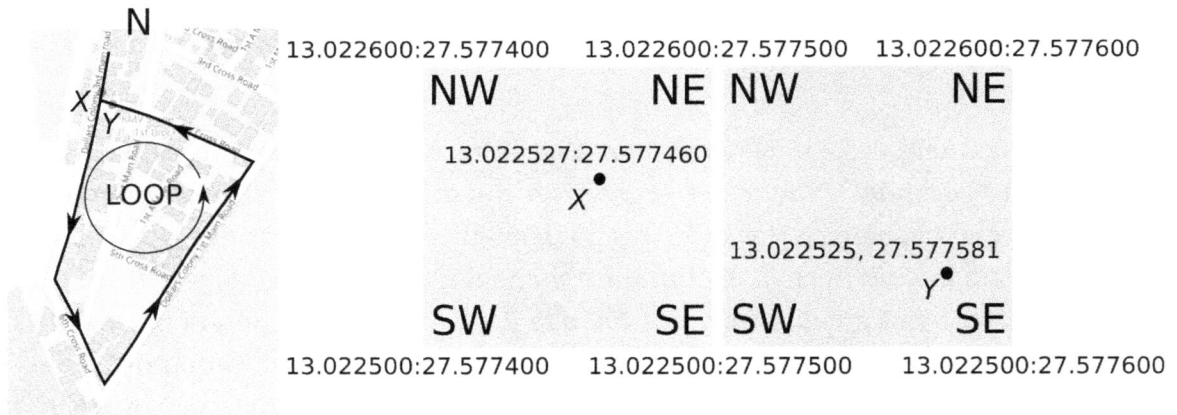

Figure 7.24: Automatically detecting segments on a trip using overlapping rounded up latitude and longitude

A loop may begin anywhere on the trip and in Figure 7.24 the locations X and Y are visited at the beginning and end of the loop respectively. The latitude and longitude of locations X and Y are rounded up to fourth decimal place from six decimal places. A location with latitude and longitude 13.022527 and 27.577460 respectively is rounded up to four locations to the north west, north east, south west, and south east. All 100×100 or $10,000$ locations in this square block will

260

be rounded up to the same four locations. An overlap between two locations X and Y is detected when one or more rounded up locations are common.

In Figure 7.24 the NE and SE corners of X are common with the NW and SW corner of Y. The distance between locations X and Y is 13 meters. The maximum possible distance between two overlapping locations is 20 meters. Associated with each corner location is a timestamp to indicate when the location was visited. The purpose of using corners to find overlapping locations is to limit the computation required to calculate distances between all possible location pairs and instead use a collection of corner locations as keys with their associated timestamps as values. The size of the collection for long trips with over a thousand readings is not excessive.

Detecting a U-Turn   A trip with a U-turn will have an onward and a return leg on the same route. If locations on each leg are not far apart, then the tracks of the legs will overlap making it difficult to analyze each leg separately. GPS Trip Analyzer creates a separate segment for each leg.

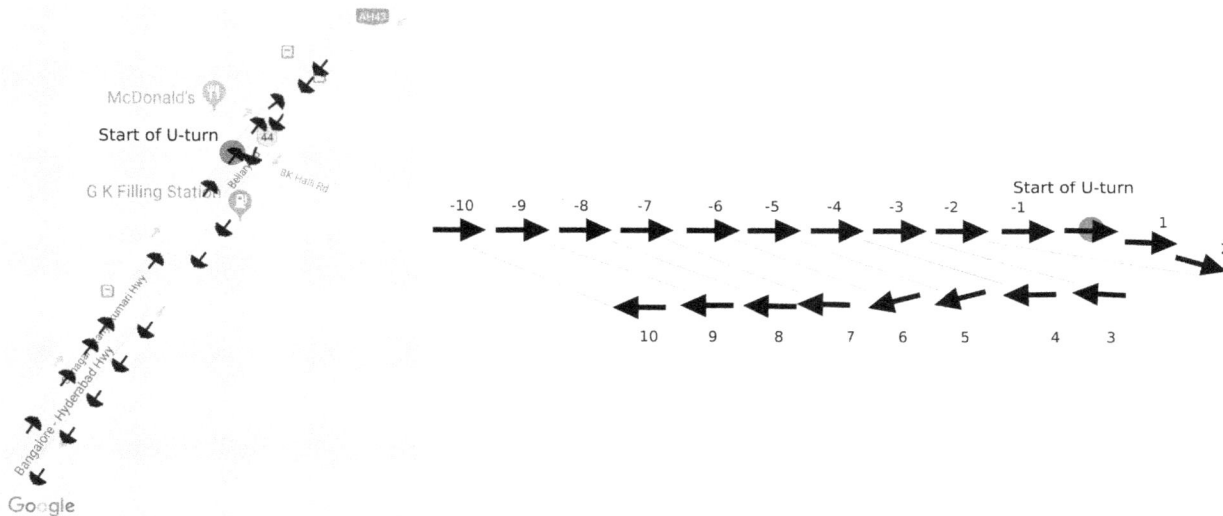

Figure 7.25: Detecting an U-turn using 10 Readings before and after a reading

Ten readings before and after the start of a U-turn are matched to verify that a change in direction of the trip represents a U-turn. The difference in the bearings between the ith reading before and after the start of a U-turn is computed where i ranges from 1 to 10. A bearing difference close to 180° is given a higher score with a higher weight for larger values of i (away from the turn). This method is not certain to identify a legitimate U-turn and may also erroneously flag a U-turn.

Export/Import Trips and API Keys   There are a number of file formats to save GPS readings from a trip. GPS Trip Analyzer handles only GPX and XML formats. Both Google and Open-WeatherMap require API keys to access their services. GPS Trip Analyzer uses the Google *Snap*

to Road and *Altitude* APIs to correct locations and altitudes of readings. The OpenWeatherMap API is used to collect the wind degree, wind magnitude, temperature, and relative humidity at each reading. The API keys can be imported from a file in the JSON format shown below

```
{ "google_api_key": "AIzaSyCmN2L4fDW9Jc5TdunCt9gQh????????",
  "openweather_api_key": "c5d14b4cafe6af9c3e1606b7????????" }
```

For users with limited requirements, access is generally free. However Google and OpenWeatherMap can change their terms of service at any time.

Commit Period   On a long ride of 10 hours or more, a large set of GPS readings will be collected and there is a risk of losing these readings if the battery of the device loses power. The commit period in the settings is the maximum time before the current readings of a trip are saved to a database. For a commit period of four hours, a long ride of 12 hours will be split into three separate trips of four hours each. At the end of the ride, the three rides can be combined into a single ride. The purpose of periodically committing GPS readings to a database is to limit the loss of collected data, if the battery or device fails.

Battery Management   With many apps running in the background, battery power can be quickly drained. Some versions of Android do have options for battery optimization. An app is usually by default set for battery optimization and will be halted when not in use. But on a long ride, it is not desirable to have a tracking app automatically switched off. There are many tips on the Net that you could follow to reduce battery usage including -

- Switching WiFi or Mobile Data off

- Using adaptive battery management or optimization for apps that will run in the foreground

- Using adaptive screen brightness, if possible

7.5   File Formats

There are many different file formats to store data from a GPS device or mobile phone. The GPX format is one of the popular formats and is usually supported by most applications. The GPS data that must be stored includes summary information such as the name of the trip, waypoints, and other information that is common for the entire trip. The detail information for each reading will often include the latitude, longitude, timestamp, and elevation. In a GPX format, this data is stored in a plain text file with XML elements to describe the type of data. The summary data which follows a standard header for a sample file is shown below -

```
<metadata>
```

```
        <name>Trip 1620</name>
        <category>Cycling</category>
        <start_time>1566481614338</start_time>
        <end_time>1566483560577</end_time>
        <distance>5309.7207</distance>
        ...

    </metadata>
```

Trip information such as the name, category, start/end times, and distance are saved in the metadata. Following the metadata, is information for each reading saved as -

```
<trk>
    <trkseg>
        <trkpt lat="23.023514" lon="77.577603">
            <d_timestamp>2019-08-22T19:16:54.338+0530</d_timestamp>
            <d_altitude>923.0</d_altitude>
            <d_speed>10.443276</d_speed>
            <d_bearing>191</d_bearing>
            ...
        </trkpt>
        <trkpt lat="23.02339" lon="77.577579">
            ...
        </trkpt>
    </trkseg>
</trk>
```

The set of readings is saved as a series of trkpts, each of which is associated with a latitude and longitude. The series of trkpts are usually listed in ascending order by time. Each trkpt also includes related information as the altitude, speed, and bearing. A track, trk, is made up of one or more track segments, trksegs, each of which has an associated list of trkpts.

The standard names for XML tags with the names like time and ele have been replaced d_timestamp and d_altitude as shown in the example. GPS Trip Analyzer stores other detail or trkpt information and a consistent naming convention was used for convenience.

The GPX file format is simple and easy to parse. Another popular format is the TCX format from Garmin that is also an XML based file format. The names of the XML tags in a TCX file are more inclusive and descriptive. A TCX file will include information such as heart rate, distance covered, and state of a sensor.

Besides these two file formats, there are a number of other file formats as well. GPSBabel is a tool to convert from other file formats to GPX and vice versa. A FIT file is a more concise version of a TCX file and includes sensor specific information for any type of physical activity including cycling.

References

[1] GPS World Staff. The origins of GPS and the pioneers who launched the system, 2010. URL `http://bit.ly/31IhdDO`.

[2] Per Enge and Frank van Diggelen. GPS: An introduction to satellite navigation, with an interactive worldwide laboratory using smartphones (Coursera MOOC), 2014. URL `https://stanford.io/2WPDTxz`.

[3] Peter Ranacher, Richard Brunauer, Wolfgang Trutschnig, Stefan Van der Spek, and Siegfried Reich. Why GPS makes distances bigger than they are. *International Journal of Geographical Information Science*, (30:2):316–333, 2016.

[4] Ishveena Singh. How accurate is the altimeter in a GPS watch?, 2017. URL `http://bit.ly/2XzOOqH`.

[5] Heartmonitors.com. Heart rate training zone calculator, 2019. URL `http://bit.ly/30dIo7n`.

[6] Osman Isvan. Wind speed, wind yaw and the aerodynamic drag acting on a bicycle and rider. *Journal of Science and Cycling,*, 4(1):42–50, 2014.

[7] Compendium of Physical Activities. Unit conversions, 2019. URL `http://bit.ly/2LPPkV1`.

[8] Andrew R. Coggan. Training and racing using a power meter: an introduction, 2003. URL `http://bit.ly/2GHEkVA`.

[9] Omnicalculator.com. Air density calculator, 2019. URL `http://bit.ly/2LnHD8c`.

[10] Peter Gray, Dimension Data Sports. Seeing data as a competitive advantage, 2019. URL `http://bit.ly/2MRZOOy`.

[11] Joe Lindsey. Why strava is getting more social than ever, 2019. URL `http://bit.ly/2MFWUkE`.

[12] Manu Konchady. GPS Trip Analyzer Android app, 2020. URL `http://bit.ly/2MZBsr1`.

8 Riding Safely

While most cyclists do not spend a lot of time calculating the risks of riding a bicycle on each ride, it is worth the effort to lower the risks as far as possible. Some of the reasons for a cycling accident include road rage, dealing with an aggressive dog, road conditions, or a mechanical failure of the bicycle. Listing all possible reasons or causes for an accident is impractical and the best that a rider can do to avoid an accident, is to ride in a manner which will minimize the risk to the largest extent possible.

8.1 How safe is cycling?

If you have been riding for a year or more, you may assume that accidents are what happens to "other" riders, until you have an accident. On the other hand, there are long time cyclists who have never been involved in a serious accident. While both cases are likely, cycling does have a measurable risk.

Figure 8.1: Is Cycling Risky? (Reprinted with permission of Taj Mihelich)

Ronald Howard defined the "micromort" to compare the riskiness of activities. An activity with one micromort of risk is an activity with a one in a million chance of death. For example, skydiving has a 8 micromort risk per jump based on the prior 413 deaths from 48.6 million jumps while climbing Mount Everest has a 75,000 micromort risk (about 300 deaths from 4000 attempts). We can use a similar measure to evaluate the risk of cycling. If there are x fatalities from y million rides, then cycling has a risk of $\frac{x}{y}$ micromorts. However, a cycle ride can last much longer than a skydive and therefore the riskiness of a ride will also depend on the duration of the ride.

The U.S. Department of Transportation's National Highway Traffic Safety Administration (NHTSA) reported 630 cyclist deaths from about 4.1 billion rides in 2009 or a risk of 6.5 micromorts [1]. Oddly enough, most of the cycling fatalities (about 70%) occurred at non-intersections and during the daytime. Almost three fourths of the fatal accidents occured in urban areas and close to 90% of the riders involved in these accidents were males.

Besides the odds of being involved in a fatal accident, a cyclist has a much higher chance of being injured in an accident. The same study from NHTSA reported 51,000 accidents with injuries or about 80 times the number of fatal accidents. Accidents with injuries may not be fatal but can vary with severity from the simple bruise to a major fracture or worse. Nonetheless, the risk of being involved in any type of accident is of concern to all cyclists.

8.1.1 Is Cycling a Risky Mode of Transportation?

One of the reasons that cycling appears risky is that a rider appears vulnerable and does not have any type of protection in an accident unlike other vehicles (see Figure 8.1). Without a protective steel cage to protect a cyclist, the impact from a fall or collision will be absorbed by some part of the body leading to an injury or worse.

A slightly older study [2] of the risks in all modes of transportation found that cycling was not as risky as some other modes of transporation. The study estimated the number of fatalities in terms of the number of deaths per 100 million person trips and defined six primary travel modes: passenger vehicle (passenger car, sport utility vehicle, van, or light truck), motorcycle, walking, bicycle, bus, and all other vehicles (e.g., large truck, motor home, taxi, limousine, or hotel/airport shuttle). Motorcyclists had the highest number of fatalities (500+) per 100 million person trips followed by other vehicle occupants (28), bicyclists (21), pedestrians (14), passenger vehicle occupants (9) and bus riders (0.4).

While riding a bus is clearly much safer than riding a bicycle, the risk of cycling is much lower than the risk of riding a motorcycle. You would assume that most of the serious cycling accidents (involving surgery and requiring intensive care) are due to collisions with larger vehicles on the road. Suprisingly, a study [3] of 2500 serious cycling accidents at San Francisco General Hospital found that over half of these types of accidents did not involve a car, but instead were possibly due to a cyclist avoiding a pedestrian, a pothole, or some other road obstacle. Still, a collision with a car or some other heavier vehicle always resulted in a fairly serious injury. Some observers also claim that not all bicycling accidents are reported and therefore cycling is actually more risky than what statistics reveal. With no clear trends and insufficient data, any conclusion is premature and the risks of cycling will remain.

One of the frequently cited deterrents for using cycling as a mode of transportation is safety [4]. These concerns are legitimate since the accident rate for cycling is still relatively high. However in countries like the Netherlands and Denmark with an extensive infrastructure and a relatively large number of cyclists, the accident rate is lower than in countries where cycling is not as popular. This is a catch-22 situation where it is likely that cycling will become safer with larger numbers of cyclists, but few ride bicycles because it is perceived as unsafe.

Measuring Safety When comparing the risks of different modes of transportation, a common measure is the number of fatalities per million or billion kilometers travelled. Air, train, and bus travel are among the safest modes of transportation using this measure at 0.05, 0.4, and 0.6

deaths per billion kilometers travelled respectively [5]. Motorists, cyclists, pedestrians, and motorcyclists have a much higher risk at 3.1, 44, 54, and 109 deaths per billion kilometers travelled.

However, a more fair comparison would use the time spent travelling instead of distance covered, since both pedestrians and cyclists cover much smaller distances than a plane or train. The time spent commuting on a road maybe the period of highest risk during a day spent at an office or at home. The accident rate for a mode of transportation can be defined as -

$$Accident\ Rate = \frac{No.\ of\ fatalities\ or\ injuries}{Exposure\ (kilometers\ or\ hours)}$$

If the accident rate is described in terms of the number of fatalities per billion hours travelled, then the bus is safest mode of transportation followed by plane, train, car, walk, cycle and motorcycle. The time spent travelling is computed based on the average speed of a mode of transportation. While accidents involving a fatality are likely to be reported accurately, minor accidents are frequently under reported. Without accurate records of all types of bicycle accidents, measuring cycle safety will be inexact.

While many cities worldwide have policies to encourage cycling, the perception that cycling is an unsafe mode of transportation will hinder such objectives. If citizens do not feel cycling is safe, then few riders will take on the additional risk. At the same time, there is an inverse relationship between the number of cyclists and the number of accidents. The European Cyclists' Federation claims that countries with the highest number of cyclists tend to have the best safety record. Encouraging cycling in a city with few cyclists is a quandary out of which there appears no easy solution.

Staying Balanced   One of the fears of a cyclist is losing balance and falling on the road. An asphalt or cement road is hard and the impact of any part of the body with the road is painful. Cycling with elbow pads, knee pads, and a helmet is obviously safer, but few riders will take the trouble of wearing such pads, besides the appearance of being a newbie cyclist. Losing control of a bicycle is not very hard. If you brake hard or a stick is jammed in the spokes of your front wheel, you will most likely fall. Without the freedom of the front wheel to swivel, it is extremely difficult to stay balanced.

A collision with a larger vehicle is almost always more harmful for the cyclist than the vehicle occupant. Years of experience riding on a bicycle will not help you to stay balanced if you are bumped by a vehicle with much larger momentum. Consider a car that weighs about 1500 kgs. at 40 kmph. colliding with a 100 kg. bicycle plus rider at 25 kmph. in the same direction. Following the collision, if the car loses just 5 kmph. and continues at 35 kmph., the speed of the cyclist would briefly rise to about 80 kmph. if kinetic energy is conserved. In reality some energy would be lost to sound, heat, and friction, yet even an experienced rider would still find it hard to stay balanced following the collision.

8.2 Accidents

Frequently, bicycle accidents in mixed traffic conditions are due to the impatience of a driver in a faster motorized vehicle. A bicyclist holding up the movement of traffic is annoying and it is not uncommon for a motorist to squeeze in a gap to pass a cyclist as quickly as possible. The risks and injuries in such accidents are almost always serious for the cyclist and not the motorist.

Although the initial reaction of a cyclist is to blame the driver of the larger vehicle for an accident, a significant number of accidents are due to the carelessness of the cyclist. When cyclists ignore red lights and break traffic laws, it does lead to the perception that cyclists are reckless and therefore more likely to have accidents. A survey [6] of over 3000 cyclists found that roughly 50%, 47%, and 3% of all cyclists obey traffic laws all the time, usually, and occasionally respectively. While the cyclists who obeyed traffic laws usually or all the time met with roughly the same number of accidents, the rate for cyclists who occasionally obeyed traffic laws had a 36% higher accident rate. Cyclists who rode in the night or in the rain did not have a higher accident rate than cyclists who rode in fair weather and the daytime. One explanation maybe that cyclists are more conscious of the risks of riding in the night and bad weather and therefore take more precautions while riding.

8.2.1 Is Bicycling Safer than Driving?

The accident rate is a measure of the risk involved in using a mode of transportation and is useful in comparing cycling with driving or other modes of transportation. One way of computing the accident rate is to estimate the average number of kilometers a cyclist would have to ride before an accident and the chance of an accident in a lifetime.

Non-Fatal Accidents If an average US cyclist rides about 1000 kilometers in a year (see Section 8.6.3), then the total distance travelled by about 45 million cyclists (about 14% of a population of 320 million) is 45 billion kilometers. With approximately 45,000 non-fatal accidents in 2015 [7], the estimated accident rate is one accident for every million kilometers travelled. If an average cyclist rides about 50,000 kilometers in a lifetime, then there is a 5% chance of having an accident at some point in a cyclist's life. This rate does appear low and is somewhat contrary to the large number of anecdotal stories of cycling accidents. Using data provided by the US National Highway Traffic Safety Administration [8], an average cyclist rode about 666 kilometers per year, while Kaplan's survey [6] from the 70s estimated the annual distance per cyclist at a higher 2000 miles or 3200 kilometers.

In Table 8.1 the chance of an accident in a lifetime of riding a cycle and driving a car is computed as 5% and 33% respectively. The total distance is annual distance × number of cyclists (45 million). The number of accidents per kilometer is the $\frac{45,000}{total\ distance}$ where $45,000$ is the number of non-fatal cycle accidents in 2015. The lifetime distance is $50\times$ annual distance and the chance of an accident is $lifetime\ distance \times accidents/million\ kms$.

Mode	Ann. Distance	Tot. Distance	Acc. / million kms.	Lifetime Distance	Accident%
Cycle	666 kms.	30 billion kms.	1.500	0.033 million kms.	5
Cycle	1000 kms.	45 billion kms.	1.000	0.050 million kms.	5
Car	20,666 kms.	4960 billion kms.	0.342	1.033 million kms.	33

Table 8.1: Chance of an Accident for Cycles and Cars

Data from the US Federal Reserve [9] estimate that US motorists drove 4960 billion kilometers in 2015 and were involved in 1.7 million non-fatal accidents [7]. If three fourths of the population of 320 million drive a motor vehicle, then the annual distance travelled per motorist is about 20,666 kilometers ($\sim$ 13,000 miles). The number of accidents per kilometer is $\frac{1.7\ million}{4960\ billion}$ or 0.342 accidents per million kilometers. The lifetime distance for a motorist is $50 \times 20,666 = 1.033$ million kms. and therefore the chance of a motorist having an accident in a lifetime is $\frac{0.342}{1.033} = 33\%$.

Fatal Accidents Fortunately, fatal accidents make up a small fraction of all bicycle and motor vehicle accidents. In 2015, there were about 800 and 32,000 fatal bicycle and motor vehicle fatal accidents respectively. For cyclists who annually ride about 1000 kilometers for a total of 45 billion kilometers, the accident rate was $\frac{800}{45\ billion}$ or 0.017 fatalities per million kilometers. With a lifetime distance of 50,000 kilometers, the chance of a fatality for a cyclist is $\frac{50,000 \times 0.017}{1\ million}$ or 0.09%. For motorists, the accident rate was $\frac{32,000}{4960\ billion}$ or 0.006 fatalities per million kilometers. Assuming a lifetime distance of 1.033 million kilometers, the chance of a fatality for a motorist is $\frac{1.033\ million \times 0.006}{1\ million}$ or 0.62%.

Why does bicycling appear safer than driving? Although the number of accidents / million kms. travelled by a motorist is about a third or less than the number of accidents / million kms. travelled by a cyclist, the lifetime distance for a motorist is more than 20 times the lifetime distance for a cyclist. On average, a cyclist spends fewer hours riding than a motorist who may drive for much longer durations. Therefore, the cumulative risk of driving for a motorist is greater than the risk of riding for a cyclist.

8.2.2 Sharing Space

Most shared road systems have not been designed with cyclists in mind and cater to motor vehicles that are both much heavier and travel at significantly higher speeds than a bicycle. Cyclists occupy much less space on a road than a motorized vehicle which does reduce traffic density but also reduces visibility. In an accident, lack of visibility is an often cited excuse by drivers of motor vehicles. While reflective clothing, red lights, and blinkers do help, this problem is worse in the night.

Unfortunately accidents involving cyclists and motorized vehicles will happen, since someone will be careless and make a mistake. There is a perception that society is somewhat indifferent to accidents involving cyclists, while pedestrian fatalities are taken more seriously and prosecuted. Often a motorist in an accident will claim that a cyclist was not visible on the road and as a result is acquitted or fined a minimal penalty. Frequently, the real reason is that a motorist is driving relatively fast and only sees the cyclist when it is too late. Besides, cycling is also viewed as a risky activity and therefore accidents with cyclists are not unexpected. It is discouraging for a cyclist when law enforcement tends to side with motorists in an accident and are more inclined to blame the former for an accident.

In the first few seconds following an accident, a large part of the kinetic energy of a bicycle and rider is dissipated, when the frame or tyre crumples or due to friction from a slide on the side of a road. Even though the bicycle comes to a stop, the cyclist continues to move either forwards over the handlebars or tumbles sideways to the ground.

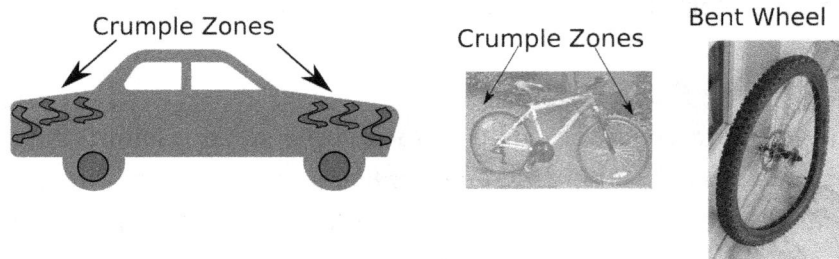

Figure 8.2: Crumple Zones in a Car and a Bicycle

Cars use a crumple zone in the front and rear of the vehicle to absorb a large fraction of the energy from a head-on or rear collision (see Figure 8.2). The occupants in a car are also fairly well protected from severe injuries inside the car and with air bags. The equivalent of automobile crumple zones for a cyclist is simply the front and rear tyres. In a head on collision, the front tyre will be bent following a collision, but a cyclist will be far more likely to have more injuries following the collision than the driver of a car.

While accidents do occur in many forms, there are a few common types of bicycle accidents. You would assume that a large fraction of bicycle accidents are due to collisions with motor vehicles. Suprisingly, 70% of the reported bicycle injuries treated in a hospital did not involve a motor vehicle, which included accidents that occurred on a road [10].

8.2.3 Single-bicycle Accidents

Most bicycle safety studies focus on accidents involving a motorized vehicle and a bicycle. However, a number of bicycle accidents occur due to other infrastructure related reasons such as a pot holed road. Roughly half [11] of all bicycle accidents did not involve another vehicle. A single bicycle accident is an accident where the poor condition of the road, a mechanical problem, or a bicyclist error was the contributing factor. The road surface may have been slippery, cracked, or uneven leading to the accident. Some of the causes of single bicycle accidents include -

- Collision with an obstacle on the road such as a bollard (used to prevent motorized traffic on the pavement) or a parked vehicle

- Being forced off the road when riding too close to the kerb, riding over debris or sand on the shoulder.

- Skidding of a wheel due to oil, mud, gravel, or wet leaves on the road surface. Other common incidents included getting a relatively thin bicycle tyre stuck in the gaps in the road between iron or concrete plates.

- Losing balance when one foot slips off the pedal, a shoelace gets entangled, or a heavy load sways the bike excessively in one direction.

- Braking errors: not braking quickly enough to stop within a reasonable distance or braking excessively with skidding or the cyclist flying over the handlebars

One of the conclusions of the study [11] was that cyclists who do not ride as often (less than once a week) are more likely to be involved in a single bicycle accident that is linked to a rider's cycling skill and strength. Accident statistics that involve a cyclist and a motorist are usually more easily available and accurate.

8.2.4   Fatal Bicycle Accidents

The Fatality Analysis Reporting System (FARS) is a database [12] maintained by the US National Highway Traffic Safety Administration containing information on all types of accidents that involve a fatality and occur on a public roadway. The FARS data includes all fatal accidents that involve a cyclist and a motorist on a public roadway.

Every accident has an associated report that describes data such as the location of the accident, an accident timestamp, type of collision, and the weather. FARS uses over 60 different codes to categorize a bicycle accident. Besides the type of accident, FARS also describes the direction and position of the cyclist. Table 8.2 shows the percentage of fatal bicycle accidents by direction, position, and location that occurred in 2017. In some cases, there was insufficient information about the accident and therefore the percentages in the table do not cover all accidents.

Bicycle Direction		Bicycle Position			Bicycle Location	
With Traffic	Facing Traffic	Traffic Lane	Sidewalk	Bike Lane	Non-Inter.	Inter.
67%	15%	80%	10%	8%	63%	36%

Table 8.2: Percentage of Bicycle Accidents based on Direction, Position, and Location in 2017

Type	Description
Bicyclist / Motorist Lost Control	Common reasons for the loss of control include - flat tyre, brake failure, not braking, stalled engine, over steering, riding too fast, alcohol impairment, bad road conditions (sand, debris, or potholes), falling asleep, or an illness.
Bicyclist / Motorist failed to yield	Either the cyclist or motorist failed to yield at an intersection and proceeded straight through an intersection leading to a collision. A bicycle or car did not stop at a signed intersection or red light. These accidents can also happen at a circle or roundabout.
Bicyclist / Motorist Turning Error	A bad left or right turn at an intersection or driveway where the cyclist or motorist cut a corner or swung too wide and entered the opposing traffic lane. These errors also include a collision with a bicyclist or motorist entering or exiting a driveway.
Bicyclist Failed to Clear	A bicyclist entered an intersection when the light was green just before turning red and could not cross the junction before the light turned green for cross traffic.
Bicyclist / Motorist Crossed Paths	A collision occurred at an intersection with a sign or light, but further information for the cause is not available.
Bicyclist / Motorist Left or Right Turn	Either a motorist or bicyclist made a left or right turn and entered in front of the path of the other vehicle in the same or opposite direction
Motorist Overtaking	A collision occurred because a motorist overtaking a bicycle did not have enough space, the bicycle swerved into the path of the motorist, or could not be detected in time.
Bicyclist Overtaking	A bicyclist had a collision with a parked vehicle, car door, or when overtaking from the left or right.
Motorist / Bicyclist on Wrong Side	A collision when the bicyclist or motorist was travelling in the wrong direction on a two way or one way road.
Bicyclist / Motorist Ride-Out	A bicyclist or motorist rode or drove out of a residential / commercial driveway into the street into the path of the other vehicle.
Other types	A motorist backs into a bicyclist, a collision in a parking lot or driveway, or there is insufficient information to understand the reason for the accident.

Table 8.3: A List of 11 different types of FARS Accidents

In a small fraction of the accidents (15%) the cyclist is riding against traffic. However, this maybe legal if the rider is on the sidewalk and not the traffic lane. Clearly, a majority (80%) of the fatal accidents occur in a traffic lane and not on the sidewalk or bike lane. An initial guess would be that many accidents occur at an intersection, but surprisingly only about one third of all accidents do occur at an intersection.

Types of Accidents Although FARS does use over 60 codes [13] to categorize bicycle accidents, they can be combined into about 11 groups. Broadly, all accidents are categorized into two types - either a motorist or a cyclist was responsible for the accident. Table 8.3 includes a list of the different types of FARS accident types and a short description of each type.

Clearly, there are many reasons for a collision and they largely have to do with the sharing of common space on the roadway. While rules for stopping, giving way, and overtaking do exist, the differences in size and speed between a bicycle and motor vehicle do lead to judgment errors and miscalculations.

Many accidents are preventable with some extra vigilance and often occur due to carelessness, distraction, or fatigue. Statistics from the FARS database reveal that both cyclist and motorists make mistakes. The list in Table 8.3 covers almost all possible errors that a motorist or cyclist could make. The reasons for less than 15% of all accidents are unknown and are typically due to lack of sufficient information.

Assigning the right code and creating a report for an accident is the responsiblility of a local official who then transmits the information to the FARS database. Every accident with a bicyclist fatality is assigned a case number and includes a report with the type of road (highway, county road, or local street), the type of intersection, the time of day, the weather conditions, and latitude / longitude of the accident location.

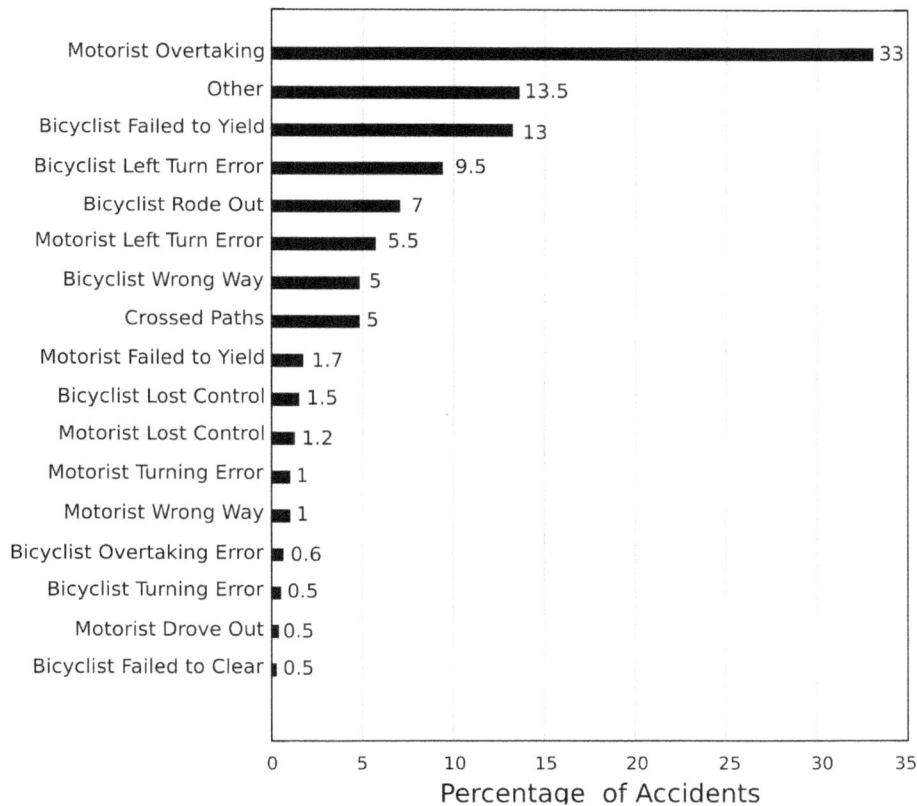

Figure 8.3: Percentage of Bicycle Accidents by Type from FARS 2017 data

Some types of accidents occur more often than others and therefore it is of interest to limit the number of such accidents. In 2017, there were about 800 bicycle related fatalities in the US. The frequency of accident types is skewed with a handful of accident types accounting for most of the accidents (see Figure 8.3).

At 33% a **motorist overtaking** a cyclist is the most frequent reason for an accident. There are several ways in which a motorist overtaking may collide with a cyclist. A bicyclist on the roadway was undetected or sighted too late by a motorist to avoid a collision. The driver may have been distracted / inattentive, the bicyclist was wearing dark clothes and therefore not visible, or the bicycle and road did not have sufficient lighting to be detected by the motorist are some of the reasons for the accident. A motorist may also have mis-judged the space required to overtake leading to a collision.

In about half the cases with a motorist overtaking, the specific circumstances leading to the collision are unknown or not clear. About a third of the cases involved a motorist who did not detect the cyclist. The remaining cases were due to a motorist mis-judging (15%) the space to overtake and the least frequent case (4%) was due a bicyclist swerving into the path of the motorist.

The next leading cause of accidents (**Other**) includes a number of types of accidents where there is insufficient information. About 41% of such accidents include unknown approach paths by the motorist and cyclist leading to a collision. In another 39% of accidents, the bicyclist and motorist were travelling in parallel paths, either in the same or opposite directions. The remaining 20% of accidents included causes such as uncertain location of collision between motorist and cyclist, a parking lot accident where a motor vehicle was entering or leaving a parking space, either the bicyclist or motorist was travelling the wrong way, or some other unusual circumstances.

Almost equally frequent (13%) are accidents where the **bicyclist failed to yield**. These accidents occur at an intersection where a sign or a signal light is used to control the flow of traffic. In both a signed and signalized intersection, the bicyclist may or may not have to come to a stop before crossing an intersection where the motorist had the right of way. A motorist's view of the bicyclist maybe blocked by other traffic at the intersection before the collision.

The fourth most frequent reason for an accident is the bicyclist making a **left turn error**. This is an accident where the cyclist turned or merged left in front of motorist who was travelling in the same or opposite direction. For countries like Britain or India where traffic flows on the left, this type of accident would be the same as a right turn error.

The top four types of accidents account for over two thirds of all accidents with over a dozen reasons for the remaining one third. Among them include a bicyclist riding out of a driveway directly into the path of a motorist driving straight ahead on a roadway, a motorist turning left in front of a bicyclist, and a bicyclist riding the wrong way or on the wrong side of the roadway.

In about 85% of the accidents, the motorist was driving a car, light truck, SUV, pickup, or van. A bicycle accident with a bus was rare at about 1% of all accidents. The remaining accidents were

Type	2014	2015	2016	2017
Motorist Overtaking	27	28	31	33
Other	15	17	15	14
Bicyclist failed to yield	14	13	15	13
Bicyclist left turn error	11	9	8	9
Total	**67**	**67**	**69**	**69**

Table 8.4: Percentage of Top Four Fatal Accident Types from 2014-2017

collisions with a large truck or some other vehicle. The most dangerous vehicle for a cyclist was the light truck that accounted for over 40% of all accidents. The front of the vehicle was involved in the collision in most cases.

The FARS data for a four year period from 2014-17 show a consistent pattern for the same top four accident types (see Table 8.4). In all four years, these accident types account for close to 70% of all accidents, with the average number of fatalities just over 800 per year. While FARS data indicate that motorists overtaking were responsible for the most frequent type of fatal accident, in many of the cases where a motorist was overtaking the specific circumstances were not clear, i.e. a motorist could not be clearly blamed for mis-judging the space to overtake or not detecting the bicyclist.

8.2.5 Other Accidents

Although FARS provides an accurate count of fatalities, there is no equivalent database for injuries due to a bicycle accident[16]. You would expect the number of non-fatal accidents to be significantly higher than the number of fatal accidents. In most cases, accidents with minor injuries will not be reported and therefore an accurate count of all bicycle accidents is not possible.

Even though most accidents do occur on urban roads, the percentage of accidents that result in a fatality is higher in rural roads [14]. On a rural road that is not divided, speeds are higher than on a city road and an accident is more likely to be fatal. Similarly, the number of accidents that occur at a non-intersection is higher on a rural road than on an urban road.

About 75% of all accidents occurred at intersections, were intersection related, or were at a junction / driveway [15]. Though this type (crossing paths) of accident accounts for a majority of all accidents, the accidents with parallel paths (in the same or opposing directions) resulted in more severe injuries. Speeds of both cyclists and motorists are often lower at an intersection and therefore accidents at an intersection are less likely to be fatal.

Even though an accident is not fatal, the injuries sustained in such accidents can be painful. Scraping an elbow or knee on a hard road is definitely not something a rider looks forward to on a ride. Falling over the handlebars is another painful accident with a high risk of a head or

---

[16]NHTSA estimated that 45,000 cyclists were injured in 2015 compared to 831 fatalities or over 50 times as many cyclists were injured compared to the number of fatalities.

rib injuries. Both head and rib injuries can be far more serious than the minor bruises from a sideways fall.

A study [6] surveyed over 3000 cyclists in 1974 to get an estimate of the number of cyclists who had a collision or a fall in a year of riding. Close to 80% of the respondents did not have any type of accident during the year while about 18% had more than one accident. This could be due to the negligence of the cyclist or a bike-unfriendly area. The accident rate was dependent on the number of years of experience. After a year of experience, the accident rate (measured in accidents per million miles) declined with more years of experience, albeit at a slower pace. Further, the accident rate was the lowest for cyclists riding to work or exercise on a familiar route and higher for touring or utility rides on unfamiliar roads. The advantage of riding on a familiar route is that a rider knows when to pay more attention on some parts of the route that may include poor road conditions, intersections, or heavy traffic.

All non-fatal accidents are not similar and such accidents can be broadly assigned four categories -

- No injuries to the rider and some damage to the bicycle alone

- Minor scrapes and bruises

- Moderate injuries requiring some medical treatment

- Major injury requiring hospitalization.

The study found that close to three fourths of all accidents were of the first and second category types and a small (6.7%) percentage of the accidents required hospitalization.

Accidents Reasons   Although, the number of bicycle accidents ($\sim 45,000$) may not be as large as the number of motor vehicle accidents, it is large enough for lawyers to take an interest in pursuing such cases. Unlike government statistics which are fairly detailed, websites from law firms simply list the common reasons for an accident. The reasons for accidents are often due to a fall or a collision with a motor vehicle or some other object. The fall may occur due to a pothole, debris on the road, metal grates, sudden braking, or losing balance. A fraction (roughly 50%) of accidents are due to cyclist error. There are a number of reasons for accidents where the cyclist is not at fault. Among the list of common reasons mentioned in legal websites include -

- A motorist fails to yield the right-of-way to a cyclist and causes a collision. These types of accidents happen at an intersection (signed or with a traffic light) and at the junction of a driveway with a roadway (see Figure 8.4 ❷, ❹, ❺, ❽, and ❿).

- The door of a parked car was unexpectedly opened in the path of a cyclist (see Figure 8.5).

- A motorist was busy texting, speeding, careless, or driving drunk at the time of the accident.

- A turning accident (see Figure 8.4 ❶, ❸, and ❼) where the motorist failed to let the bicyclist pass, was not far enough ahead to safely make the turn, or the bicyclist was in the blind spot of the motorist.

- A sideswipe accident (see Figure 8.4 ❻ and ❾) where the motorist was driving in the same or opposite direction as the cyclist and drove on a path to make contact with the cyclist leading to a collision.

A motorist is not always to blame for a collision. Cyclists breaking traffic rules, making sudden changes in path, or riding in an aggressive manner can also be the cause of an accident.

8.2.6 Avoiding an Accident

Cyclists are unique in that they can both ride on a roadway with cars and walk a bicycle on the sidewalk along with pedestrians. The rules that a cyclist must follow on the roadway are different from the rules on a sidewalk. On a sidewalk with sufficient space, you can travel in both directions.

While a cyclist does have the flexibility of riding on the sidewalk or road as needed, following the traffic rules on the sidewalk or road will reduce the chance of an accident. Riding on the sidewalk is generally discouraged, since you may have to transition to the roadway at a junction. There is a reasonable possibility that a motorist, who is focused on the roadway, will not notice you moving from the sidewalk to the roadway in time to avoid a collision.

Intersections If you ride in the city, you will inevitably encounter an intersection. A majority (about 75%) of motor vehicle - bicycle collisions occur at intersections or junctions [15]. Collisions at intersections also account for about a third of all fatal bicycle accidents. As a cyclist approaches an intersection, it becomes more important to be aware of traffic conditions at the intersection. Motorists often fail to notice a cyclist because of an obstruction or simply because other motor vehicles are more visible and obvious.

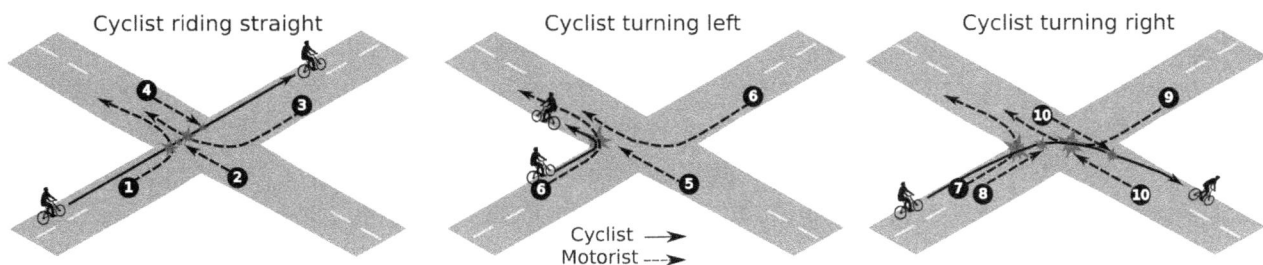

Figure 8.4: Cyclist riding straight, turning left, and turning right at an intersection

A cyclist riding on the left hand side of the road approaching an intersection can either ride straight, turn left, or turn right (see Figure 8.4). The path of the motorist is shown as a dashed line and intersects with a continuous line for the path of the cyclist.

Riding Straight   There are four potential collision areas when a cyclist rides straight through the intersection. The first collision is at ❶ where a motorist makes a left turn in front of the cyclist. The second and third collisions at ❷ and ❹ respectively are due to motorist driving straight from the left and right through the intersection. The final collision at ❸ is a motorist making a right from the opposite direction.

Turning Left   The safest of the three routes at an intersection is a left turn with three sources of collisions from motorists driving straight from the right ❺, a motorist making a right from the opposite direction ❻, or a motorist making a left in the same direction as the cyclist ❻. Since a cyclist could be in the blind spot of a driver, the cyclist is cut off by the motor vehicle and will not have enough space to make the turn.

Turning Right   The right turn is more accident prone than the left turn. A cyclist has to change lanes from the edge of the road to the center of the road just before the intersection. If you change lanes too early, then you will have to ride with fast moving traffic in the center of the road till the intersection. If you change lanes too late, then you risk a collision with motorists at ❼ and ❽. Finally, motorists driving straight at ❿ and turning right at ❾ are another source of collisions.

Vehicle / Bicycle	Straight	Left	Right
Straight	❷, ❹	❶	❸
Left	❺	❻	❻
Right	❽, ❿	❼	❾

Table 8.5: Collision Paths for Motor Vehicles and Bicycles

Preventing Collisions   Given all the possibilities for a collision at an intersection, it is best to slow down before the intersection to check traffic conditions, traffic lights, and signs. There are a large number of variations to the ten listed collisions and more detailed guides of the different types of collisions and methods to avoid collisions have been published [15]. A cyclist can have either a cross path or parallel path collision. In cross path and parallel path collisions, the cyclist and motorist are travelling in different and the same directions respectively before the collision. The reasons for a cross path collision maybe lack of visibility or bad judgment of speed and time to cross the intersection. Trying to outrace a motor vehicle is often futile for a cyclist and it is best to slow down and pick up speed after an intersection.

Cars do have blind spots and can make turns without any signal. It is best to stay in the front or the rear of a car and be prepared for a vehicle to make an unexpected turn. One of the frequently cited reasons by motorists for an accident is that a cyclist was not visible. Figure 8.5 shows the three most common types of collisions with motor vehicles.

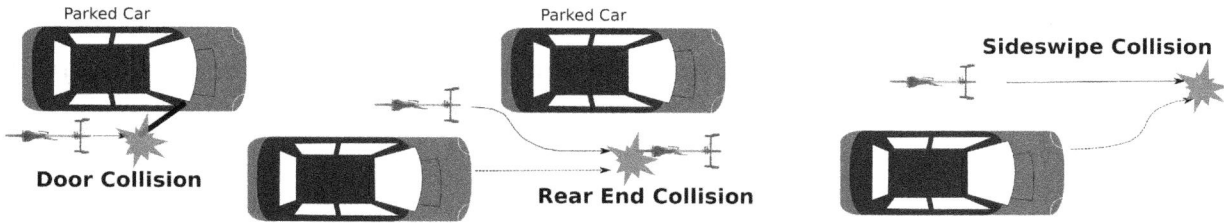

Figure 8.5: Three Types of Collisions with Motor Vehicles

The first type of collision is when the door of a parked car is unexpectedly opened in the path of a cyclist who does not have enough space to pass without a collision. In the second type of collision, a cyclist passing a parked car is hit from the rear by a motorist who does not slow down in time to allow the cyclist to pass the parked car. The third type of collision is a side swipe collision where a motor vehicle swerves unexpectedly towards the cyclist.

8.2.7   Forces in a Collision

In all collisions, a cyclist is almost certain to fall and sustain injuries. Simulation software like Madymo developed in the Netherlands gives realistic analysis of a collision between a cyclist and a motor vehicle. However, even a simplistic theoretical analysis can reveal the approximate forces you would experience in a collision.

The two equations to analyze a collision are based on the conservation of momentum and the calculation of the coefficient of restitution (C_r). The principle of conservation of momentum states that momentum before is the same as the momentum after the collision, given that there are no external forces in the collision. If m_b (mass of the bicycle), m_c (mass of the car), v_{b_before} (velocity of the bicycle before the collision), v_{b_after} (velocity of the bicycle after the collision), v_{c_before} (velocity of the car before the collision), and v_{c_after} (velocity of the car before the collision) are parameters for the bicycle and car, then the conservation of momentum states that

$$m_b v_{b\_before} + m_c v_{c\_before} = m_b v_{b\_after} + m_c v_{c\_after}$$

The velocity, v_{b_after}, of the bicycle after the collision is not known and needed to calculate the force on the cyclist due to the impact. Since the velocity, v_{c_after}, is also unknown, a second equation using C_r is used to compute the two unknown velocities. C_r describes the "bounciness" of the collision and is defined as the ratio of the separation velocity to the approach velocity -

$$C_r = \frac{v_{sep}}{v_{app}} = -\frac{v_{b\_after|c\_after}}{v_{b\_before|c\_before}} = -\frac{v_{b\_after} - v_{c\_after}}{v_{b\_before} - v_{c\_before}}$$

If $v_{b\_after} = v_{c\_after}$, $C_r = 0$, then the collision is called a perfect plastic collision where the cycle and car stick together. In the case where $C_r = 1$, the bicycle and car bounce off each other with no loss of energy (a perfect elastic collision). The C_r for a bicycle with a rubber tyre and a steel car is approximately 0.7. Given the masses of the bicycle and car, the velocities before the collision, and the C_r, the velocities after the collision can be calculated using the equations for the calculation of C_r and the conservation of momentum.

Collision Type	v_{b_before}	v_{c_before}	v_{b_after}	v_{c_after}	Impact Force
Door	15 kmph.	0 kmph.	3 kmph.	13 kmph.	282 N
Rear End	15 kmph.	40 kmph.	54 kmph.	36 kmph.	924 N
Head on	-15 kmph.	40 kmph.	71 kmph.	32 kmph.	2034 N

Table 8.6: Velocities and Impact Force after a collision

Table 8.6 shows the velocities after a collision for the three types of collisions. The mass of the car (m_c) and bicycle + rider (m_b) are 1000 kgs. and 85 kgs. respectively. In the first case with a door collision, only the mass of the door (75 kgs.) is used instead of m_c. The time of the impact (t_i) is assumed to be one second. The impact force is the change in bicycle momentum, $m_b(v_{b\_after} - v_{b\_before})$, divided by the time of impact [17].

In a rear end collision, the impact force is over three times greater, since a vehicle that is over 10 times heavier strikes the bicycle. Finally, in the worst case head on collision, the impact force is over 2000 N or roughly the force of a person weighing 200 kgs. standing on you. Collision forces can be analyzed based on the type of collision (elastic or plastic). In an elastic collision, the bicycle and vehicle separate following the collision unlike a plastic collision where the bicycle and vehicle are a single object after the collision.

Elastic Collisions Many collisions occur at an angle and the impact force strongly depends on the angle degree. A gentle touch from a heavier vehicle may not topple you over, but an impact at an angle close to 0° or 180° will generate a much higher force. A sideswipe collision can occur when a bicycle and another vehicle are travelling in the same or opposite directions (see Figure 8.6).

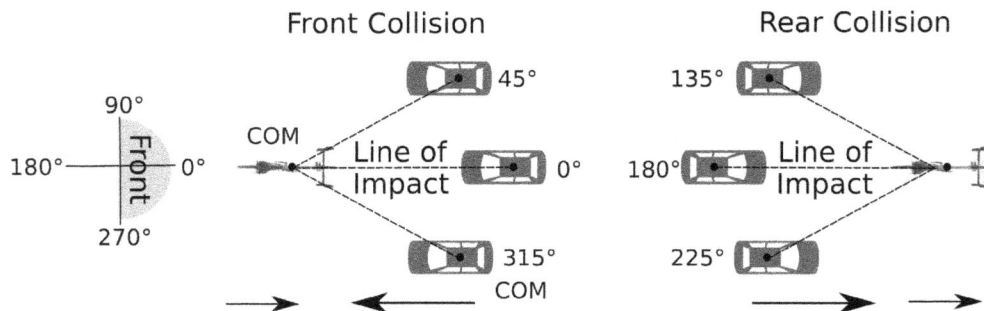

Figure 8.6: Collisions between a vehicle and a bicycle travelling in the same or opposite directions

---

[17]The velocities are converted from kmph. to meters per second in the calculation

If a bicycle and vehicle are travelling in the opposite directions as shown in the center of Figure 8.6, then a front collision is possible at angles ranging from 270° to 0° to 90°. A more oblique collision close to 270° or 90° will have less impact than a head on collision at 0°. Although a front collision is possible beyond 90°, the assumption is that the vehicle and bicycle are moving further away from each other and therefore will not collide.

A rear collision occurs when both the bicycle and vehicle are travelling in the same direction. The impact of collisions between 90° and 270° will vary with the highest impact at 180°. Like front collisions, it is assumed that there are no rear collisions between 0° to 90° and 270° to 0°, since the vehicle and bicycle are moving away from each other.

The line of impact is the line joining the centres of mass of the bicycle and the vehicle. For every collision, a plane of contact is defined perpendicular to the line of impact where physical contact occurs. The velocities of both the bicycle and vehicle are resolved along the axis of the line of impact and its perpendicular. The velocities along the line of impact will change following the collision and can be calculated using the equations for the conservation of momentum and the coefficient of restitution as before. From the change in velocity following the collision, the change in momentum and the impact force (given a one second impact time) for the bicycle can be calculated.

The collision angle in Figure 8.7 is the degree of the line of impact. Angles of 0° and 180° represent head-on and rear-end collisions respectively. A collision between a bicycle and another vehicle is possible at all angles from 0° to 360°. Figure 8.7 shows the change in impact forces for three vehicles travelling at different speeds from 0° to 360°. In all collisions, the bicycle and rider weigh 85 kgs. and speed is 20 kmph. A motorcyclist travels at 80 kmph. and the combined weight of the rider and motorcycle is 200 kgs. The impact force from an accident with a motor cycle at high speed is considerably higher than the impact force from a collision with a much heavier car at lower speeds.

A head on collision between a bicycle at 20 kmph. and a motor cycle at 80 kmph. has an impact force of over 2800 N. The impact from a rear end collision is also high at about 1700 N. A benign phrase, "throw distance", is used to describe the severity of the impact force. The higher the throw distance, the higher the impact force and the further away that the cyclist will land following the collision.

Head on collisions with a car weighing 1000 kgs. and travelling at a lower speed of 40 kmph. have a lower impact force (2200N), but are still high enough to cause a fatality or severe injuries. Even though the difference in speeds is just 20 kmph., a rear end collision has an impact force of about 740N.

Suprisingly, collisions with other cyclists of the same weight can also generate relatively high impact forces. A front end collision between one cyclist travelling at 20 kmph. and another cyclist travelling at 30 kmph. has an impact force of about 1000N. The impact force from a rear end collision at 200N is much less but maybe enough to knock a rider off the bicycle.

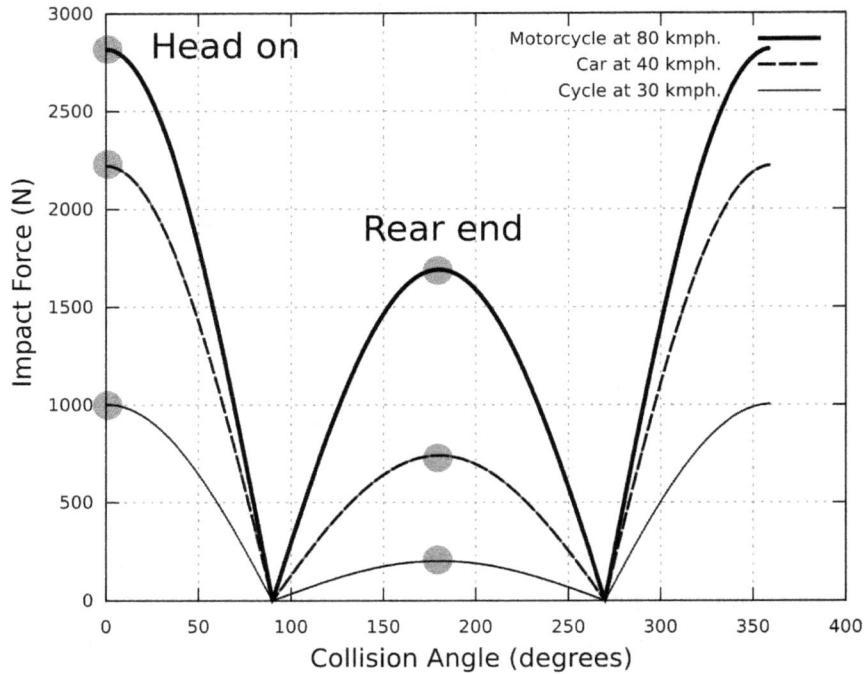

Figure 8.7: Impact Forces from a Sideswipe Collision with a Motorcycle, Car, and Bicycle at Speeds of 80 kmph, 40 kmph., and 30 kmph. respectively

The impact force that a cyclist experiences in a collision is a reactionary force. Accidents between a bicycle and another vehicle that are not in line with each other will have some part of the impact force acting perpendicular to the movement of the bicycle. When this sideways force is high enough, a cyclist can no longer maintain balance and will topple over.

Plastic Collisions The analysis here assumes that non-sideswipe collisions are inelastic, i.e. some kinetic energy will be lost in the form of heat and damage to the frame or body of a vehicle. The rightmost section of Figure 8.8 shows the collision of a vehicle at an angle θ_1 with a bicycle heading east or 0°.

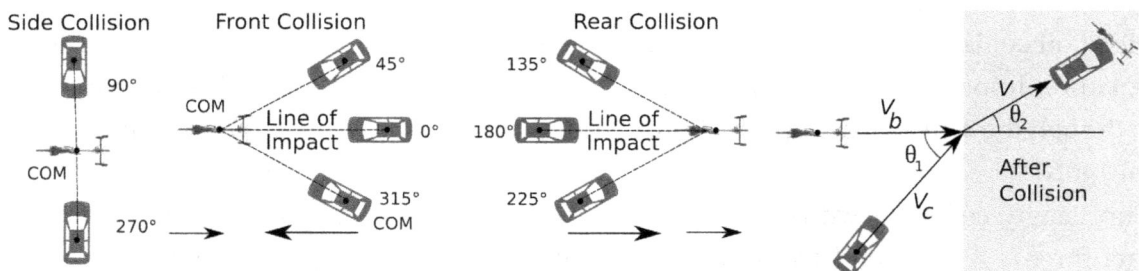

Figure 8.8: Collision of a Bicycle and a Vehicle at an angle θ_1

The velocities v_b and v_c are the velocities of the bicycle and vehicle respectively before the collision. After the collision, both the bicycle and vehicle stick and become one object moving at velocity v in the direction θ_2. The values of v_b, v_c, and θ_1 before the collision are known and the two unknowns v and θ_2 are computed using the equations for the conservation of momentum in the x and y directions.

$$m_b v_b + m_c v_c cos\theta_1 = (m_b + m_c) \times vcos\theta_2 \;\; and \;\; m_c v_c sin\theta_1 = (m_b + m_c) \times vsin\theta_2$$

The masses of the bicycle and vehicle are m_b and m_c respectively are also known. Dividing the second equation by the first equation, the value of θ_2 is computed and finally v is computed as well. The mass of a bicycle and a vehicle are estimated at 85 kgs. and 1000 kgs. respectively while the velocities of the bicycle and car are 20 kmph. and 40 kmph. respectively. The angle θ_1 is varied from 0° to 360° to reflect possible collisions from all directions. For each possible collision at a given angle θ_1, the value of v and θ_2 that represents the velocity vector after the collision, is found. The impact force is the change in momentum of the bicycle divided by the impact time (one second).

As you would expect, the impact force is the maximum near 0° that represents head-on collisions where the change in velocity is the highest and therefore the corresponding impact force is high as well (see Figure 8.9). The maximum impact force for a car weighing 1000 kgs. travelling at 40 kmph. colliding with a bicycle travelling at 20 kmph. in the opposite direction is about 1300N , which is less than the impact force in a similar sideswipe collision. In a non-sideswipe collision, more kinetic energy is dissipated when the bicycle and vehicle stick to each other and therefore the impact force is less.

A motorcycle weighing 200 kgs. travelling at 80 kmph. has a higher impact force (1650N) in a head-on collision with a bicycle. Although, the motorcycle weighs a fifth of the weight of the car, the high velocity contributes to a high change in bicycle momentum. Finally, a bicycle travelling at 30 kmph. colliding with another bicycle of the same weight travelling at 20 kmph. in the opposite direction generates an impact force of about 600N.

In a rear end collision (180°), the impact force is the least since the magnitude of the change in velocity is not as high as in collisions at other angles. Still, an impact force of just 100N is like a force of 10 kg. and can topple a cyclist.

Essential Skills to Avoid Collisions   Clearly, collisions are bad for a cyclist. Before riding on a busy road where collisions are more probable, it is best to develop some skills that it make a collision less likely. Some of these skills include

- **Riding in a straight line:** You should be able to ride along a line without too much of wobbling or swerving.

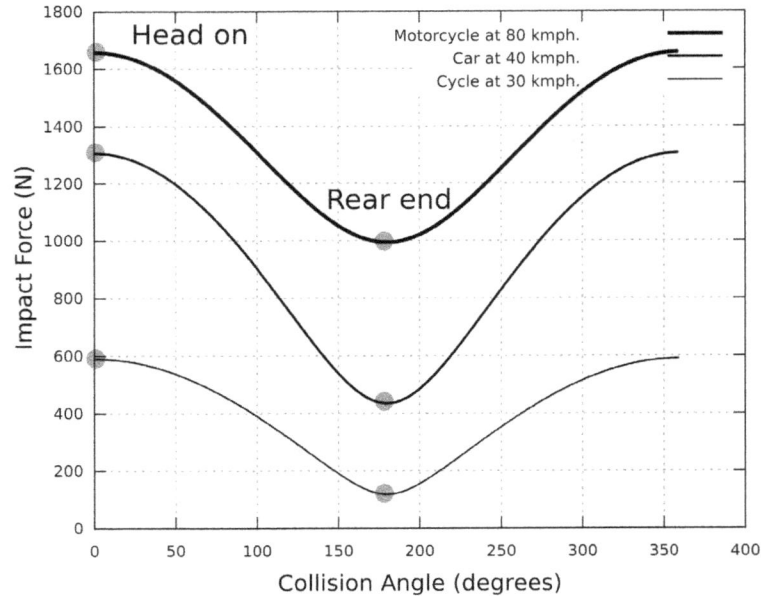

Figure 8.9: Impact Forces from a Non-sideswipe Collision with a Motorcycle, Car, and Bicycle at Speeds of 80 kmph, 40 kmph., and 30 kmph. respectively

- **Looking behind and riding:** When moving across the road to change lanes, you will need to look behind to see if traffic is approaching, while riding forward without losing balance. It only takes a momentary glance to view the traffic behind.

- **Crossing an intersection:** A significant proportion of accidents occur near an intersections, you would need to stop, look, and yield before crossing an intersection.

- **Using hand signals:** You need to able to ride with one hand while using the other hand to signal a turn.

- **Traffic rules:** Although you do not need to pass a test to ride a bicycle on the road, it is helpful to know the rules that motorists are expected to follow and use that knowledge when riding with other vehicles.

- **Avoiding sideswipe collisions:** If you ride too close to the curb or edge of the road, you will risk being overtaken by another vehicle with little separation. Fully occupying a lane and giving way to overtaking vehicles can reduce the chances of a sideswipe collision. If you ride too close to the curb, you risk losing balance since there maybe insufficient space to turn the front wheel.

- **Riding with traffic:** While it maybe legal to ride against traffic on the sidewalk, it is more risky than riding with traffic.

8.2.8 How Effective is a Bicycle Helmet?

In an accident, depending on the speed of a cyclist, the head of the cyclist could hit the ground at speeds from 15 kmph. to over 21 kmph. [16]. The US Consumer Product Safety Commission tests bicycle helmets at impact speeds of 17 kmph. to 22 kmph. Inside the helmet is a layer of polysterene foam (a type of plastic for cushioning and insulation) that will compress and spread the impact area in an accident. While the bicycle helmet can prevent severe head injuries, a cyclist can still suffer a concussion as the soft brain sloshes around the hard skull till all the energy of the impact is dissipated.

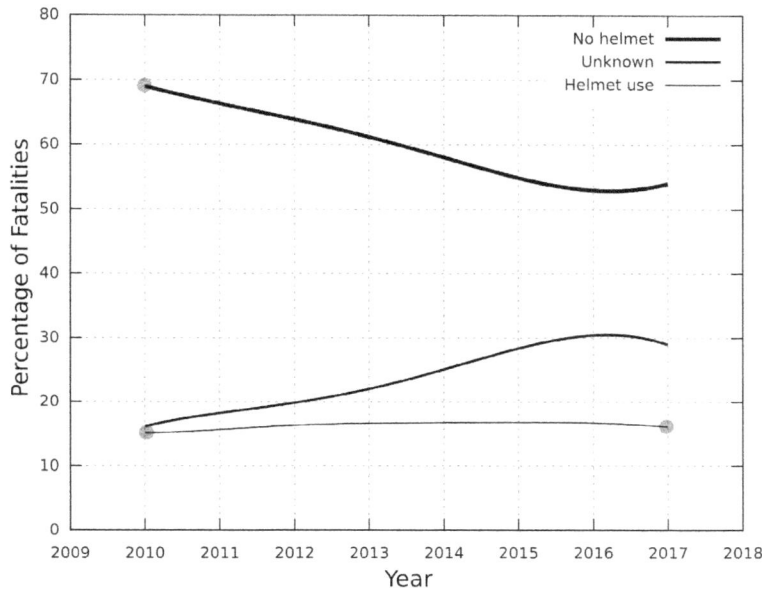

Figure 8.10: Percentage of Bicycle Fatalities based on Helmet Usage (data from iihs.org)

The Insurance Institute for Highway Safety (IIHS) published data [17] describing the number of bicycle fatalities from the year 2010 to 2017 based on the usage of a bicycle helmet or not. In all years, over half of the deaths occurred in cases where the cyclist was not wearing a helmet with a peak of close to 70% in 2010. The percentage of deaths where the rider was wearing a helmet has stayed relatively constant at between 15% and 17%.

In a fall, it may take just over half a second before the helmet hits the ground from a height over 2.0 meters. Despite the short time (0.5 seconds) it takes for the helmet to make an impact with the ground, the vertical speed at the time of impact reaches 20 kmph. or more. The vertical speed in a fall is independent of the horizontal speed and therefore slower riders will also have an impact speed of about 20 kmph. (at a slower speed, it is easier to avoid falling).

With and without a helmet, the time of impact is about six milliseconds and one milliseconds respectively [18]. However, the difference of 5 milliseconds makes a large difference in the deceleration force experienced. The g force at one millisecond and six milliseconds to decelerate from 20 kmph. to 0 kmph. is $638g$ and $106g$ respectively. The US standard for the maximum g force in a helmet is $300g$ [18].

Do you need a helmet? One argument [19] against using helmets is that there is no relationship between helmet usage and the fatality rate per billion kilometers travelled. In the U.S. and the Netherlands with about 50% and 2-3% helmet usage respectively, there were about 45 fatalities and 10 fatalities per billion kilometers travelled. Therefore you could conclude that increasing helmet usage would have no effect on the fatality rate.

However, the purpose of using a helmet is to protect a bicyclist from a head injury in an accident. It is possible that the fatality rate in the U.S. would have been even higher than 45 fatalities per billion kilometers had helmet usage been less than 10%. While a helmet cannot prevent all possible injuries, it does make it more likely that you will survive an accident with a head injury (see Figure 8.10). A slightly older study [6] from the 70s also found that safety conscious riders who used helmets were less likely to have accidents.

A more recent study [20] in the US of over 70,000 cyclists who suffered injuries from 2002-12, found that only 22% of the injured cyclists wore a helmet. The study also found that wearing a helmet reduced the severity of injuries, length of time spent in hospital and mortality for all cyclists.

Sharing the roadway with cyclists is very common in Netherlands and therefore motorists, who maybe cyclists as well, know how to react and drive with cyclists. With an emphasis on bike-friendly infrastructure and roads designed to minimize conflicts between the different modes of transportation, cycling is a relatively safe activity with or without helmets for riders who are not racing.

Types of Helmets Not all helmets are the same. A helmet maybe bulky or streamlined, light or heavy, and expensive or cheap. A budget helmet may have expanded polystrene liners (stryofoam) that is glued to a plastic shell. It is lightweight and effective in transferring the energy from an impact by collapsing and converting some of the energy to heat. However once stryofoam collapses, it does not regain its original shape and therefore must be replaced after an impact. At the same time, the helmet may appear bulky since the layer of styrofoam inside the helmet must be thick enough to absorb the energy of an impact.

The helmets in the mid-range may look more stylish and have vents for air circulation. Instead of styrofoam, a liquid foam like a elastomere polyurethane is used inside the helmet and performs the same function to absorb the energy of an impact. Unlike styrofoam, the lining in these helmets will regain its original shape. The materials used inside a budget and mid-range helmet will also feel different. While styrofoam is stiff and crushable, a liquid foam is rubbery or squishy.

As the price of helmets increases, they become more specific to the type of cyclist (a commuter, racer, or mountain bike rider). Features like a visor, a more aerodynamic shape, or better ventilation will be included. The most expensive helmets use the thinnest possible liner with same protection as the budget helmet, are extremely light, and also look appealing. However if protection is a priority, then both the budget and expensive helmets provide roughly the same performance [21].

8.3 Handling Road Rage

A motorist may get annoyed by a cyclist, but often the period of interaction is less than a minute. The cyclist is more of an irritant on the road than a source of anger. The typical motorist may have to wait for a short period of time before passing the cyclist. Still, an annoyed motorist can display anger with anyone on the road, including cyclists who are the most vulnerable. There are a number of reasons for road rage that include -

- Most motorists on the road are not out for the pleasure of driving but are probably commuting to work, transporting goods, or running errands. Any added driving time because of a cyclist in the way, becomes an annoyance.

- A cyclist is more likely to face the anger of a motorist in a bad mood than another driver. The motorist controls the period of engagement, can rave and rant, and drive off before the cyclist can react.

- On some roads motorists greatly outnumber cyclists. The lone cyclist is a minority, seen as an interloper on the road and is possibly hostile.

- Cyclists in most countries do not pay any road tax and are seen are freeloaders. However, it is simply not worth the effort for a government agency to collect road tax from cyclists since the amount of tax per cycle would be roughly $\frac{1}{50000}$ the tax of a motor vehicle (see Section 8.5).

- The available space on a road maybe scarce and some motorists find it annoying to share the limited space with cyclists. The anger of a motorist can increase sharply on a densely packed road and a cyclist is the easiest target to release the anger.

Responding to Road Rage   A cyclist on a 15 kg. bicycle is simply far outmatched by a motorist in a 1000+ kg. vehicle and will lose badly in a physical confrontation. Therefore the best option is to de-escalate the anger in the interaction with the motorist. This is hard to do in reality since the initial reaction of a cyclist who is innocent is to protest and retaliate. Even though this a typical reaction, there is good reason to stay silent and not respond. In any escalation, the cyclist is far more vulnerable and will certainly lose if the motorist stays in the vehicle. Some reasons to avoid the totally asymmetric confrontation include -

- A cyclist is not going to be able to debate, reason, and persuade an angry motorist, no matter how logical and right the argument. It is not worth the effort engaging in any sort of conversation with someone who is angry.

- It is easy to get off the road on a cycle and let the motorist continue driving ahead. In heavy traffic, a motorist cannot easily reverse to continue the confrontation.

- There is a possibility that a cyclist forced a motorist to brake or swerve to avoid a collision. The motorist will respond with some anger that may surprise a cyclist. Re-tracing your path before the interaction with the motorist can reveal who was at fault.

If you have a really aggressive driver on a two wheeler or other vehicle behind you who is honking or threatening to push you off the road, it is best to get out of the way ASAP. Bus and taxi drivers who are on the roads for long hours are usually stressed out and will become angry at a slight inconvenience due to a cyclist.

8.4 Dealing with Dogs

As you are riding, a barking dog may unexpectedly come into your path. It is an intimidating experience since you may not have foreseen the hazard. You are forced to either slow down, swerve or continue riding in the hope that you will exit the dog's territory before long. A dog keeps track of vehicles, other dogs, and people passing through its territory.

When it sees an unfamiliar vehicle or person, it makes a quick decision whether the intruder is a threat or not. If you are not a threat, then you can safely ride without any trouble. However if the dog believes you are a threat, then it will begin to bark and challenge you. Once you have been categorized as a threat, a dog will not stop barking till you are out of its territory. It is almost impossible to get a dog to change its mind and you will have to deal with being a threat.

Some dogs react to any vehicle such as a car, motorcycle, or cycle that is moving relatively fast in a predatory fashion. They may perceive a cyclist as prey to be hunted down. Unfortunately, dogs can quickly detect if you panic and show fear. A dog will typically run parallel to your path and poses a danger if it starts drifting closer to you. Rarely will a dog jump in front of you inviting a collision. Among the various recommendations to avoid an attack include -

- Stand up to appear imposing and dominant and make a loud sound. The larger appearance and noise can deter a dog from attacking or biting.

- Swerve towards the dog to show that you can be aggressive as well. This is hard to do, since your initial reaction is to get away from the dog.

- Squirt any liquid that you may have in the direction of the dog.

- Some breeds of dogs are more aggressive and go beyond simply barking. Rottweilers, German Shepherds, and Doberman Pinschers are some of the aggressive breeds that do not hesitate to attack. Recognizing these breeds will alert you to the danger, but you still have to deal with the situation. Once an attack begins, even the owners of these dogs may have difficulty controlling them.

Being able to sprint at speeds above the average speed (30 kmph.) of a dog can ensure a quick getaway. Although, this does appear counter to the advice of not showing fear by fleeing from the attack, it is sometimes the best option to get out of the way as soon as possible.

A Pack of Dogs   At the risk of being flippant, there is nothing like a pack of dogs chasing you to motivate you to pedal as fast as you can. A pack of dogs is far more dangerous than a single stray dog. Their aggressive behaviour is similar to the behaviour of a mob. Whatever fear that a single dog had as an individual is lost in a pack. While a single dog may just bark and let you pass, if a pack of dogs begin to attack, then there is no alternative but to get away as fast as possible.

Another alternative is to slow down, get off the bicycle, and walk very slowly with arms raised. While being on the bicycle may seem safer, you cannot afford to lose to your balance and fall. However, getting off the bicycle with a pack of barking dogs in the midst does take some courage. Finally, avoiding a confrontation by riding on routes that are less likely to have a lot of stray dogs is better than trying to manage the unpredictable nature of some dogs.

8.5   Right to Use the Road

Two common criticisms of cyclists are that they do not pay road tax and are not required to have licenses. In the 50s and 60s, highway engineers studied the damage done to roads by vehicles based on their weight [22]. The American Association of State Highway Officials found that damage to the roadbed is roughly proportional to the fourth power of the axle load of the vehicle. Therefore, a vehicle x with double the weight of a vehicle y will causes 16 times as much damage compared to the damage caused by y. However, few if any country follows a formula for road tax that is strictly based on the fourth power of the axle weight ratio.

The fourth power law uses the weight supported per axle instead of the total weight to account for the distribution of weight in a heavy vehicle. Faster vehicles also tend to cause more damage than slower vehicles and the amount of damage is roughly linear with speed. Calculations of road tax rarely take speed into consideration, since speed depends on a motorist's behaviour.

A mid-size car weighing 1500 kgs. with two axles causes more than 50,000 times as much damage to the roadway compared to the damage from a 100 kg. weight of a bicycle and rider $-(\frac{750}{50})^4 = 50,625$. Collecting one fifty thousandth of the road tax of a car from a bicyclist may not be worth the effort for a road transport department.

An argument can be made for a cyclist road tax based on damage caused by weather and not weight. Over a few years, harsh weather can damage a roadway, even if it is not used extensively. Yet, pedestrians do not pay a road tax to use the sidewalk or pavement.

Cycle vs. Car Area   A cyclist is physically much smaller than a car and requires roughly $\frac{1}{8}$ of the area occupied by a car. A mid-size car of dimensions $4.5m \times 1.75m$ occupies about $8m^2$ and at 20 kmph. would require a separation length of about $6m$ for a $1s$ gap and a separation width of $1m$

on either side. Therefore, the total area required for a mid-size car is about $40m^2$ (10.5×3.75). A bicycle of dimensions $1.75m \times 0.75m$ occupies about $1.3m^2$. With a separation width of $0.25m$ on either side and a separation length of $2.25m$ (stopping distance at 20 kmph. on a dry road, see Chapter 6), the total area required for a bicycle is about $5.0m^2$ (4.0×1.25) and therefore roughly 8 bicycles could use the space required for a single mid-size car (see Figure 8.11).

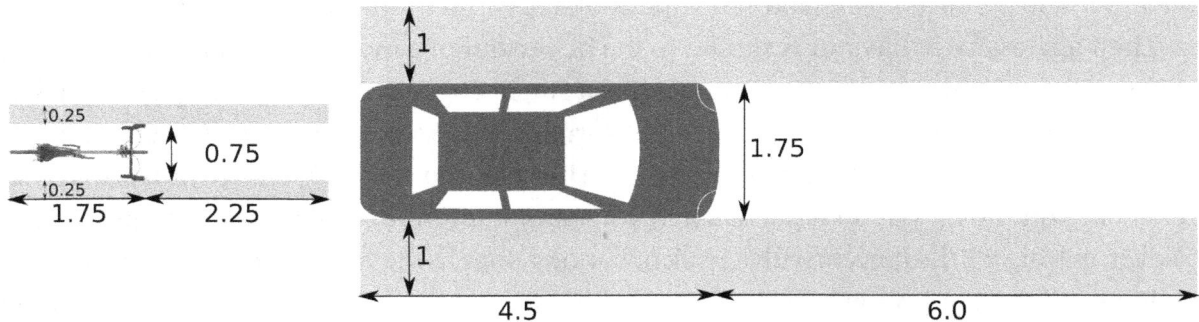

Figure 8.11: Road Areas for a Bicycle and a Car in meters at 20 kmph. on a Dry Road

8.6   Bicycle Infrastructure

Lack of safety is frequently cited as one of the primary reasons why people avoid cycling. One of main purposes of bicycle infrastructure on a roadway is to prevent collisions between faster motorists and slower cyclists and increase the level of safety for cyclists. There have been a number of studies [4] on which types of bicycle infrastructure will improve safety. In retrospect, roadways should have been designed and constructed with the assumption that they will be used by cyclists. But, often cyclists were never considered in the design of a roadway or it was simply assumed that a cyclist would find space to ride on the far edge of the road.

One of the common recommendations is that on a road where speeds are significantly different, slower vehicles should be separated from faster vehicles. Without a physical barrier separating motorists from cyclists, the space allocated for cycling can be easily occupied by other vehicles. The physical barrier comes in various forms including a series of short posts (bollards) that are close enough to prevent motorized two wheelers using the space allocated for cyclists, a series of kerbstones, or even parked cars. The terms used to describe bicycle infrastructure depend on the region, but the following terms are common -

- **Bike Lane** is a part of the roadway close to the kerb that is designated for the use of cyclists with signs and some type of marking.

- **Bikeway** is used for any type of roadway that is designated for either exclusive or shared travel by cyclists.

- **Bike Route** is a system of bikeways that maybe interconnected for cyclists to ride through a city or a scenic route.

- **Bike Path** is a physically separated bikeway that is located away from a roadway and is often meant exclusively for bicycles.

- **Bike Track** is like a bike path but is located on or next to a roadway and separated from motorists with some type of physical barrier.

- **Traffic Calming** devices are rumble strips or road humps used to slow down motorized traffic that travel alongside cyclists.

- **Shared Roadway** is a roadway that is open to the use of both cyclists and motorists.

Space for Cyclists   Designers of roadways allocate space based on the anticipated motorized traffic density. Cyclists are often taken into consideration as an afterthought and expected to squeeze into the available space on a shared roadway. A few cyclists are confident enought to navigate through and find space on roads with dense and high speed traffic. Other cyclists would like more space on a roadway to ride without fear. Slower cyclists tend to wobble a little more and therefore need at least 0.125 meters on either side. The standard recommended space for a cyclist is one meter for the rider and bicycle plus 0.125 meters on either side for a total width of 1.5 meters [23]. This is the minimum width for single lane traffic in one direction. The average length of a bicycle is roughly 1.75 meters and 2.0 meters is an average separation distance between two bicycles. Therefore, the space occupied by a cyclist on a roadway is 3.75×1.5 or 5.63 square meters.

Types of Cyclists   The US Federal Highway Administration categorized cyclists into three broad groups[23] -

- The **A**dvanced or experienced rider uses the bicycle like a motor vehicle and is comfortable sharing the roadway with motorized traffic.

- The **B**asic or less confident rider may ride shorter distances to visit or shop on roads with less traffic.

- **C**hildren who ride to school and not ride as fast and mostly use residential roads

The three cyclist groups are useful to plan the type of bicycle infrastructure that is appropriate to suit the most likely rider group on the roadway. You would expect more children to cycle near a school or park than experienced riders.

8.6.1   Bicycle Lanes vs. Bikepaths

A **bicycle lane** is often created on an existing road by limiting the number of lanes for motorized traffic. Drivers of motorized traffic may resent the use of a lane on an existing road for bicycles, since with less space, vehicle speeds are certain to be lower.

Advocates of **bikepaths** or **bikeways** claim that having riders on a completely separate path from a roadway improves the safety of the cyclists. Some argue that to the contrary, the safety record for a cyclist on a separate bikeway is worse than on a common shared roadway. This does appear odd since the likelihood of a motorist-cyclist collision is greatly diminished on a bikeway.

A study [6] of the accident rate for about 3000 cyclists on a bicycle lane, bikepath, and streets did confirm that bikepaths have the highest accident rate of all three routes. The accident rate for a bicycle lane, a minor roadway, a major roadway, and a bikepath were 25, 27, 35, and 80 accidents per million miles respectively.

A bikepath appears to be more than three times as risky for a cyclist than a bicycle lane. One explanation for this paradox is that cyclists become careless on a bikepath and the bikepath rules are relatively lax compared to the rules for a shared roadway where accidents can be more severe. Although motorists are absent on a bikepath and therefore bicyclist-motorist collisions are non-existent, a bikepath will often cross a roadway at an intersection and the same types of accidents that occur at intersections between roadways is likely.

Some of the problems with bike lanes include the accumulation of debris and obstacles such as grates, expansion joints, railway tracks, and rumble strips that can lead to falls on a bicycle with relatively thin tyres. Expansion joints and grates that are often found on flyovers or bridges make steering with a thin tyre tricky. Grates that are parallel to the roadway can trap the front wheel leading to a fall. The intersection of a roadway and bicycle lane is another potential source of accidents.

8.6.2 Making Bicycling Safe

Since safety is one of the reasons why cycling is not popular in some countries, it is worthwhile finding out which types of roads or routes are the safest for cyclists. A Canadian study [4] evaluated the accident rates for different bicycle route types.

Control and Injury Locations The main goal of the study was to identify which types of bicycle infrastructure contributed significantly to lower the risk of an accident. There are a large number of possible reasons for any accident and the study attempted to isolate infrastructure causes for an accident. The route characteristics at the location of the accident were compared with the route characteristics at a random location on the same trip which ended with an injury. The idea was to create a control location from a random location on the same route that had conditions similar to the conditions at the injury location. Any difference in route characteristics between the injury and control locations would become the primary reason for the accident. Since the control site and injury site were chosen from a single trip, characteristics such as rider behaviour, type of bicycle, use of helmet, and weather were identical for both sites. The characteristics for an accident were divided into three categories (see Table 8.7)

Some of these characteristics were more statistically significant in differentiating between an injury and control site. The study used a logistic regression model to relate the cycling environ-

Type	Characteristics
Cyclist	Age, gender, cycling experience, education, income
Trip	Purpose, timing, distance, use of helmet, injury type
Site	Intersection, street type, lighting, gradient, construction, traffic density/speed, timestamp, and weather

Table 8.7: Characteristics for an Accident

ment or characteristics to the likelihood of an injury or control site. Every control or injury site was located on one of 14 defined route types (see Table 8.8). Therefore, the model related the cycling environment with a route type that was likely to have more accidents than other route types.

Types of Routes The study estimated the accident rates on 14 different route types to identify the relatively safer routes. The study identified cyclists who were injured and hospitalized, but not fatally, to retrace the trip which led to the injury. Participants in the survey were interviewed fairly soon after the accident.

Street	Infrastructure	Parked Cars	Risk
Major	**Cycle Track**	**Yes**	**0.11**
Local	Designated bike route	Yes	0.49
Local	No bike infrastructure	Yes	0.51
Major	**Bike lane**	**Yes**	**0.54**
Off	Bike path	No	0.59
Major	**Shared lane**	**No**	**0.60**
Major	**No bike infrastucture**	**No**	**0.63**
Local	Designated bike route with traffic calming	Yes	0.66
Major	**Bike lane**	**Yes**	**0.69**
Major	**Shared lane**	**Yes**	**0.71**
Off	Multi-use path, unpaved	No	0.73
Off	Multi-use path, unpaved	No	0.79
Off	Sidewalk or pedestrian path	No	0.87
Major	**No bike infrastructure**	**Yes**	**1.00**

Table 8.8: Fourteen Route Types and Associated Risks

Table 8.8 lists the 14 route types used in the study and their associated risks[18] with 1.00 and 0.11 being the highest and lowest risk respectively. A *cycle track* is located between the roadway and the sidewalk, separated with some sort of physical barrier to protect cyclists from motorists and may also be at a different grade from the roadway. It is differentiated from a *bike lane* which is simply a part of the roadway that is on the same grade and marked for bicycles. A cycle track is equivalent to a protected or separated bike lane. In a memorandum dated March 2011 from

---

[18]The risk is actually the ratio of the odds of having an accident at a control site to the odds of having an accident at the injury site.

the office of the Mayor of New York City, Howard Wolfson claimed - "When protected bike lanes are installed, injury crashes for all road users (drivers, pedestrians, cyclists), typically drop by 40 percent and by more than 50 percent in some locations."

A *bike route* is a roadway with signs to indicate that bicyclists also use the road but without any separate lane. It is often found on local streets where there may not be enough space to create a separate bike lane. Local streets are generally safer than major streets as you would expect. Interestingly, the use of traffic calming devices such as road humps, rumble strips, and signs on a local street did not reduce risk. Confirming earlier studies that a completely off street path for bicycles does not significantly reduce risk, all the off street route types had risk values greater than 0.5.

The bicycle infrastructure or lack of on a major street or main roadway can significantly alter the risk for a cyclist from the lowest 0.11 to the highest 1.0. A major street with no bicycle infrastructure and parked cars has the highest risk, since all the space on the road is shared with all vehicles. Parked cars pose a risk for a cyclist who besides the chance of being "doored" has to squeeze through the available space between a parked car and traffic on the roadway.

Although not included as one the of route types, a bicycle lane can be one-way or two-way. A two-way bicycle lane is likely to be more accident prone, since a part of the bicycle traffic will be riding against the flow of motor vehicle traffic. Therefore, most bicycle lanes are created on the edges of the road to ensure that cyclists ride with the flow of traffic.

The two types of injuries were either a fall or a collision. Over 70% of all injuries were collisions with either a motor vehicle (33%), a surface feature like a pothole (25%), or a post, lane divider, or person (13%). Collisions with pedestrians and other bicyclists were relatively rare with the exception of areas with dense motor vehicle, bicycle, and pedestrian traffic.

8.6.3 How many kilometers does an average rider travel?

Before building bicycle infrastructure, an estimate of the number of cyclists and distances travelled will help in justifying costs and benefits. Kaplan [6] surveyed over 3000 cyclists in the U.S. and tabulated the number of miles covered in a year for 10 different ranges.

Not suprisingly, his study found that annually a few cyclists ride long (> 5000 miles) distances while over half (61%) of cyclists ride less than 2000 miles. In Figure 8.12, 35% of cyclists rode between 0 to 1000 miles, 26% rode between 1000 to 2000 miles, and so on. Many of the riders with large mileages were compiled from one or more tours. The study also found out that the annual mileage increased with more experience. A cyclist with over five years of riding experience rode roughly 25% more than a rider with less than five years of experience.

A more recent estimate [8] from the 90s estimated a range of 40 to 147 kilometers per year per person (including non-cyclists). If 14% of the US population (320 million) are cyclists, then the annual distance per cyclist ranges from 285 to 1046 kilometers or an average of 666 kilometers. The average cyclist in Netherlands rides about 1000 kilometers in a year [24]. One reason Ka-

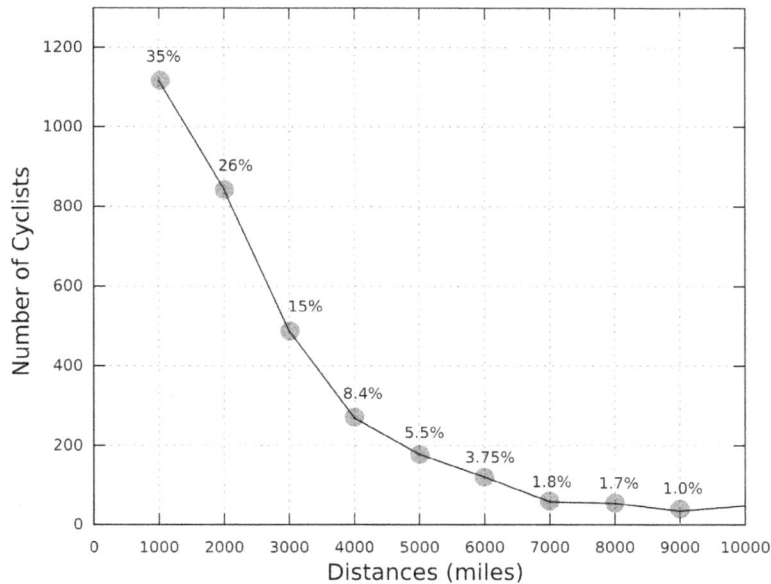

Figure 8.12: Number of Cyclists vs. Number of Miles Travelled in a Year [6]

plan's study had a higher average distance per cyclist maybe because many of the respondents to his survey were committed cyclists.

Number of Cyclists Another report [25] of the number of cyclists estimated the number of US cyclists at about 32% of the population or close to 100 million cyclists. While this estimate is high, it includes cyclists who ride just once or twice a month and make up roughly two thirds of all cyclists. A few cyclists (10% of 100 million) ride more than 100 times a year and the remainder ride between a moderate 25-100 times a year.

8.7 Cycling in Bike-Unfriendly Surroundings

Although cycling does have environment and health benefits compared to other modes of motorized transportation, a bike-unfriendly area will give preference to motorist's requirements for higher speeds and greater space. A bike lane on the roadway may not exist and a cyclist may have no choice but to ride side by side with faster, larger and more powerful motorized vehicles.

The popularity of bicycles in such conditions will be limited to a few riders willing to take some risks and who have sufficient skills to ride with a variety of other transportation modes including motorized two wheelers, cars, vans, and buses. One of the benefits of being in the minority is that a helmetted rider on a bicycle is an oddity on the road and more likely to be noticed.

Drivers of motorized vehicles have far more power in their vehicles than a humble cyclist possesses. The average cyclist can ride comfortably between 20-25 kmph on a flat road, while the average speed of a typical motorized vehicle could be double or triple a cyclists' speed. This difference in speed is exaggerated on an uphill road. This difference in power, speed, and dominant position of the motorist place the cyclist at risk. A gentle tap from a car or even a motorcycle is sufficient to topple a cyclist (see Section 8.2.7) . Often, the driver of the other vehicle may not notice the collision and stop to check if a cyclist was injured. Although the law may eventually catch up with the other vehicle's driver, the risk of a serious injury is high.

8.7.1 Traffic Patterns

It may appear challenging to ride on bike-unfriendly roads, but there are behavioural patterns that if you keep in mind will made riding easier and predictable. Some of these patterns include -

- A vehicle will overtake you at high speed and then brake in front of you. The driver may have saved a few seconds by accelerating and braking, but this behaviour is repeated and predictable. Driving a powerful vehicle behind a relatively slow moving bicycle seems illogical and therefore the overtake and brake pattern is one that a cyclist should expect.

- If a bus overtakes you, it may happen just before a bus stop. This is very similar to the previous pattern. A bus driver may prefer to accelerate and brake than simply slow down near a bus stop. The sound of a bus behind you and an approaching bus stop is a warning of the need to slow down and let the bus overtake you.

- Even if you are riding at a reasonable speed, the driver of the vehicle behind you may honk incessantly and then suddenly decide to overtake at high speed.

- Just when you have picked up speed, a vehicle in front of you will brake for no apparent reason. You have no choice but to slow down and pick up speed again. Similarly, a motor vehicle turning will overtake and brake in front of you, forcing you to slow down. While the safer option for the motorist would be to slow down and let a cyclist go ahead and then turn, it is extremely unlikely on bike unfriendly roads. As a consequence, you will lose momentum and it is often pointless to try and accelerate near a turn to force the motorist to slow down and turn behind you.

- On a main road, a vehicle approaching from a side road will reach a junction at the same time as you do and will not give way. The driver of the larger and more powerful vehicle will assume higher preference even though you are riding on the main road.

- It is not worth accelerating if the motorist behind you honks. No matter what speed you reach, it will not be sufficient to keep up with the speed of a motorized vehicle. Besides, the driver does not really care how fast you can ride and is simply in a hurry to move ahead.

- Sometimes there may not be sufficient space for a vehicle to overtake you and a vehicle behind you may still honk repeatedly. Since you cannot vanish into thin air to give enough room to overtake, your best bet is to wait for the next opportunity and move to the side of the road. These situations are tricky since you don't want to move at the wrong moment to the side of the road, when there is insufficient space to overtake. An impatient driver will try and drive through the narrow gap with possibly disastrous consequences for you. When you ride in your space on the road, you want to make sure that it is obvious to the driver behind whether there is enough to space to overtake or not. Impatient drivers will take risks and overtake you, even if it means squeezing you between a car and the side of the road.

- Changing lanes to make a turn along side fast moving vehicles at a junction can be treacherous. Even though it takes longer, it is often better to make a turn in the same direction as traffic and then make a U-turn.

8.7.2 Tips

Although a cyclist cannot ride as fast as a motorized vehicle, the safest option is to ride as if you are on a motor vehicle and follow the same traffic laws. Some of the tips to ride safely include -

- **Staying at least 0.25m from the edge of the road:** While it may appear to be safe to ride as close to the edge of the road as possible to give enough room for other vehicles, it is actually more risky. The drivers of other vehicles can mis-calculate the space to overtake and risk a collision. However, you do need enough room to steer and stay balanced.

- **Claiming your space:** Even though a bicycle and rider take about $\frac{1}{8}$ the space of an average car, you still have to occupy that limited space firmly (i.e. in the center of your $\sim 1.25m$ wide lane) to ensure that another vehicle does not encroach and force you to the edge.

- **Making your presence known:** Establishing eye contact with a driver of another vehicle is essential if your paths cross. Giving a hand signal in addition to looking at the driver of the other vehicle to indicate your intentions should ensure that a motorist has noticed your presence. Besides wearing bright fluorescent clothing and waving your hands aka bio-motion, there is nothing more you can do to attract attention.

- **Behaving consistently and obeying traffic rules:** Riding through a red light is an invitation for an accident and confirms the perceptions of motorists that cyclists are reckless. A traffic light may turn from green to red just as you approach it. While a car may speed up and just make it past a traffic light without running a red light, it is very hard for a cyclist to attempt the same idea. You will end up in the middle of a junction blocking traffic from the other road.

- **Avoiding the blind spot:** If you are riding on the side of a large vehicle like a bus or lorry, you could be in the blind spot and not visible to the driver. Also riding in the direction of the sun, a motorist may not be able to see you. Besides watching for motorists, you also have to look out for pedestrians who can change directions suddenly and appear in your path.

- **Handling an accident:** If you have an accident, you may not get much sympathy. Since the cycling community is limited, in an accident the police, lawyers, judges, and other motorists are unlikely to be fellow cyclists and therefore cannot see the cyclist's point of view. The common refrain in an accident is that the cyclist was invisible and therefore the motorist is not liable. Since proving visibility is not possible, most motorists who are guilty in an accident can get away with a minimal penalty.

- **Riding Uphill:** When traffic comes to a standstill on a slope and begins to move, it will be hard to start riding uphill without wobbling. One option is to switch to a lower gear before the climb.

Tips for Indian Roads   On Indian roads, the number of motorized two wheelers (scooters and motorcycles) is much higher and a greater threat to a cyclist than a car or a bus. In 2017, motorized two wheelers accounted for over a third of all road accidents [26]. It does not take much time to accelerate on a motorized two wheeler, like a motorcycle, and the power can lead to aggressive behaviour on the road. Since a cyclist's power cannot match the power of a motorcycle, the riders of motorized two wheelers are more likely to become impatient with slower moving cyclists. Like bicycles, motorized two wheelers occupy a fraction of the space for a car, and therefore riders can navigate between gaps on the road that would be impossible for a car. Given the challenges of riding with a higher number of motorized two wheelers, the list of tips below can help in riding on Indian roads -

- A vehicle may signal a left turn, but continue straight or even turn right. The turn indicator may have been switched on for an earlier turn or some other reason and you cannot always assume that a vehicle will move in the direction of the turn indicator.

- Vehicles will be parked on the side of the road forcing you to move towards the center of the road. If you don't move towards the center of the road quickly enough, you may find ourself trapped behind a vehicle waiting for a break in traffic. Being able to accelerate is a great help in these situations to avoid being cut off.

- The same idea applies when you are behind an autorickshaw driver who is looking for passengers. They drive very slowly and may unexpectedly stop. When you are trapped behind a slow moving autorickshaw, you have to move to the center of the road without interrupting the flow of traffic.

- Pedestrians will cross the road anywhere and come in your path without any warning. On a footpath, pedestrians have the right of way and it is best to wheel the bicycle along. You also have to be prepared for the pedestrian who may suddenly get off a bus that has stopped.

- Honking is a habit and drivers often do it for several reasons. It maybe out of frustration with traffic congestion, to communicate that a vehicle is behind you, or to indicate that a red traffic light will shortly become green. The result is a cacophony of noise on the roads. When a traffic light changes to green, it is like the start of a formula one race and you have to get away as quickly as possible.

The chances of an accident can be minimized despite the lack of bicycle infrastructure to separate cyclist traffic from motorized traffic by being predictable and keeping out of the way of faster vehicles. Also, a cyclist with greater experience is less likely to have an accident [6].

References

[1] U.S. Department of Transportation National Highway Traffic Safety Administration. Traffic safety facts 2009 data, 2009. URL `http://bit.ly/2jUfmu3`.

[2] Laurie F. Beck, Ann M. Dellinger, and Mary E. O'Neil. Motor Vehicle Crash Injury Rates by Mode of Travel, United States: Using Exposure-Based Methods to Quantify Differences. *American Journal of Epidemiology*, 2007.

[3] Gina Kolata. How safe is cycling? it's hard to say, 2013. URL `https://nyti.ms/2lWs8sA`.

[4] Kay Teschke et al. Route Infrastructure and the Risk of Ijuries to Bicyclists: A Case-Crossover Study. *American Journal of Public Health*, 102(12), 2012.

[5] Wikipedia. Aviation safety, 2020. URL `http://bit.ly/2lHu4F0`.

[6] Jerrold A. Kaplan. Characteristics of the Regular Adult Bicycle User, University of Maryland, Master's thesis, 1975.

[7] U.S. Department of Transportation Federal Highway Administration. Traffic Safety Facts 2015: A Compilation of Motor Vehicle Crash Data from the Fatality Analysis Reporting System and the General Estimates System, 2020. URL `http://bit.ly/2kPSsod`.

[8] U.S. Department of Transportation Federal Highway Administration. The Environmental Benefits of Bicycling and Walking, 1993.

[9] Federal Reserve Economic Data. Moving 12-Month Total Vehicle Miles Traveled, 2020. URL `http://bit.ly/2m142gq`.

[10] U.S. Department of Transportation Federal Highway Administration. Injury to Pedestrians and Bicyclists - An Analysis based on Hospital Emergency Department Data, 1999.

[11] Paul Schepers and Karin Klein Wolt. Single-bicycle crash types and characteristics. *Cycling Research International*, Vol. 2, 2012.

[12] US National Highway Traffic Safety Administration. Fatality Analysis Reporting System, 2018. URL http://bit.ly/2kDEBRv.

[13] U.S. Department of Transportation Federal Highway Administration. 2014 FARS NASS GES Pedestrian Bicyclists Manual, 1999.

[14] Dan Nabors, Elissa Goughnour, Libby Thomas, William DeSantis, and Michael Sawyer. *Bicycle Road Safety Audit Guidelines and Prompt Lists*. U.S. Federal Highway Administration, 2012.

[15] National Academies of Sciences, Engineering, and Medicine. *A Guide for Reducing Collisions Involving Bicycles*. The National Academies Press, 2008. URL https://doi.org/10.17226/13897.

[16] Allen St. John. Anatomy of a bike accident, 2016. URL http://bit.ly/2kuYQAT.

[17] Insurance Institute for Highway Safety. Fatality facts 2017 bicyclists, 2017. URL http://bit.ly/2k2Ft1P.

[18] Bicycle Helmet Safety Institute. Bicycle Helmet Liners, Foam and Other Materials, 2019. URL http://bit.ly/2me6jVI.

[19] Angie Schmitt. Why Helmets are not the Answer to Bike Safety: In One Chart, 2016. URL http://bit.ly/21UEdOM.

[20] Sciencedaily.com. Many injured US adult cyclists not wearing a helmet, 2019. URL http://bit.ly/2msQsT9.

[21] Bicycle Helmet Safety Institute. Cheap or Expensive Bicycle Helmets, 2019. URL http://bit.ly/21wObGm.

[22] Richard Masoner. The Fourth Power Rule, 2014. URL http://bit.ly/2kCTvrt.

[23] American Association of State Highway and Transportation Officials. *Guide for the Development of Bicycle Facilities*, 1999.

[24] Bicycle Dutch. Dutch cycling figures, 2018. URL http://bit.ly/2m7JKSj.

[25] PeopleforBikes. U.S. Bicycling Participation Report, 2019. URL http://bit.ly/21e621g.

[26] Ministry of Road Transport and Highways. Road Accidents in India - 2017, Report. URL http://bit.ly/2KzdD7g.

9 Training

Even though a bicycle is very efficent at converting the force you apply on the pedals to a propulsive force, the human body is not as efficient in converting the chemical energy in your body to mechanical energy. Roughly 70% or more of the energy generated by your muscles while pedalling is dissipated in the form of heat energy. The aim of training is to increase the force generated by muscles and to recover faster following a workout. Even a short uphill ride at a reasonable intensity will increase your heart rate as well as body temperature. Since body temperature is regulated, you will begin sweating to keep your body temperature within a safe range. In addition to maintaining temperature, you will also need to prevent your heart rate from exceeding a recommended limit based on your age.

9.1 Heart Rate and Power

The heart rate for an average adult is between 60 to 100 beats per minute (BPM) or 86,400 to 144,000 beats per day or roughly 2.6 billion beats in a lifetime of 70 years. There is no rest for the heart and only a recovery period when the demand for energy is low. Any activity which demands energy will require the heart to pump more blood and correspondingly increase the heart rate. Keeping the rate of increase low, yet pumping more blood is key to riding long distances with a sustainable level of power. Since your heart rate will fluctuate during the day depending on your level of activity, the "resting heart rate" (HR_{rest}) is one heart rate to evaluate (the other is the maximum heart rate, HR_{max}).

9.1.1 Resting Heart Rate

The HR_{rest} is the heart rate when a person is resting and has not exercised for at least 1-2 hours earlier. HR_{rest} will vary depending on your age and level of fitness. In general, a younger person will have a lower HR_{rest} than an elder person and a more fit individual will have a lower HR_{rest} than someone who is sedentary.

If you have measured your HR_{rest}, then you can identify a category for the condition of your heart from a table, depending on your age and sex [1]. When your HR_{rest} exceeds the base rate (50 for male and 55 for female), then a rough calculation for your heart condition is $x = floor(\frac{HR_{rest} - Base\ Rate}{5})$ where x defines the condition (0 - Athlete, 1 - Excellent, 2 - Good, 3 - Above average, 4 - Average, 5 - Below average, and 6 - Poor). A fitter individual has a stronger and bigger heart which can pump more blood per beat, compared to the volume of blood pumped per beat from a sedentary individual's heart.

Your HR_{rest} is just one of the factors to consider when training. However it is a key factor, since it is an indicator of how well your heart muscle is functioning. If the muscle is weak, you will find it harder to ride long distances that a cyclist with a stronger and larger heart can. Since the volume of blood per stroke is larger for an athlete, the number of strokes (beats) will be less and therefore the heart does not have to work as hard.

9.1.2 Maximum Heart Rate

While your HR_{rest} is good to know, your maximum heart rate (HR_{max}) is more important when training. This is the heart rate that you can sustain for a brief period (minutes) before exhaustion. An old formula to calculate HR_{max} was $220-age$. So for a 40 year old individual (male or female), HR_{max} was 180. This calculation was viewed as too simplistic and an alternate calculation [2] was suggested from experimental data collected from over 3000 healthy men and women, which took into account the difference between male and female heart rates. If x is the heart rate, the HR_{max} for a male and female was computed as $213 - 0.65 \times x$ and $210 - 0.62 \times x$ respectively.

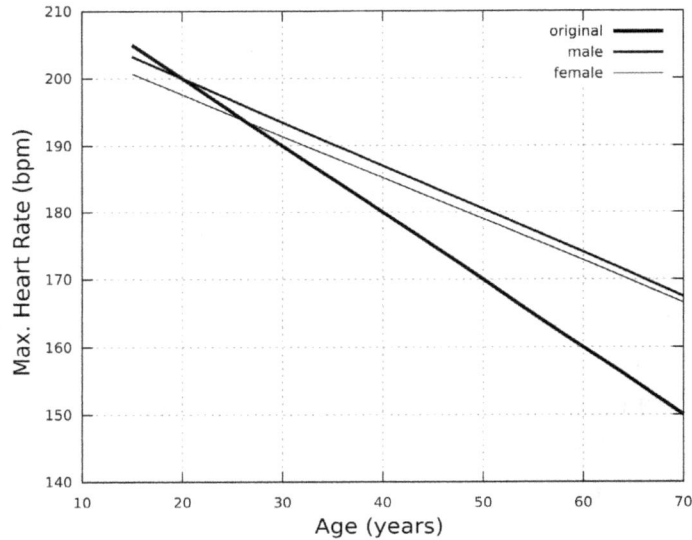

Figure 9.1: Maximum Heart Rate vs. Age [2]

Figure 9.1 shows the HR_{max} for different ages using the old and new formulas. The old formula had a slightly higher HR_{max} for the ages 15-25 and then deviated substantially lower at higher ages. The HR_{max} for healthy elder individuals using the new formula was about 10 bpm higher than the HR_{max} using the old formula.

However, HR_{max} is not necessarily an indicator of fitness. A person with a lower HR_{max} maybe able to maintain the lower rate for a longer period than a person with a higher HR_{max}. Your heart rate and power generated are related. If your power increases by some percentage, then your heart rate will also increase by a similar percentage. However, your heart rate may increase further to sustain the higher power output. Alternatively, your heart rate can also stay constant, even though your power output maybe reducing.

9.1.3 Heart Rate Monitors

Regularly working out at your HR_{max} is not recommended [3]. If you monitor your heart rate during a workout, you will know if you are over exerting yourself. When you increase the intensity of your workout (by generating more power), your heart rate will increase as well. When your exercise intensity is too high, there will be other signs such as rapid breathing besides a high heart rate.

Currently, there are a wide range of heart rate monitors from the ones that you strap on your wrist to the band with a sensor on your chest. All monitors either display the real time heart rate and optionally also record the heart rate on a ride. Watching your real time heart rate can indicate if you are over exerting or whether you can accelerate. If you can record your heart rate, then post-ride you can view how your heart rate fluctuated during the ride. You could see how your heart rate changed with change in the gradient or a head wind.

The wrist based heart monitors are sufficiently accurate and look identical to wrist watches. You can check your heart rate by looking at your wrist just as you would check the current time. They work by detecting your heart rate from pulses of light (see Figure 9.2). An emitter sends a beam of light towards an artery, some of which is absorbed by the blood in the artery. In the left side of the figure, before the pulse wave reaches a location where the beam of light from the emitter will be directed, most of the light is reflected back to a detector. In the right side, when the pulse wave interacts with the beam of light, less light is reflected since more blood in the pulse wave absorbs light. The device counts the rate at which the detector receives less reflected light from the artery to compute the real time heart rate. The differences between the magnitude of transmitted and reflected light is exaggerated in Figure 9.2.

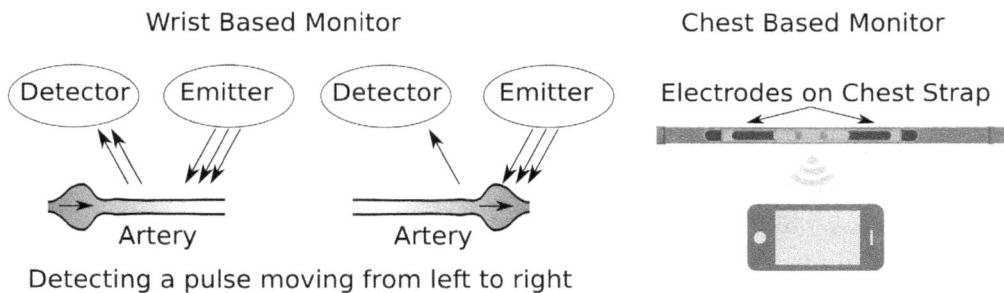

Figure 9.2: Optical based and Chest Strap Heart Monitor

Many manufacturers of optical based heart rate monitors state that their devices are meant for recreational and not medical purposes. For a runner or a cyclist, a wrist based monitor will have physical movements that can make measurements inaccurate. A study [4] has shown that optical based heart monitors will have some error when used in physical activities, but the low cost and convenience of these devices still make them appealing for the recreational cyclist.

A more accurate gadget to monitor heart rate, is the chest based sensor device (see Figure 9.2). This device has two parts - a strap around the chest with embedded sensors (electrodes) to detect electrical activity as the heart beats and a receiver (smart phone) to collect the data transmitted by the strap sensors. As the heart beats, pulsating electric waves are transmitted that are detected by sensors embedded in the strap. The sensors then transmit the data to a monitoring device or smart phone that can store or display the real time heart rate. This method of monitoring the heart rate is more accurate than the optical based method and usually more expensive as well.

Heart Rate Zones   The American Heart Association defines moderate exercise intensity as 50%-70% of HR_{max} and vigorous exercise intensity as 70%-85% of HR_{max}. The formula to calculate HR_{max} based on age may not be appropriate for everyone, since the experiment to derive the formula was conducted with a population of healthy men and women from a particular region. To find your unique HR_{max}, you would have to ride at maximum intensity on a long uphill till exhaustion. The assumption being that you will reach your HR_{max} at the end of this tough test.

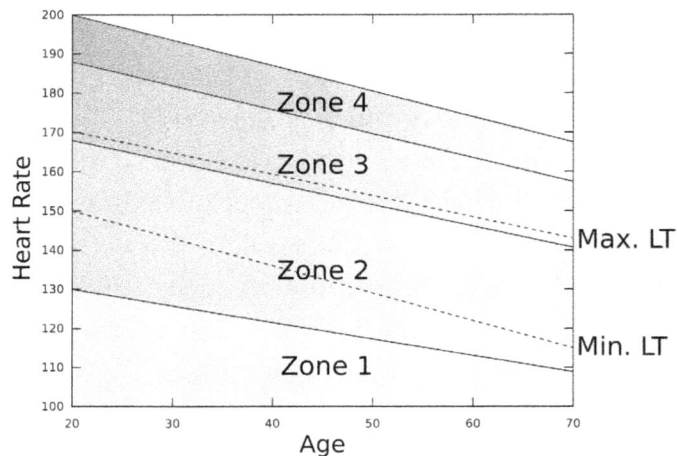

Figure 9.3: Heart Rate Zones and Minimum / Maximum Lactate Threshold vs. Age

Burke and Pavelka [5] define four zones based on HR_{max}. The first zone ends at roughly two thirds your HR_{max} - in this zone, you will be pedalling with ease and you are recovering or warming up. Fat is the primary fuel to generate energy in zone 1. The second zone is where your heart rate is between 66% and 85% of HR_{max}. At the lower end of zone 2, fat is still the primary fuel, with glycogen[19] as the secondary fuel. The quantity of glycogen stored in the liver and muscles is significantly less than the quantity of fat in the body and can be quickly (a few hours) used up. Heart rates at zone 3 cannot be sustained for long ($\simeq 1\ hour$) and heart rates at zone 4 are possible for an even shorter duration (in minutes or seconds).

As the demand for power increases with an associated increase in heart rate, the amount of energy produced aerobically (with oxygen) from fat exclusively is not sufficient to generate all the required energy. Even though you maybe breathing more heavily and inhaling more oxygen, the body cannot generate enough energy and supplements aerobic energy with additional anaerobic energy from glycogen.

---

[19]Glucose and carbohydrates are stored in the form of glycogen.

Lactate Threshold However, since glycogen stores are limited compared to fat, they can be depleted relatively quickly. A by-product from anaerobic energy generation is lactic acid or lactate. The lactate in the body is recycled by the liver, however if the production of lactate is high when energy demand is high, then the liver cannot recycle the lactate fast enough and it begins to accumulate in the blood. When the concentration of lactate in the blood is high, you have reached your lactate threshold (LT). The lactate concentration levels increase rapidly at the threshold level and you often have to stop and rest. Once you have crossed the LT, it does not take long before the body stops producing sufficient energy to continue at the same level of intensity and you are forced to stop.

If you could raise your lactate threshold towards the maximum level (see Figure 9.3), then you could ride with a higher heart rate for a longer duration. With a higher lactate threshold, at the higher end of zone 2 your heart rate is close to 85% of HR_{max}, but you are not yet exhausted.

9.1.4 Power Meters

A power meter fitted on a bicycle measures the power output at any given instant. Power can be evaluated at the pedals where you apply force to generate torque to turn the pedals. Most power meters use a strain gauge to measure the applied force (f in newtons). The electric resistance of a strain gauge reduces in proportion to the applied force. A cadence sensor attached to the power meter measures the angular velocity (ω in radians per second) and the radius (r in meters) of the crank is known. The product $f \times \omega \times r$ gives the power generated in watts. If you apply a force f of 100 N with a cadence is 70 rpm ($\omega = \frac{2 \times \pi \times 70}{60}$ radians / second) on a crank of radius r 0.17 m, then you would generate about 124 watts. This represents generate the power generated on one side and the total power generated is simply twice the value on one side or 248 watts (assuming that you generate the same power with each leg).

More expensive power meters may have sensors in multiple locations including both pedals. With multiple sensors, you can detect if one leg is generating more power than the other leg. Most power meters communicate with a device like a smart phone to transmit the measured data. You can also measure power at the hub of the rear wheel. The power measured here is assumed to be more accurate since it reflects the power transmitted to the rim of the wheel and losses in the drive train are not considered.

Power meters are still relatively expensive for the average rider and one way to check if you really need a power meter is to rent one. Although, you would want to use it multiple times to check if the power that you generate is changing (increasing or decreasing) over time. Smart phone apps that use a wind sensor, GPS location, and altitude will not be as accurate as a power meter on the hub. However, these apps and low cost wind sensors are an alternative for the cost conscious cyclist to get a reasonable estimate of power generated.

Power Zones The heart rate from resting to peak was divided into four zones with an increasing heart rate at higher zones. The power zones are similarly defined with a lower power at the initial power zones and the maximum power at the highest power zone. Like the heart rate zones, the high power zones are not sustainable for long. The lactate threshold level was defined in a range between 75% to 85% of HR_{max}.

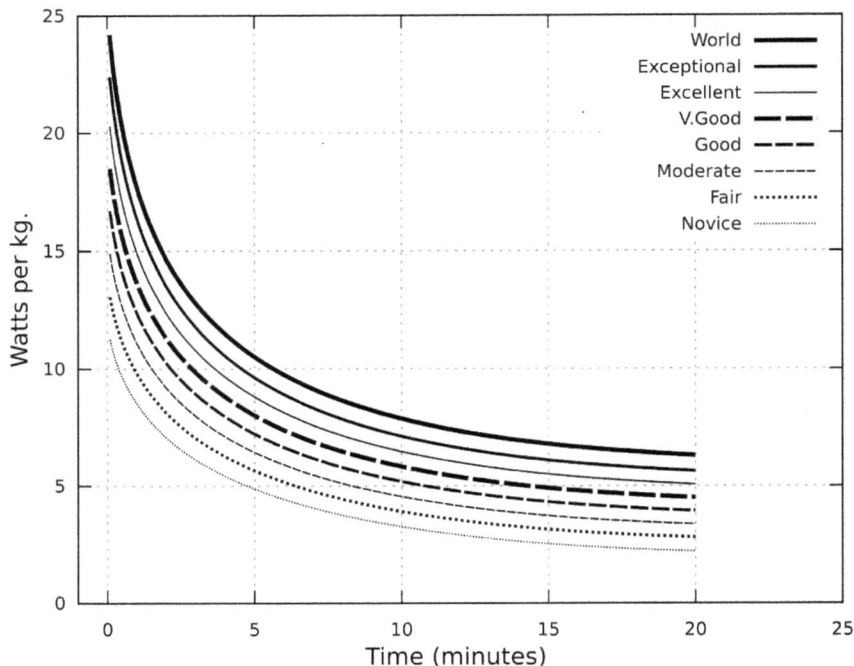

Figure 9.4: Watts per kg. vs. Time in minutes for different categories of cyclists

Instead of comparing raw power values, the power per kg. is used to compare performance. A heavy cyclist maybe capable of generating higher power than a lighter cyclist, but may still ride slower than the lighter cyclist, since the rolling resistance and slope resistance for the heavy cyclist will be greater. An average cyclist weighing 70 kgs. and generating 210 watts will be producing power at 3 watts per kg. A highly trained cyclist will have a much higher (5-6) watts per kg. than a novice cyclist. The watts per kg. of a cyclist will depend on the duration of power generation. All cyclists can generate a relatively high power for a very short period (seconds) which quickly diminishes for longer periods that range from minutes to hours.

Figure 9.4 shows the watts per kg. vs. time for eight categories of cyclists from world class to novice (from Dr. Andrew Coggan's data[6]). In all categories, the watts per kg. drops quickly in the first few minutes and stabilizes after 15 minutes. This is consistent with other sports like running where athletes can generate high power for short distances in the range of 100-400 meters. The power sustainable for a few seconds can range in thousands of watts [20].

The lactate threshold described as a percent (75%-85%) of HR_{max} has a power analog, functional threshold power (FTP), described as watts per kg. The change in heart rate and power

---

[20]Usain Bolt's power in the first 1-2 seconds of the 100 meters dash has been estimated at over 2600 watts.

generated are not always synchronized. Your heart rate may increase even though you are still generating the same power, since some of your energy maybe used to cool the body.

Your FTP is measured from the maximum power you generate for an hour. A FTP test may reduce the time to 20 minutes (due to the difficulty of finding a one hour route without any stops) instead of an hour. On a long uphill ride, you can ride as hard as possible to generate power for a period of roughly 20 minutes. After the test, you can find your average power for the period and divide by your weight in kilograms to get watts per kg. This value is reduced by 5% [7] for the FTP. If tests over a period of months show that your FTP value is rising, then you are either losing weight but still generating the same power or you are generating more power with the same weight. This is of course hard to accomplish. A trained cyclist will have a higher FTP similar to the higher lactate threshold described earlier.

Zones are described as a percent of FTP in a similar manner to the heart rate zones described as a percent of HR_{max}. The seven zones in Table 9.1 from Allen and Cogan [8] cover rides of all durations from the really short intense ride to the long endurance ride.

Zone	% of FTP	Ride Type
1	< 55%	Active recovery
2	56%-75%	Long endurance rides
3	76%-90%	Shorter tempo rides of 30 minutes or more
4	91%-105%	Threshold rides of 8-30 minutes
5	106%-120%	3-8 minutes at VO_2 max level
6	121%-150%	30 second - 3 minute effort to improve anaerobic capacity
7	151%+	< 30 seconds for neuro-muscular power

Table 9.1: Training Zones based on FTP Threshold

The seven zones of Table 9.1 reflect the curves in Figure 9.4 where a high percent of FTP can only be generated for very short durations. Since FTP is the average power over 20 minutes or longer, you can generate over 100% of FTP for periods less than 20 minutes. Training in zones 3 and 4 may increase your FTP while in zone 2, you will be riding at a steady pace to build a base. Although, there maybe many ways to increase FTP, each person may have a physical limit based on age and build.

Anaerobic vs. Aerobic   The anaerobic system by definition does not need oxygen to generate energy and is sustainable for short bursts of energy. The effort to extract energy from the anaerobic system is high and your heart rate rises to the zones 3 and 4 (see Figure 9.3). The anaerobic system can only be used for a short period (see Table 9.2 for the features of the two energy sources).

It would seem logical to train the energy source depending on the type of ride. So, an endurance ride and a sprint would mean training the aerobic and anaerobic energy sources respectively. Although the two types of energy sources, anaerobic and aerobic, are clearly defined, training in one or the other will have benefits for the other type. You will need energy from both types regardless of the type of ride and you cannot train exclusively for one source. Instead, you can prioritize one source over the other.

Source	Duration	Need Oxygen	Effort	Ride Type
Anaerobic	Minutes	No	90%	Sprint or Climb
Aerobic	Hours	Yes	60%-80%	Endurance

Table 9.2: Features of Energy Sources

To get the most of your training, Doyle [9] suggests mixing training intensities (see Section 9.1.6). Besides training both energy sources effectively, an added benefit is the shorter training duration. In a lab test (Wingate) of high intensity for about 30 seconds, while the anaerobic energy was the primary source, aerobic energy was also used.

However for an endurance rider, it may not seem to be necessary to train the anaerobic energy source, since it is largely not used. A study [9] of 174 cyclists over 30 years showed that while interval training could not prevent the loss of anaerobic power, aerobic power did not diminish as much. The steady aerobic energy source was sustainable for many years and explains why many cyclists in their 50s or 60s can still ride long distances.

One way of detecting whether you are using aerobic or anaerobic energy sources is based on your cadence. If you are pedalling at over 75 rpm without a lot of effort on each stroke, then you are using your aerobic energy sources. Once cadence drops below 60 rpm, you will most likely be generating more force for each stroke using anaerobic energy sources.

9.1.5 $VO_2\,Max$

While your body needs oxygen at all times, the need is greater when you are exercising. Muscles need more oxygen when they are generating power and $VO_2\,Max$ is a measure of how efficiently you can get oxygen from the air to your muscles. Even though a higher $VO_2\,Max$ is an indicator of superior aerobic capacity, it does not necessarily mean that the rider with a higher $VO_2\,Max$ will ride faster than a rider with a lower $VO_2\,Max$. A cyclist with a slightly lower $VO_2\,Max$ maybe able to sustain a higher level of power for a longer duration. Still, you would expect a cyclist in the world class, exceptional, or excellent categories to have a higher than average $VO_2\,Max$.

Measuring $VO_2\,Max$ is harder than measuring other parameters such as speed, power, or cadence. It is usually measured in a lab with a treadmill or a stationary bicycle. The required effort is steadily increased and oxygen intake is monitored. $VO_2\,Max$ is measured in milliliters per kilogram per minute. A cyclist with values of about 50 ml. / kg-min. would be considered *trained*, while any value above 70 ml. / kg-min. indicates a *highly trained* cyclist.

There are several formulas to measure $VO_2\,Max$ [10]. One formula is $Q \times VO_{2\_diff}$ where is Q is the cardiac output and VO_{2_diff} is the fraction of oxygen absorbed. The cardiac output Q (measured in liters / minute) is the product $HR_{max} \times SV_{max}$ where SV_{max} is volume of blood pumped by the heart in a single beat (measured in liters per beat). The fraction of oxygen absorbed VO_{2_diff} can increase from 0.05 at rest to 0.16 during an intense ride. To maintain uniformity and compare results between cyclists of different weights, the product $Q \times VO_{2\_diff}$ is divided by the weight of the cyclist.

The values for this $VO_2 Max$ formula are necessarily evaluated in a lab and an alternate simpler formula uses just HR_{max} and HR_{rest}. It is approximately $\frac{HR_{max}}{HR_{rest}} \times 15.3$. If your HR_{max} is 175 beats / minute and HR_{rest} is 54 beats / minute, then your $VO_2 Max$ is about 50 ml. / kg-min. The idea behind this formula being that if your HR_{rest} is low, then your cardiac output Q must be high and therefore $VO_2 Max$ is also higher. While $VO_2 max$ can predict overall physical fitness, it cannot predict the outcome of a race.

Increasing $VO_2 max$ A rider with high $VO_2 max$ will use oxygen more efficiently and therefore be able to generate more power than another rider with low $VO_2 max$. Short bursts of intense riding in intensity level training (see Section 9.1.6) increase your breathing rate to supply more oxygen to blood in the lungs that eventually is transported to the muscles that convert oxygen and fat into energy.

With training, the respiratory and leg muscles become more efficient and stronger to handle the higher power demands during long rides and short sprints. Suprisingly, it does not take many years of training to notice an improvement and weeks of training are sufficient to build stronger lungs.

Do you need a heart rate monitor or power meter?   The purpose of your training may presumably be to ride faster, become more fit, lose weight, or relieve stress. In any case, to get the most of your training, you need to measure the intensity of your training and not necessarily the duration or distance covered. Clearly, if your heart rate during your workout is in zones 3 and 4 in Figure 9.3, the intensity is also high.

A heart rate monitor is not necessary to realize that your heart rate is in the top two zones, since you will be sweating, breathing heavily, and able to say a few words at a time. This level of intensity is not sustainable for more than 5-10 minutes. However, a heart rate monitor will give you a more precise indication (numeric values) of the duration and heart rate. Besides, you can also evaluate how your heart rate is changing over time with a monitor.

You may notice that your speed varies on rides over the same route due to changes in weather conditions or fitness levels. A power meter will indicate if you are generating more power, even though your average speed may fluctuate. The data from a power meter is not subject to changes in weather. While smart phone apps will give an estimate of your power over a half hour to an hour, a power meter will give more accurate readings at shorter time periods (in seconds). These accurate readings are necessary to build a plot like the one in Figure 9.4 to identify a cyclist category.

Even though a ride may have taken longer due to a head wind, the results from a power meter may show that you are still generating power at the same or higher rate than before. Knowing your average power over some duration, you can use a power calculator to estimate the time to cover some given distance with an average gradient and head wind [11]. An alternative low cost but less accurate method is to use a smart phone to calculate power generation based on changes in the gradient, instantaneous speed, and wind direction (see Chapter 7).

9.1.6 Steady State or High Intensity Interval Training

While many advocate high intensity interval training [5] to raise the lactate threshold, cycling at a steady rate of 60%-70% of HR_{max} does have its benefits as well. You can ride at a steady pace which is a manageable, but still tiring over a period of an hour or more. The steady state form of training does not have the stress of keeping track of intervals or intensity and can even be relaxing. This maybe appropriate if you are not interested in generating more power or building muscle.

Not everyone has the time to ride at a steady rate over a long period, and studies [12] have shown that short intense workouts can be just as effective as a long slower workout. If you are riding to loose weight, then high intensity interval training (HIIT) was clearly better than steady state. However, HIIT is more intensive and does impose an additional strain on the body. Pushing yourself to exercise repeatedly at the top two heart rate zones can take a toll and you may lose interest in continuing to exercise at the same level of intensity.

Choosing which form of training – steady state or HIIT – may depend on your HR_{rest}. If your HR_{rest} is above 60 bpm, then steady state training is suggested over HIIT [12]. The idea being that a high HR_{rest} is associated with relatively poor aerobic function and therefore first building up aerobic performance through steady state training is preferred before HIIT to generate power. HIIT improves anaerobic performance since higher intensity levels of exercise generate energy from both carbohydrates and fat.

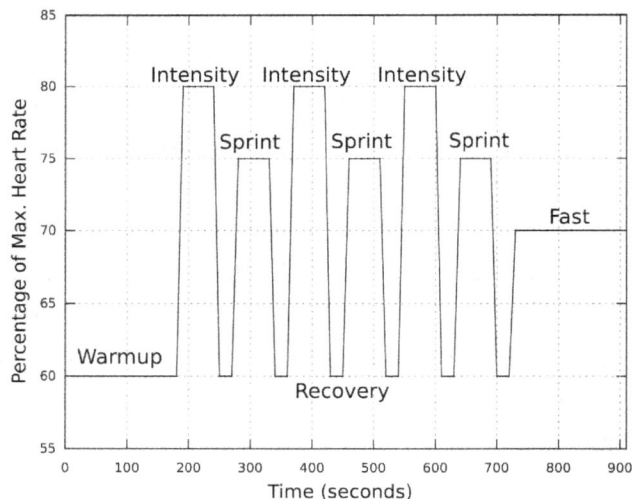

Figure 9.5: A High Intensity Interval Training Schedule

While the shorter duration of HIIT to gain fitness is probably very appealing, there is no universal formula for the length of the steady state and HIIT periods and how you should alternate between the two. If you spend too much time exercising at a high intensity, then you will be stressing your body excessively and it may lead to fatigue or joint pains. At the same time, if you do not exercise at a high intensity long enough, you will not get the benefits of HIIT.

In Figure 9.5 a schedule [13] for HIIT over 15 minutes shows one schedule for alternating between intense and recovery periods during a ride. The ride starts with a three minute warmup period followed by three high intensity periods of one minute each with a 30 second recovery and a minute sprint between. In the 15 minute workout, a total of three minutes is spent in each of the five types of rides - warmup, intensity, recovery, sprint, and fast.

Following a rigid schedule like the one shown in Figure 9.5 is possible on a stationary bicycle, but you may not find a route that will fit the schedule, given traffic and roads in a city. Still, the purpose of the workout can be achieved by using the schedule as a guideline without adhering strictly to the times specified.

Why does interval training work? A repeated schedule of high intensity work followed by recovery is better than simply trying to ride at the fastest possible pace for a half hour or more [14]. You actually spend more time riding hard on a HIIT schedule. The schedule shown in Figure 9.5 is a sample schedule to build muscles for sprints over relatively short distances. Over short distances, the fast twitch muscle fibers generate the power to ride with short bursts of speed, while over long distances, the slow twitch muscle fibers generate less but sustainable power for longer durations. The HIIT schedule for endurance training besides having an overall longer period, will also have intensities that last for a few minutes instead of a minute or less for sprints.

Aside from riding more efficiently, HIIT also has an added benefit of reducing cancer cell growth [15]. Although, increased physical activity has been associated with extending the lives of cancer patients, this study [15] reported the changes in colon cancer cell count before and after HIIT. Following a HIIT session, researchers found a significant reduction in the number of colon cancer cells compared to the number before exercise. The study suggests that the acute nature of HIIT is more beneficial than a long term low intensity exercise routine. However, the benefits of the reduced number of cancer cells following HIIT did not continue beyond two hours. Even though the advantages of HIIT did not last very long, repeated HIIT sessions may limit the growth of cancer cells. Further, there are few negative side effects from physical activity compared to medication.

Length of a Training Ride Marathon runners use the 20 mile limit as a marker of preparedness for the run. This would amount to slightly over 75% of the total distance. While completing 75% of a 200 km. ride will definitely give you the confidence to complete the entire distance, it is not necessary to complete a 150 km. training ride before attempting a 200 km. ride.

One reason that a cyclist may not attempt a long training ride is the time to finish the ride. Carmichael [16] points out that long training rides are a small component of riding faster. Improving fitness and power output is possible with shorter training rides based on HIIT.

However, the experience of riding at least 1 or 2 long rides will identify sources of problems such as saddle pain, hydration, and improper clothing. On a long ride, sitting on a saddle for 10-12 hours without any back support can be an ordeal besides dealing with other issues such as nutrition and fatigue. Carmichael suggests not worrying too much about covering long distances, since beyond the experience, they add little in overall fitness.

Over Training A little motivation to train is good, but training excessively can lead to injury and may end up doing more harm than good [17]. The reasoning behind over training is that if riding 50 kilometers is good, then riding 100 kilometers must be twice as good. However, the benefits of riding longer and longer distances may not leave enough time for the body to recover.

Some riders are quickly able to cope with the higher demands from longer distances, while others take longer to recover and therefore never fully recover from a long ride. The fatigue from one long ride accumulates in the next long ride and over time this tiredness builds up to the point that the motivation to ride any distance is depleted.

One way to detect if you are over training is to monitor your heart rate at the beginning of every day. If it stays elevated compared to your heart rate before training, then your body is not recovering fast enough to resume training at the same intensity. Other symptoms of over training include poor sleep, constant muscle soreness, and lack of appetite.

9.1.7 Tips to Ride Fast

A search on the Web for tips to ride fast reveals a number of methods, plans, and training techniques. Many of these techniques will improve the speed of a novice rider, but you may have physical limitations beyond which you cannot ride faster, no matter how hard you train. Since power is directly proportional to speed, you can only increase your speed by generating more power (ignoring other factors such as a lighter and more aerodynamic bicycle).

While HIIT is appealing to the impatient cyclist, since it is does not take much time, a strong base is recommended before riding fast. One schedule [18] is based on three types of rides in a week. The first ride is a base building ride that lasts about 2-3 hours. It is a steady paced ride in zone 2. These rides are aerobic rides that are mainly used to train muscle fibers to burn fat and preserve glycogen for a ride in zone 3. Your body will also become better at processing lactate that accumulates during a ride. If the lactate is removed at a rate faster than the rate at which accumulates in the blood, then you will not approach the lactate threshold.

The main purpose of the base building ride or long slow ride is to train the body to use fat as the energy source and conserve high energy sources like carbohydrates for situations where high power is needed. A ride in zone 2 is not necessarily an easy ride, since your average heart rate should be about 75% of HR_{max}. Without a heart monitor, you can identify your zone based on your cadence and pedalling effort. If your cadence is high (> 75 rpm), every pedal stroke does not take a lot of effort, and you are not out of breath, then you are most likely in zone 2.

The second type of ride is the cruise control ride or a ride that is faster than the base building ride, but not sustainable for long (more than hour). This ride alternates between sub-rides of 10 minutes or more of higher intensity (80% of HR_{max} or near zone 3) riding and 10 minutes of easy pedalling in zone 2.

The last type of ride is more intense than the second type of ride, but of a shorter duration. The ride alternates between a maximum power effort for a minute or so and a few minutes of recovery at a relaxed pace. The purpose of this ride is to raise your lactate threshold, the heart rate at which lactate begins to accumulate in the blood, beyond which riding becomes very hard.

9.1.8 Climbing Fast

A climb slows down all riders and it takes more power to ride uphill than on a flat at the same speed. If you ride at about 25 kmph on a flat with 200 watts of power, then your speed will drop down to 15 kmph. on a uphill of gradient 0.04 with the same power (see Chapter 3). There is plenty of advice on the Web on climbing faster and what works for you may not work for others and vice versa. For some riders, climbing comes naturally. You can see them gliding uphill almost effortlessly, while the average rider struggles to keep pace and rocks the bicycle back and forth to gain some distance. Watching an efficient climber pedal smoothly and methodically, an uphill ride may even seem deceptively easy.

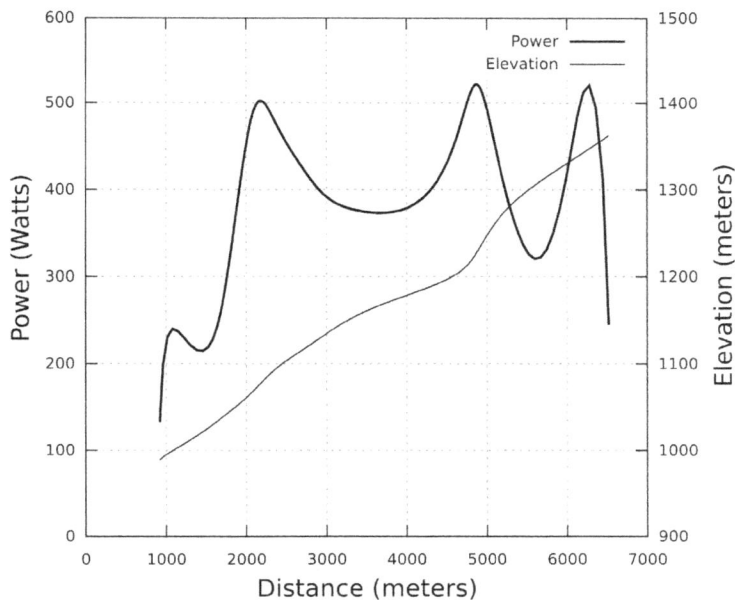

Figure 9.6: Power in watts and Elevation in meters vs. Distance in meters

You can climb faster on a lighter bicycle, but the increase in speed is not as much as you would expect. If the weight of the bicycle and rider are 10 kgs. and 70 kgs. respectively and

313

you can generate about 172 watts of power on a slope of gradient 0.04, then your speed would increase from 15 kmph. to about 15.75 kmph. on a bicycle that weighs 5 kgs. for the same 172 watts of power. To sustain the same speed of 15 kmph., the required power on a 5 kg. bicycle would be reduced by about 10 watts compared to a 10 kg. bicycle from 172 watts to 162 watts.

Most riders will not be severely challenged with a short climb of a few hundred meters (horizontal) that is fairly common in a city. A long climb of several kilometers will test any rider, since there is no recovery period for at least 10 or more minutes. Riding uphill at a steady pace for a half hour or more is hard. Accelerating uphill is even harder and is possible when you are riding at a high cadence of 80 rpm or more.

How much power do you need on a climb? At the start of a climb, it maybe tempting to ride in a high gear to build up speed. Inevitably, if the climb is longer than a few kilometers, it becomes difficult to sustain the pace. Instead, the recommended method is to start a climb in a relatively low gear at 70 rpm and then gradually work up to a higher intensity near the end of the climb.

The power required for a climb does not uniformly increase with time and distance (see Figure 9.6). A 6 km. climb with an average gradient of 0.05 at 10 kmph from an elevation of 1000 meters to 1360 meters requires power in the range from 150 to over 500 watts. The calculated power values is smoothed to show a pattern of changing power demands. Power of 400-500 watts is significantly above the power that an average rider can generate. Needless to say, sustaining this level of power generation for more than several kilometers is hard work.

Measuring Ascent After a few rides and estimates of speed, you will have a rough idea of the power that you can generate. On a climb, you will be riding in the both the horizontal and vertical dimensions. The gradient is the ratio of the vertical (rise) to the horizontal (run). A gradient of 4% would mean that you would ride 100 meters horizontally for a 4 meter climb vertically. A gradient can be negative (for a descent) or positive (for an ascent).

Horizontal speed is measured in kmph. (or meters per second) and the equivalent vertical speed is similarly measured in vertical meters ascended per hour or VAM (an abbreviation for an Italian term that translates to average ascent speed). A gradient factor [19] proportional to the grade percent is computed as $2 + \frac{grade\%}{10}$, so a 6% gradient translates to a 2.6 gradient factor. The VAM for a given rider is then calculated as the product $pw \times gradient\ factor \times 100$ where pw is the power-to-weight ratio (see Chapter 3).

The figure 9.7 shows the VAM for the power generated by a 64 kg. rider. The gradient ranges from a low of 2% to a high of 8% and power output ranges from 100 watts to 400 watts. The slopes of the 2% and 8% gradients appear to be deceptively small and not significantly different. It takes about 97 watts for a 64 kg. rider on a 10 kg. bicycle to ride up a 2% slope at 15 kmph. However, if the slope is increased to 8%, the same rider has to generate 288 watts to sustain the same speed of 15 kmph. The required power to ride at the same horizontal speed increases by 3

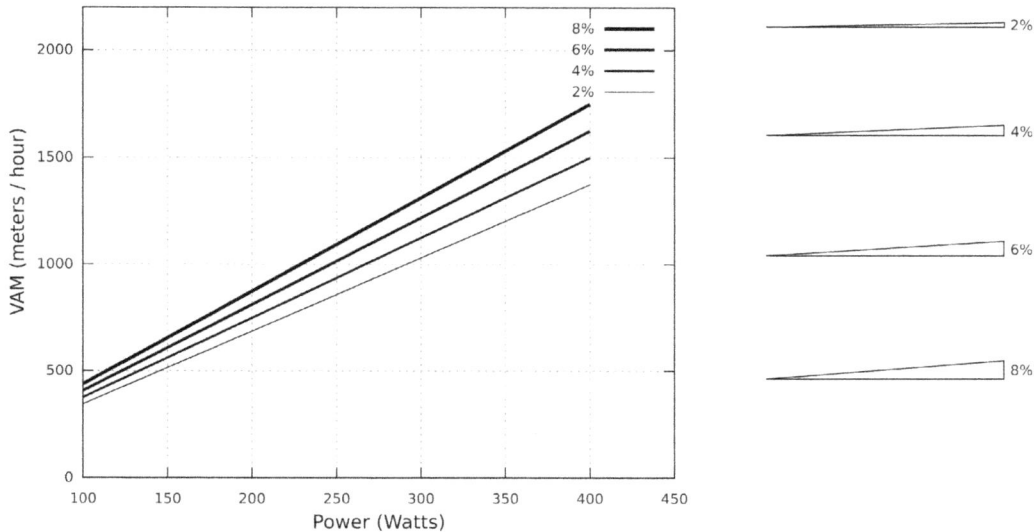

Figure 9.7: Power in watts vs. VAM in meters

times for a 2% gradient compared to an 8% gradient, while VAM increases fourfold from 300 to 1200 vertical meters per hour.

Gradient and Power Usage As the gradient increases, more power is used to climb uphill than to ride horizontally. Therefore VAM increases for the same power when the gradient is higher. A cyclist generating 200 watts of power and with a total bicycle and rider weight of 80 kgs. will cover different vertical and horizontal distances depending on the gradient.

At 200 watts with a zero percent gradient, a rider will cover about 32 kilometers in an hour assuming no head or tail wind and an aerodynamic factor of 0.23 (see Chapter 3). When the gradient increases to 2%, the horizontal distance covered in a hour drops from 32 kilometers to 23.3 kilometers and the vertical distance covered increases from 0 to 467 meters. At a gradient of 8%, an even larger fraction of the 200 watts is used to climb 800 meters vertically and the horizontal distance covered decreases to just 10 kilometers (see Figure 9.8).

Why is a higher cadence better? A common mistake that some riders make on a long climb is starting the climb at a high intensity with a relatively low heart rate. Riding at a high level of intensity means that lactate will begin to accumulate at a faster rate than the body can remove it. Once you have crossed your lactate threshold, you will soon have to slow down to a crawl and lower the intensity substantially.

The idea behind riding at a high cadence (> 70 rpm) is that you can sustain a lower level of power for a longer time, given that you are riding at a lower gear that corresponds to the power required. At a low cadence, you are probably riding at a high gear and applying more force. The

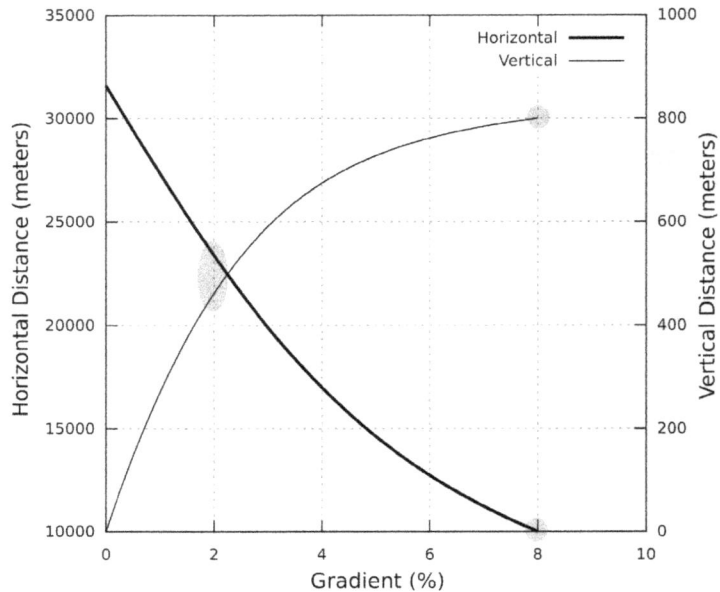

Figure 9.8: Gradient Percent vs. Horizontal and Vertical Distance

fast twitch muscle fibers must supply a larger fraction of the energy required for the higher force and quickly get depleted of the needed glycogen in the muscles.

A study [20] showed that fast twitch muscle fibers lost about 50% of their glycogen at 50 rpm and just 33% at 100 rpm. As fast twitch muscle fibers lose glycogen, they become less efficient at contracting to generate the required force. Therefore more muscle cells are needed leading to more oxygen consumption and lower efficiency. This assumes that a lower rpm corresponds to riding at a higher gear.

Finding the right gear and cadence also depends on the gradient. On a 4% gradient, if you can generate 200 watts, then you can ride at about 75 rpm with a gear ratio of 1.79 (see Figure 9.9). At a gear ratio of 3.12, your rpm would drop to less than 45. The figure also shows that power required increases more rapidly with cadence for higher gear ratios.

When the gradient increases from 4% to 8% the required power to sustain 75 rpm with a gear ratio of 1.79 increases by about 50 watts to 250 watts. At higher gear ratios, the required power increases by a larger amount with an increase in gradient. If you can generate 400 watts on a 4% gradient with a gear ratio 3.12 at 70 rpm, then you would need 500 watts on a 8% gradient at the same gear ratio and rpm.

The "Optimal" Cadence   The benefits of riding at a high cadence (> 100 rpm) will diminish past some threshold. You maybe spending more energy moving your legs up and down than applying power to the pedals. The ideal cadence for a cyclist will depend on the power generated and the gradient. Knowing your unique optimal cadence can prevent situations where you experience exhaustion and have to slow down or stop.

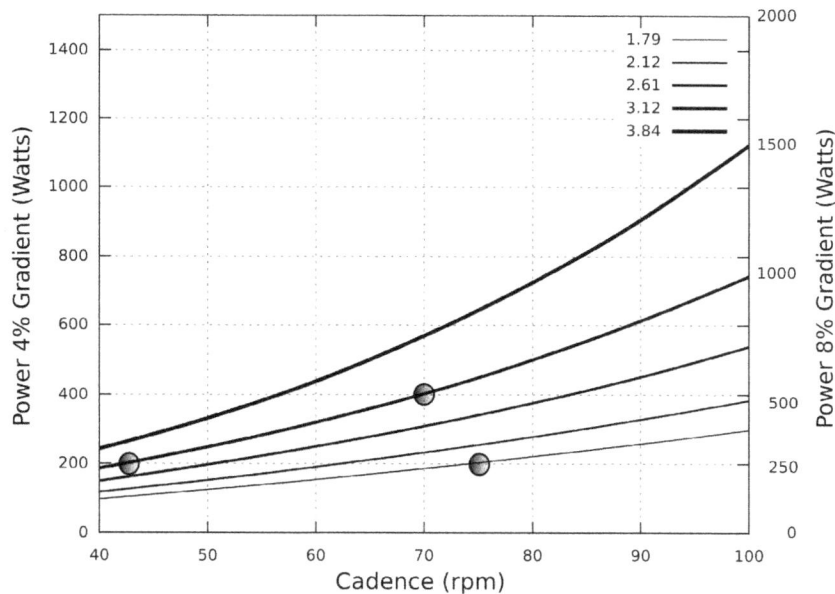

Figure 9.9: Cadence (rpm) vs. Power (watts) for a 4% and 8% gradient

A study [21] evaluated a range of cadences with the corresponding neuro-muscular fatigue and found that the minimum fatigue threshold was in the cadence range of 80-90 rpm. However, one factor that may limit the average cyclist from riding at that cadence range is $VO_2\ Max$ (see Section 9.1.5). The same study found that the oxygen uptake or $VO_2\ Max$ was the least for cadence in the range of 60-70 rpm. Since professionals have a higher $VO_2\ Max$, it seems logical that a professional with a large oxygen uptake can sustain a high cadence for a long duration. In general, professional cyclists ride at a higher cadence than the average cyclist. One explanation is that professionals can generate more power per pedal stroke and therefore can ride with both a high gear ratio and cadence.

Finally, the best cadence for you maybe the one you find through trial and error. The three parameters that a cyclist can control are the gear ratio, cadence, and power output. One method would be to start at a low gear and high cadence and gradually increase to a higher gear till the power to sustain the cadence is excessive. The optimal cadence may also depend on the duration of the ride. On an endurance ride of more than an hour, you would ride at a lower cadence than on a short duration ride. Similarly on a climb of several kilometers, your cadence would be lower than on a short sprint of half kilometer.

9.2   Endurance Cycling

If you ride a bicycle for fitness or to commute and *don't* enjoy it, then you will quickly seek an alternative. But if riding a bicycle has been fun, then the next step would be ride to longer distances. At first, riding distances of 25, 50, and 100 kilometers will seem arduous and not within an average rider's capability. But as you ride further you will notice that your limits grow along with confidence in your abilities.

This does not imply that endurance cycling is easy. Actually, it does become harder as the distances increase. One of the reasons endurance cycling is popular is because riding further than what seemed possible is a challenge. For some riders, the challenge is not the distance but covering a relatively long distance (100+ kms.) in a shorter duration compared to the time to cover the same distance on a prior ride.

The first time you cover 100 kms., it may appear hard. By the second or third time you ride the same route, it no longer appears as hard. Your body is adapting to riding long distances. Still, it is not likely that you will ride faster and faster each time you cover the same distance. Although riding distances of 100, 200, or 300 kilometers may seem daunting, it is easier to cover distances on a highway than to ride the same distance in a city. You don't have to brake as often on a highway and it takes less effort to maintain a steady speed than to repeatedly start and stop riding on city roads.

Wind can be your friend or foe depending on its direction. A tail wind may initially lead you to wonder why riding suddenly seems much easier than on other days. You may also be deceived into thinking that your fitness level has improved. When the wind or you change direction, the truth will be obvious. The opposite, a head wind, will slow you down and unfortunately on a highway there is little that can be done to counter a head wind. It may not be feasible to change directions and there maybe few obstacles to block the wind.

Some cyclists have a natural talent for riding fast and no matter how hard you try, you may not be able to duplicate the speeds and distances of a professional. Training wisely is key to getting the most of you physical capabilities. The two main purposes of training are -

- to build a base to supply fuel at a steady rate from your aerobic energy sources (mainly fat) that will last many hours

- and to experience riding a bicycle for many hours which will stress parts of your body far more than on a shorter ride like a commute

9.2.1 Fuels

The two main sources of fuel to generate muscular energy are fat and carbohydrate. In an endurance ride, different proportions of these fuels are used depending on the intensity of the ride. If you ride at a high intensity (see Figure 9.10), most of your fuel source is from stored carbohydrate (glycogen) and the remainder from fat. When your intensity is low to medium, the source of fuel is reversed and a large fuel fraction is obtained from fat with much less from carbohydrates (see Figure 9.11).

Figure 9.10 shows the number of stored calories of fat and carbohydrates for an average person. Although the number of stored calories is unique to each individual, all riders will have a much smaller (about 3%) store of carbohydrates compared to fat [5].

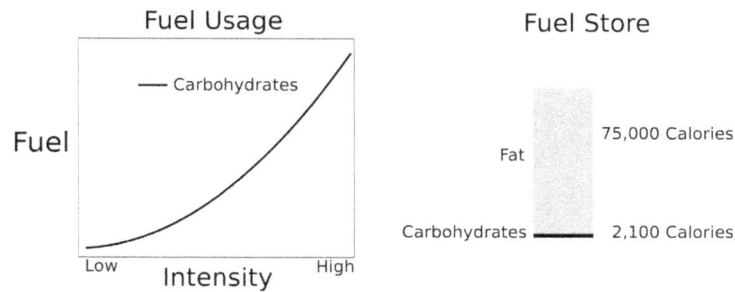

Figure 9.10: Fuel Store and Usage in a Long Distance Ride

Burning Fat   Since your store of fat is more than enough for a long ride, you can increase your glycogen or stored carbohydrate level by "cabrohydrate loading", a few days before you begin the ride. There are several formulas on the Web for "carbohydrate loading" and the essential idea is to make close to 70% of consumed calories from carbohydrates alone. One caveat to this formula is that if you consume carbohydrates at the same rate as a professional cyclist and don't ride the same distances, then any excess carbohydrate will be converted to fat, which is already in abundant supply.

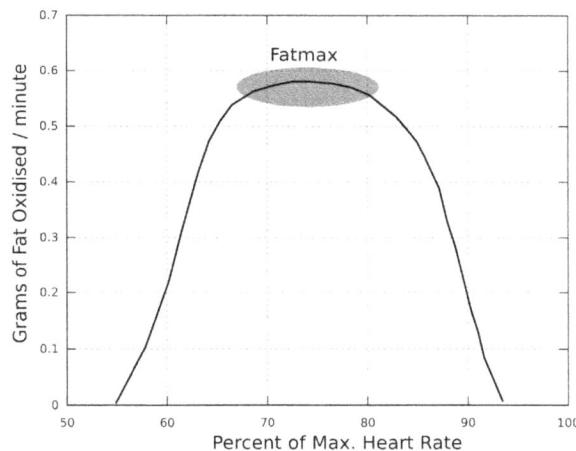

Figure 9.11: Percent of HR_{max} vs. Grams of Fat Oxidised per minute

Since fat is the most abundant fuel in the body, you would like to use it over a limited supply of carbohydrate. However, the usage of fat is dependent on the intensity of the ride, that we can measure as a percentage of maximum heart rate. Initially, you use more fat as your intensity and heart rate increases (see Figure 9.11) [21]. When your heart rate increases to over 80% of HR_{max}, fat is used much less and is almost zero at very high intensities.

The number of grams of fat that you oxidise reaches a plateau at about 0.6 grams per minute when your heart rate is between 65% to 85% of HR_{max} (or heart rate zone 2). This region of

---

[21]The figure also shows why it is hard to lose fat through exercise alone. If the oxidisation rate is 0.5 grams per minute, then it takes about 33 hours of exercise to lose one kilogram of fat.

maximal fat oxidisation called "Fatmax" [22] is the optimal intensity to use fat as the fuel for cycling. The heart rate corresponding to this region is not identical for all riders. A trained rider may have a "Fatmax" region in a higher range with the peak at a higher percent of HR_{max}.

Bonking Running out of fuel is something a cyclist would not like to experience and it is the physiological equivalent of a flat tyre. You are forced to stop riding. Although, you can replace the tube in a flat tyre fairly quickly and continue riding, it takes longer to build up your carbohydrate levels to start riding again.

This situation maybe appear to odd, since it is very unlikely that you will have exhausted your fat fuel supply. Even though you have more than enough fat to continue riding, it is not possible to burn anymore fat. A study [23] proposed an explanation for this paradox when the body is unable to tap into the abundant fat supply in the absence of carbohydrates.

The authors claim that current training methods emphasize carbohydrate usage over fat, before and during an endurance ride which inhibits the use of fat. Instead they suggest the opposite. They recommend lowering carbohydrate and increasing fat intake over a period of several weeks to better utilize fat. The study mentions that the shift from fat as a major part of human diet to a mix of carbohydrates and other food types followed the advent of agriculture. The human body had adapted to fat as the primary fuel over a period of a million or more years and the transition to a mixed diet occurred more recently.

One example the authors cite is the observation of Arctic explorers who travelled a few thousand miles in the company of Inuit families, living on a fat diet alone. While walking in cold weather is not the same as endurance cycling, it still requires a lot of energy that must come from the fat store of the body. The explorers noted that their bodies adapted over a few weeks to an almost exclusive fat diet and did not suffer any nutritional deficiency. The authors conclude that after a month on a zero carbohydrate diet, the body adapts to using fat as the exclusive fuel for exercise.

The debate over which diet (low or high carbohydrate) is better will continue. Experienced endurance riders [5] continue to recommend a high carbohydrate diet and a cyclist could experiment with each diet to see which is better. Further, current sports drinks have far more carbohydrate than fat along with other elements.

9.2.2 Stress

Riding a bicycle for more than 2-3 hours does stress the body and the minor aches / pains on a shorter ride will become far more noticeable and could limit your ability to ride. The first step is to ensure that the bike fits you (see Chapter 2). Primarily, the frame size should NOT be larger than necessary, forcing you to stretch your arms to reach the handle bar. The effort to reach the handle bar on a large frame can cause a back ache which may become more painful on longer rides.

Secondly, you want to ride with the saddle at a height that is most comfortable. When the saddle is too low, you will stress your knees and when the saddle is too high, you will be rocking back and forth on the saddle. Both of these issues - the wrong frame size and saddle height - are easily detected before starting a long ride.

On a long ride, there are a number of places in your body where you may notice pain. Figure 9.12 shows some of the common complaints from a survey of over 600 riders [24]. The top three problems faced are - the saddle, upper back (including shoulders and neck), and hands.

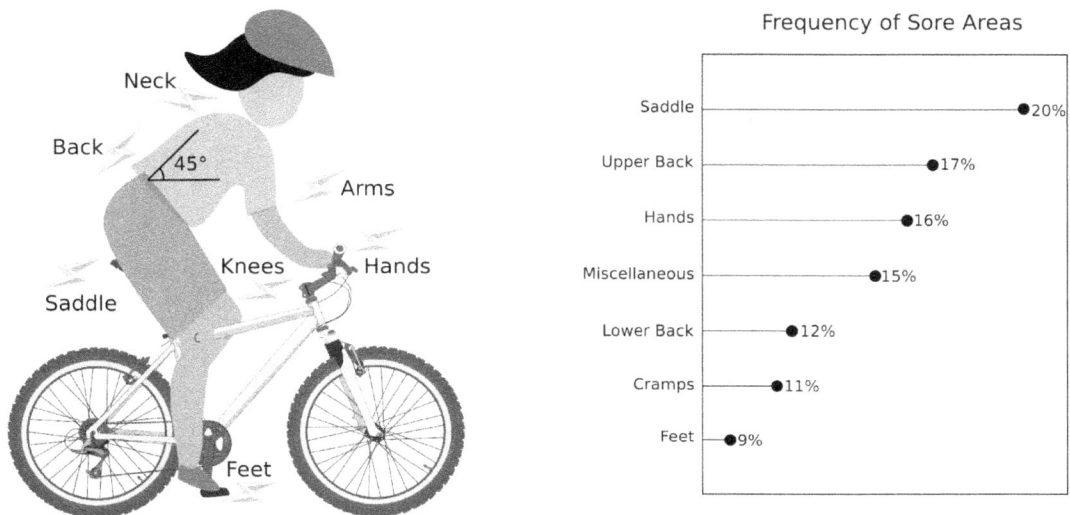

Figure 9.12: Contact and Pain Points on a Long Ride

9.2.3 Soft or Hard Saddle?

Two sit bones support your weight when you sit on the saddle. The other contact points on the handle bar and pedals may support some of your weight, but the saddle where the "sit bones" make contact must support the majority of your weight. The pressure due to your weight can reduce blood circulation in the skin leading to pain.

A first reaction to solve this problem would be to use a soft saddle with a lot of padding. Unfortunately, this does not work. A thin layer of padding will absorb some of the vibrations and cushion the "sit bones". If the padding is excessive, then the sit bones will sink into the seat and most of the pressure will be distributed to the crotch (the area between the "sit bones") leading to more discomfort. For short rides this discomfort may not be noticeable, but on a longer ride it does become a painful issue.

When deciding on a saddle, another issue to consider is whether a narrow or wide saddle is better. In general, road bicycles have narrower saddles than touring bicycles. A narrow saddle is appropriate if you will be leaning forward most of the time and your sit bones are not as wide.

If your sit bones are wide, then you are likely to feel the pressure of your weight in your crotch (see Figure 9.13). Some saddle makers add a relief channel in the center of the saddle to lower the pressure felt in the crotch. The idea behind this type of saddle is that the area around the sit bones is less sensitive to pressure than the area around the crotch.

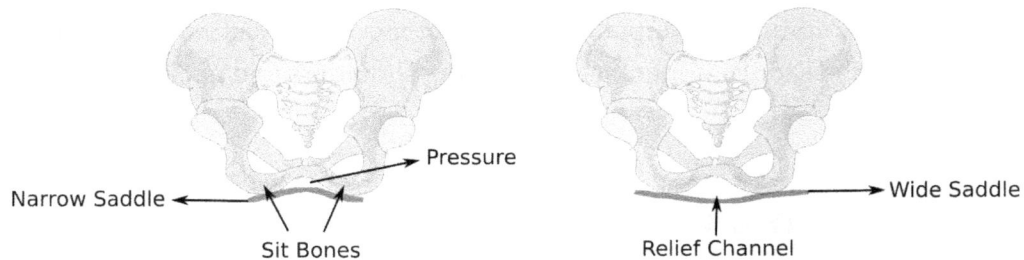

Narrow Saddle ← | Pressure | Sit Bones | Relief Channel | → Wide Saddle

Figure 9.13: Wide vs. Narrow Saddle

The other causes of pain can be the constant rubbing of your thighs and the saddle, a pedalling style with the heel too low, or the saddle tilted up / down. A saddle sore will form as a result of the increase in temperature, friction from the constant pedalling motion, moisture from sweat, and reduced blood flow. What begins as a small sore can quickly grow unless treated. Managing pain due to saddle discomfort is a complex issue since riders comes in all sizes and shapes and there is no single unique saddle that is the best for all riders.

A saddle sore is also similar to a bed sore that forms after prolonged pressure on a certain part of skin. Several suggestions to avoid saddle sores include periodically moving back and forth on the saddle, occasionally standing and pedalling, and sitting slightly above the saddle when riding over bumps.

Adding springs to the bottom of the saddle is another option to improve comfort. When you sit upright, the springs will absorb some of the road vibrations and this type of saddle is most appropriate for touring bicycles and leisure riding. On a road bicycle, a rider tends to lean forward and distributes more of the weight (see Figure 9.14). The springs that are typically located in the rear of saddle will sink and the front of the saddle will press against the crotch. This of course leads to the same problems as the soft saddle.

Issues with the saddle are the most frequent problem that riders face. With every pedal stroke, there is friction as your leg moves up and down, besides the pressure on your crotch. There is a lot of advice [5] on how to avoid saddle sores on the Web and you may have to find the best saddle for you through some trial and error.

One recommendation for an appropriate saddle is based on your riding style [25]. For a commuter or a long distance rider, a wider saddle is recommended and you will be sitting more towards the rear of the saddle. In contrast, a racer will be positioned on the forward half of the saddle. Saddles come in a variety of sizes and shapes, but in general the wider and narrower saddles are designed to minimize pressure zones for the commuter and racer respectively.

Commuter   Relaxed   Aggressive   Racer

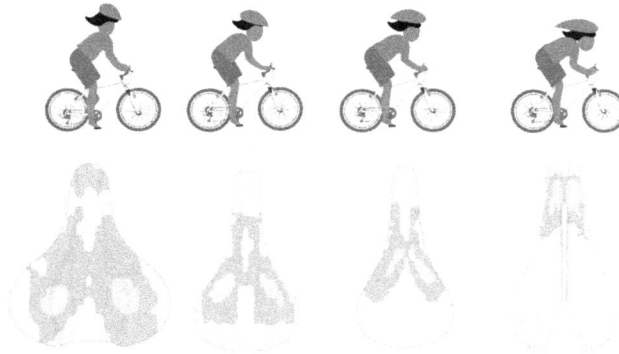

Figure 9.14: Pressure Zones on the Saddle based on Riding Style (courtesy Bontrager)

9.2.4 Neck and Back

As you tire, you will tend to let your head droop which is risky when you have to watch the road ahead. Also, keeping your head at the same angle for many hours will make your neck muscles stiff. Suggestions [5] include tilting your head from side to side to stretch and relax neck muscles.

You will be riding without any back support for many hours and unless you have trained, it is likely that at some point your back may begin to ache. A back position at about 45° (see Figure 9.12) is considered an efficient angle [5]. One benefit is that some of your weight is shifted forward and this decreases the pressure on the bottom muscles. The other benefit is better usage of the leg muscles.

However the more upright you sit, the larger the air drag (see Chapter 3) which increases with the square of the velocity. If speed is a priority, then clearly you would want to reduce air drag, which would mean lowering your back to a smaller angle than 45° to reduce your frontal area. If back pain becomes a real problem, then recumbent bicycles are an alternative. There is much less pressure on the back and your entire back has some support. Sitting reclined reduces the pressure on the discs in the spine compared to the more aggressive posture on a road bicycle.

Unfortunately with fatigue, it becomes harder for your muscles to maintain an optimum spine posture. Strengthening your core muscles before a long ride can reduce the likelihood back pain. The core muscles are a group of muscles in your mid-section including the front, back and sides. There are many ways to build your core muscles and one of the popular methods is called the plank [26]. If you ride at a gear that is relatively high for an endurance ride, you maybe riding fast but you will be using your lower back muscles to keep your mid-section steady. Without a strong core, the lower back muscles will begin to ache after a few hours.

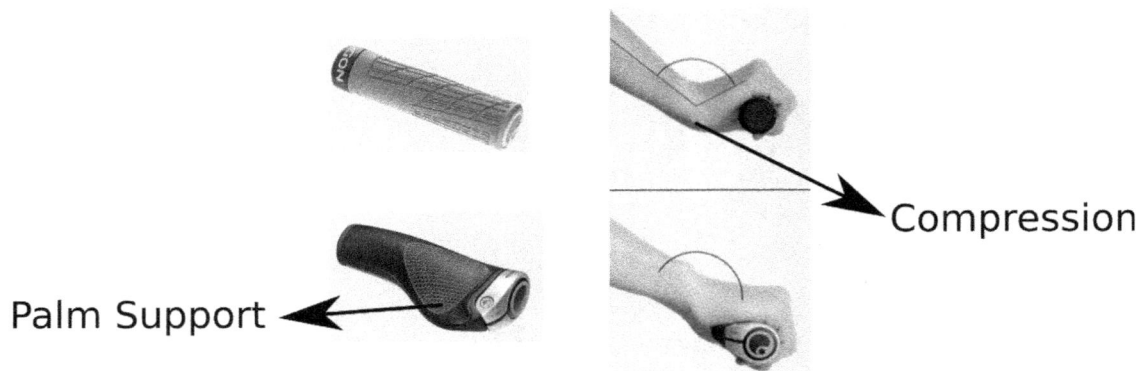

Figure 9.15: Grips and Hand Angles to Avoid Numbness (courtesy Ergonbike.com)

9.2.5 Arms and Hands

Since you have to steer your bicycle for many hours, your arms and hands can ache as well. Occasionally, you maybe able to ride hands-free, but on a long ride you will be spending many hours holding the handle bar. If your elbows and arms are stiff, then not only will your arms absorb the shock from bumps on the road, you will overuse your arm muscles.

When you hold the handlebar for many hours, you may end up with calluses on your hands from the repeated contact and pressure. A simple solution is to use gloves which reduce the friction between your hands and the handlebar and also absorb some of the vibrations. The padding in some gloves can reduce road vibrations. Another reason to use gloves is get a better grip of the handlebar, if your hands sweat.

Other options include using a winged grip on the handlebar. The pressure from holding the handlebar is distributed across a larger area over the palm and fingers (see Figure 9.15). This may prevent a nerve from being pinched. The hand has little fat and muscle to protect nerves from the pressure due to a part of your weight and the grip on the handle bar.

Another injury is due the angle of the wrist as shown in the upper image. The nerves in a bent wrist are compressed and over time the pressure surrounding the nerve will cause a loss of sensation or numbness. A winged grip can avoid this problem by supporting your palm to keep the wrist in line with your forearm and hand.

Other problems with holding the handlebar, include pressure on the wrists and palms of your hand. When the nerves in your palms are pinched, you will feel numbness in your hand. In normal circumstances, you could massage the numb area to restore blood circulation. On a bicycle this is a little harder, since you do need to hold the handlebar with at least one hand. When you need to brake, numbness in your hand will make it a little harder to control the bicycle.

9.2.6 Cramps

If you've experienced cramps on a bicycle, then you know how it suddenly appears when you have been riding for a few hours. The sharp intense pain forces you to slow down or stop. Although cramps do not last very long, the experience is a cautionary signal from your body that you are pushing the limits of your muscles. A cramp occurs when a muscle suddenly contracts and you will feel a lump at the location of the pain.

Causes of Cramps   One theory to explain why cramps occur is that on a long ride you have lost fluid and associated chemicals through sweat and therefore you need to replenish both to maintain a chemical balance in your body. This seems logical since sweat is clearly visible and you can even see crusted salt on the faces of some riders. Besides sodium, other elements such as pottasium, calcium, and magnesium are present in sweat in much smaller quantities.

Research results from sports drink companies showing that a sports drink can prevent cramps should be taken with a grain of salt. Clearly a sports drink containing water, electrolytes and carbohydrates is refreshing and useful for a tired rider. But whether a sports drink alone can prevent cramps is doubtful. A study [27] showed that cyclists who suffered cramps and those who did not, had similar levels of dehydration and blood minerals. Still, replenishing lost electrolytes does help in preventing other issues such as mental confusion and dizziness.

Another reason to doubt that cramps are caused mainly by a chemical imbalance is that sweat has more water than electrolytes, i.e. the concentration of electrolytes in less in sweat than body fluid. Therefore as you sweat, you do lose water, but the concentration of electrolytes in your blood will increase. Since these changes in electrolytes and water depletion occur in the entire body, any muscle in the body should be equally likely to cramp. But the muscles most likely to cramp are the ones that you have being used excessively during a long ride and not the muscles in other parts of the body. This may happen if your body is not able to supply enough blood to your leg muscles.

Recovery   Prevention is the best option to handle cramps. Sometimes you will get a warning with a mild tightening of the muscle just before the full blown cramp occurs. If you can avoid or limit using the particular muscle for some time, there is a good possibility that you will not experience a cramp. Clearly once you suffer a cramp in your leg muscles, you want to alleviate the pain as soon as possible. Suggestions to recover from a cramp include gently massaging the cramped muscle, stretching your leg, and staying hydrated.

If you continue to ride after getting cramps and recover, the likelihood of multiple occurrences of a cramp in the same muscle will increase. Since a few major leg muscles are the ones used most often in cycling, you have little choice but to continue using the same muscles with lesser power. Stopping and stretching is an effective method to relieve the cramp, but there is still a possibility of the cramp re-appearing later in the ride.

Incremental training can also prepare your muscles for what will be expected on a long ride. One recommended technique is to increase the mileage during training per week by a maximum of 10%. If you experience a cramp in a 100 km. ride, then you will almost certainly have multiple cramps on a 200 km. ride. Training by riding upto 75% of the distance of an upcoming long

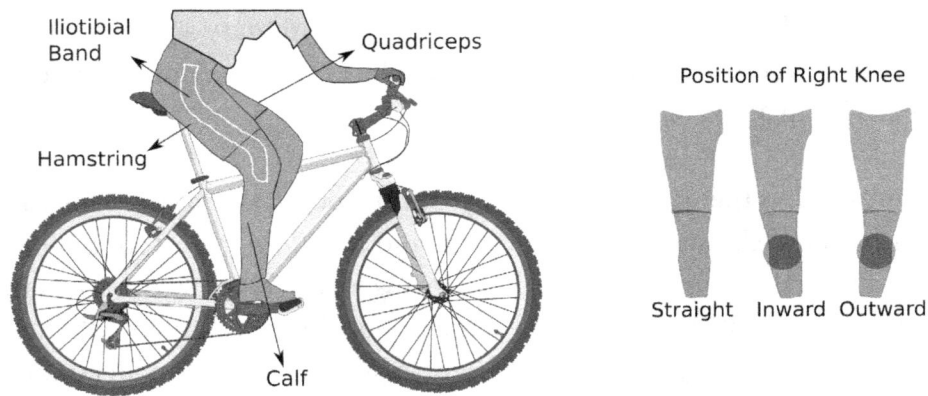

Figure 9.16: Upper Leg and Knee Muscles

ride should be sufficient to prevent cramps. You can train for long rides using a schedule from *Bicycling* magazine (see Section 9.2.11).

9.2.7 Knees and Feet

While a cramp is a sharp pain that you cannot ignore, a mild pain in the knees may start without hindering your ride as much. Often, the knee pain vanishes when you stop riding. Although cycling is a low impact activity for your knees compared to running, on a long ride the repetitive nature of thousands of flexes of the knees can lead to injury. The knee is a complicated joint and there are many things that can go wrong on a long ride.

If you pedal with your knees facing inwards or outwards (see Figure 9.16), you are likely to be less efficient and also encounter knee pain. Since force must be transmitted to the pedal vertically, some of the force will be directed inwards or outwards when your knees are not in the same plane as your thighs and calf. If the distance between your hips is wider or narrower than the distance between the pedals, then it may not be possible or even painful to get a perfect vertical alignment of your upper and lower leg. While the ideal position is straight, it may not be possible for all cyclists to ride in this position.

Saddle Height   Two simple recommendations to avoid knee pain are - maintaining the "right" saddle height and gear. When your saddle is too **high**, your leg extension is excessive during a pedal stroke. The repeated extensions stretch and compress the iliotibial band (a collection of fibers that runs along the side of the leg connecting the hip to the tibia – see Figure 9.16). The nerves located behind the kneecap between the iliotibial band and the femur bone below are compressed, causing the pain felt behind the knee.

When the saddle is too **low**, not only are you generating less power than you could, you are also bending your knee excessively. The muscles near the kneecap and quadriceps (a group of four muscles around the thigh) are pushed against the femur. The quadriceps are one of the main muscles used to generate power in the downward pedal stroke. Instead of being applied to the pedals, some of the generated power will be used to pull the kneecap inwards, leading to pain in the tendons around the knee.

Since the saddle has limited horizontal movement (forwards and backwards) compared to vertical movement, most problems with the saddle are due to an incorrect saddle height. However, riding with a saddle that is too far back can also stress the iliotibial band and similarly a saddle that is too forward will stress the front of the knee.

High Gear   If you ride in the city, you may have to stop many times due to traffic or stoplights. When you start moving again, you have to generate substantial power depending on your gear. This is because not only do you have to overcome drag forces, you also have to accelerate to gain speed. In a high gear, it takes a lot of effort to apply power on the pedals, which increases the forces on your knees. Therefore on a higher gear, the time to regain speed is longer and can lead to knee pain due to the repetitive high forces on the knees. On a lower gear, the force on the pedal is lower and correspondingly the forces on the knees are lower as well. You can also accelerate faster by simply spinning the pedals at a higher cadence.

Cadence   On an endurance ride, you want to plan the gear ratio and a cadence range for the long duration of the ride. The risk of injury from repetitive strain is high in a long ride of 200 kilometers or more. If your gear ratio is 2.6 on a road bicycle with 700×28 tyres, then it would take about 36,000 pedal strokes to cover a distance of 200 kilometers. At an average cadence of 60 rpm, the time for the ride would be 10 hours. Changing to a lower gear of 2.1 on the same bicycle, it would take about 88,000 pedal strokes to cover 400 kilometers and the ride would take over 24 hours. These are upper limits of the number of pedal strokes and in reality, you would spend some fraction of the time coasting.

Since using energy as efficiently as possible is key to finishing an endurance ride, you would like to use the most optimal gear and cadence. A study [28] showed that riding on a low gear (low power of 50 watts) at 110 (high) rpm was actually only transmitting 40% of the generated power to the crank while the remaining 60% was used to move the legs up and down.

Yet, professional cyclists ride at close to 100 rpm, which would appear to be inefficient. However, one major distinction between a professional cyclist and an amateur is the difference in power. A professional may generate over 300 watts of power and ride at a much higher gear, while the amateur cannot sustain that level of power for very long. According to physiologists, muscle efficiency is based on the speed at which your muscles can contract. "If you choose a gear and cadence that allows your muscles to contract at one third of their maximum velocity, you'll maximise your power output" [28].

The idea behind riding at a high cadence is that your muscles will contract more frequently, but not as intensely. Although, the force applied to the pedals decreases, your power generated will not reduce, since the period between the applied force on the pedal also decreases. A study [29] found the heart rate of non-professional cyclists increases with higher cadence. With a high heart rate, it becomes difficult to consume enough oxygen. There appeared to be no benefit for a recreational cyclist to increase cadence beyond 90 rpm.

The ideal cadence is in a range of 70-100 rpm and on most rides your cadence will vary more or less than some average value. Still this range is a higher cadence than the range of a typical cyclist who may prefer to ride at close to 60 rpm, which is close to the walking cadence (in steps per minute).

Feet   Finally, all the power generated by the leg muscles is transmitted to the pedals through the feet. Therefore, this is another part of the body that can afflict a cyclist. Like saddle issues due to friction, the same problems arise with feet. Repeated pressure on some part of the foot can cause blisters.

A study [30] of about 400 cyclists found that roughly half experienced some foot pain while riding. Most of the pain was located in the upper region of the foot near the toenails, toes, and balls of the foot. The study also found that riders with cleated shoes had a higher risk of pain. This is not unexpected, since with cleats the area of the foot that is used to apply pressure on the pedal is localised compared to the larger area of the foot that can be used on flat pedals.

A loosely fitting shoe means that your feet will be repeatedly rubbing the insole, upper, and side of the shoe. A tight shoe also has problems that include - discomfort from pressure on the foot, restricted blood flow, and numbness. While riding your feet may swell due to the additional blood flow to the muscles of the foot, therefore a shoe that was fitting before a ride may become tight during a ride. A loosely fitting shoe can be made more snug with thick socks, but a tight shoe cannot be remedied.

Standing vs. sitting   You may have observed that on an uphill, many professional riders stand and pedal. One obvious advantage of standing and pedalling is that you will be able to apply a greater force on the pedals than when you were sitting. The part of your weight that was supported by the saddle must now be supported by the pedals as well as your upper body. Typically, the decision to stand and pedal is made when the cadence rate falls below what seems efficient (< 60 rpm).

While you will be able to use additional muscles to generate power when standing, you also use some energy in staying balanced. Standing and pedalling will increase your heart rate by a few beats per minute or more. This alone may not be a factor, but when you are riding close to your lactate threshold, it may lead to exhaustion. One study [31] estimates that you can generate about 37% more power while standing than sitting with a corresponding increase of eight heart beats per minute. Of course this additional power does not come for free. While your muscles still have to generate this power, efficiency is not compromised.

Since you are generating about 37% more power when standing than sitting, you will ride faster and therefore save time. Assume a 70 kg. rider on a 10 kg. bicycle with no wind and a commuter aerodynamic factor, riding three kilometers uphill. The rider with the least power on

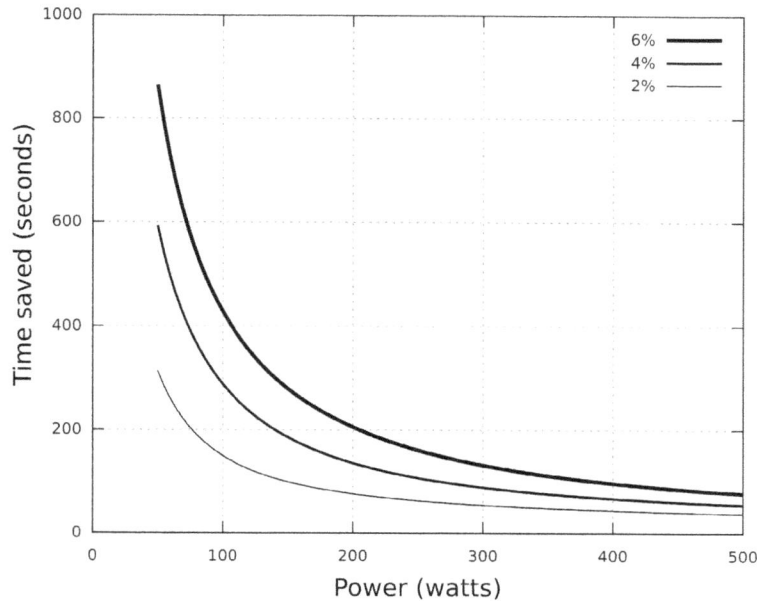

Figure 9.17: Time saved in seconds when standing and pedalling vs. Power in Watts for three gradients on a three kilometer uphill ride

the steepest gradient gains the most time when standing and pedalling (see Figure 9.17). This is intuitive since the change in speed is much larger when power is low.

If the generated power when sitting is 50 watts, then your speed on a 2% gradient would be 8.2 kmph. Generated power increases to 68.5 watts when standing and speed increases to 10.7 kmph (an increase of about 30%). When the generated power while sitting is 300 watts under the same conditions, speed is 29.5 kmph. Generated power increases to 411 watts when standing and the corresponding speed becomes 34.6 kmph (an increase of about 17% compared to sitting).

The steeper the gradient, the larger the slope resistance and more power makes a larger difference in time saved. For a 2% gradient at 200 watts, you would ride at 23.4 kmph. and 28.0 kmph. while sitting and standing respectively (an increase of about 20%). When the gradient is 6% at 200 watts, your sitting and stand speeds would be 12.7 kmph. and 16.7 kmph. respectively (an increase of about 32%).

A disadvantage with standing is that your frontal area is larger when standing and therefore air drag is larger. However, most riders stand and pedal on a climb at speeds that are not very high. On a sprint, riders maybe off the saddle, but in a very aggressive aerodynamic position. Some riders lean forward while standing. This does help in positioning your feet over the pedals at a slight angle when your feet are at zero degrees or the top dead center. However, if you lean forward too much then you may push the front wheel into the pavement and the rear wheel may lose traction.

When one of the pedals is at the bottom dead center (180°), most of your weight is on one side of the bicycle and therefore you may have to tilt the bicycle to the other side to maintain balance. Most riders rock the bicycle a few degrees while standing and pedalling, but rocking too much will be a waste of energy.

Additionally, your center of mass (COM) will be located closer to the front of the bicycle when standing. This is an issue when braking (see Chapter 6) since the normal force on the rear wheel will be less than the normal force when sitting, and therefore more likely to skid when you brake. However since more riders stand and pedal on an uphill than on a flat, braking forces are minimal and unlikely to cause a skid.

9.2.8 Do you really want to suffer?

A 1-2 hour bicycle ride is very different from a 10+ hour ride. The mental and physical challenges in long rides can lead you wonder if it is worth it. Among the reasons a cyclist may ride several hundred kilometers include - confronting the mental and physical challenges of a long ride, bragging rights, and simply getting away from it all. Although you may assume that long hours on a bicycle become monotonous and dull, you cannot afford to lose focus. In the back of your mind you have to manage your thoughts, monitor traffic on the road, sense the tiredness of your body, and safely get to the next destination in the ride.

A long bicycle ride is like an escape from your day-to-day activities and when the road is open and relatively free from traffic, you are free to linger on thoughts and ideas. The psychological explanation for why anyone would stress their body, when most of the time we seek to minimize effort, is that the experience of accomplishing a hard task is valued highly and remembered for a long time.

The outcome of a long round trip ride of several hundred kilometers may seem pointless to an observer, since it appears as if you expended a lot of energy to end up where you began your ride. However, the reward from the completion of a long ride compensates for the hard work and effort of the ride. The primary experience that a rider recalls of a long ride is often not the difficulties faced during the ride, but the happiness at the end of the ride.

Finishing a long ride Even though you may have trained sufficiently, on the day of a long ride you maybe tempted to give up as the day progresses and the workload becomes more intense. One or more parts of your body maybe aching and the temptation to stop increases. If you are not familiar with the route, then riding on an unfamiliar road will increase your anxiety. Fear will make your muscles tense when you should actually be breathing deeply to stay relaxed. The combination of stress, anxiety, and fear of riding on an unfamiliar route can overwhelm a rider leading to a mental block.

A common suggested method [32] to overcome the mental block is to break the goal of completing a long ride into sub-goals of smaller sections of the ride. If you are currently tired and realize that you have to cover another 100 kilometers to complete a ride, it may not appear achievable. Instead, you can focus on completing the next 25 kilometers alone and then set another goal of completing the following 25 kilometers and so on. Of course, this may mean that you will have to slow down and it takes some patience to complete these sub-goals when you are riding at below your normal pace.

Another suggestion [5] is to visualize success and think positively. Visualizing yourself completing the ride and achieving your goal is a proven way to overcome a mental block. If you can imaging yourself climbing uphill at moderate speed, some of the inherent anxiety and tension will reduce. Once you have climbed one hill, you can repeat the process at the next one. All of this assumes that you are hydrated and have eaten sufficiently.

When to stop riding   Cyclists do not begin a ride thinking that they will not finish the ride. Despite positive thinking and training, there maybe a point in your ride when you should stop riding. If you have a fever or are otherwise unwell, it is not worth risking permanent damage to your body. Your heart rate is elevated when you have a fever and the stress of riding besides staying cool will increase your heart rate further, increasing your body temperature to the danger zone.

On a long ride, you maybe so tired that you simply fall asleep while riding. Needless to say, this is highly dangerous and other vehicles on the road have no clue that you have fallen asleep. Anything can happen and it is simply not worth the risk of riding, if you are likely to fall asleep.

9.2.9   Staying cool

No matter how hard you try to stay cool when cycling, your body cannot convert all the chemical energy in your muscles to mechanical energy and a substantial fraction (75% or more) is converted to heat energy. Only 25% of your effort is converted to force on the pedals with the remaining 75% released as heat energy. If this heat is not released, you risk becoming overheated and your performance will drop.

When the external air temperature is much less than your body temperature, you will not notice the heat you generate as much, since it is quickly dissipated into the surrounding air. However if the weather is hot, then the generated heat adds to the ambient temperature and you end up sweating more than you would normally. While you do sweat in cold weather, your sweat rate increases substantially in hot and humid weather. It is also obvious when you are sweating profusely that you will need to replace the lost fluid at some point to stay hydrated.

Practically all cyclists will agree that it is harder to ride when the weather is hot (> 30 °C) than on a pleasant cool day. The two ways that you can lower your body temperature are -

- **Convection**: Even on a still day without any wind, resistance from air flow is a large drag force. Although air drag slows you down, it does help in keeping you cool. The flow of air removes heat from the surface of your body.

- **Evaporation**: When sweat on the surface of your body evaporates, it absorbs some of the surrounding heat and adds to the moisture in the air. If the air is humid, then rate of evaporation is lower. Therefore less sweat evaporates and the remaining sweat is simply lost to the ground or some other surface.

Convection   A study [33] of the heat transfer and drag from different body segments, found that some parts of the body have a much higher rate of heat transfer than others. The same study also compared different cycling positions. While drag can be substantially reduced when a cyclist changes from an upright position to an aggressive position (hands on the drops of the handle bar), the change in positions does not change the rate of heat transfer as much.

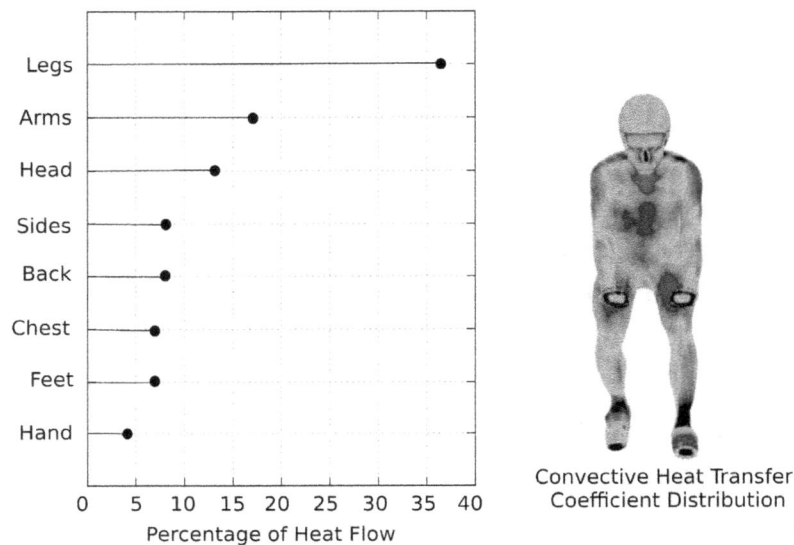

Figure 9.18: Percentage of Heat Flow and Convective Heat Transfer Coefficient Distribution for a Cyclist [33]

The loss of heat due to convection is measured by Newton's law of cooling which states that the loss of heat from a body is proportional to the difference in temperature between the body and its surroundings in the presence of a wind or $Q \propto (T_s - T_a)$ where Q is the heat loss per unit time, T_s is the temperature of the body surface, and T_a is the ambient temperature of the surroundings. Introducing a constant of proportionality, the equation becomes $Q = CHTC \times A \times (T_s - T_a)$ where $CHTC$ is the convective heat transfer coefficient and A is the area. The units for the constant $CHTC$ are expressed in $\frac{watts}{m^2 \times K}$.

If $CHTC$ is higher for some parts of the body than others, then more heat will be lost due to convection in those parts. High $CHTC$s were found in the legs, arms, hands, and feet (see Figure 9.18, the colour image in the study shows differences in $CHTC$ with red and blue colours for high and low $CHTC$s respectively). The less exposed parts of the body like the back had lower $CHTC$s, but since the area of the back is relatively large, heat loss from the back accounts for about 8% of the total heat lost. Conversely the heat lost from your hands is relatively low, even though the $CHTC$ for hands is high, the hand area is much smaller compared to the rest of the body.

The average $CHTC$ was $90\frac{watts}{m^2\times K}$ and $88\frac{watts}{m^2\times K}$ for the upright and aggressive positions respectively. There appears to be no heat loss benefit from riding in an aggressive position. Some assumptions in the study include an uniform body surface temperature and a relatively high speed of 60 kmph. On a hot day, a black shirt is likely to have a higher surface temperature than other parts of the body.

Evaporation   The ambient temperature is often not the real temperature that a rider feels. The heat index published by meteorologists usually only takes into account the relative humidity. The Australian Apparent Temperature formula [34] for the actual temperature is a function of the ambient temperature, the relative humidity, and the wind speed (published by Robert G. Steadman in 1994). For a cyclist, the wind speed is a significant factor in calculating the apparent temperature. The formula to calculate the apparent temperature is defined as

$$T_a + 0.33 \times \rho - 0.7 \times ws - 4.00$$

where ws is the wind speed in kmph. and T_a is the ambient temperature in centigrade. The value of ρ is defined as $rh \times 6.105 \times e^x$ where rh is the relative humidity and x is defined as $\frac{17.27\times T_a}{237.7+T_a}$. The relative humidity is measured as a percentage and represents the fraction of water vapour in the air compared to fully saturated air. When the air is dry, it is easier to stay cool since sweat evaporates more easily than in humid air and therefore the apparent temperature of humid air is higher than the ambient temperature.

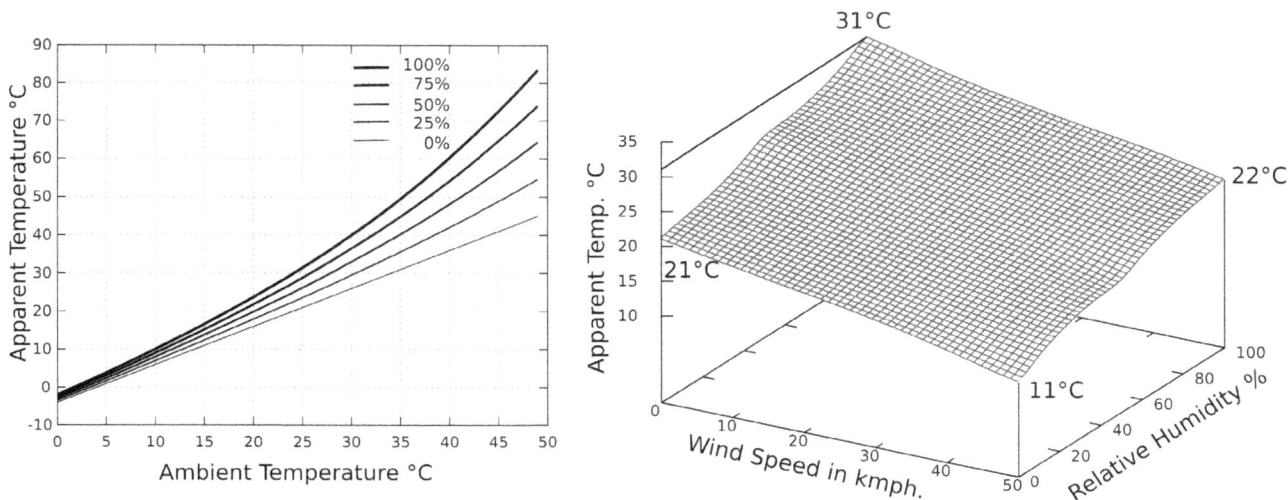

Figure 9.19: Effects of Humidity alone on Ambient Temperature and the Effects of Wind + Relative Humidity on Ambient Temperature of 25°C

The left chart in Figure 9.19 shows that an increase in relative humidity is accompanied by an increase in the apparent temperature. At low ambient temperatures (< 10°C), the effects of relative humidity on ambient temperature are not noticeable (in fact the apparent temperature is lower than the ambient temperature). At higher ambient temperatures, the effects on the apparent temperature due to humidity are more pronounced. For relative humidities of 25%, 50%, 75%, and 100%, the apparent temperature is 9%, 26%, 43%, and 61% respectively higher than the ambient temperature of 45 °C.

The right chart shows the combined effects of wind speed and relative humidity on the apparent temperature given a fixed ambient temperature of 25 °C. At 0% relative humidity, as wind speed increases from 0 kmph. to 50 kmph., the apparent temperature drops from 21 °C to 11 °C. For every 10 kmph. increase in wind speed, the apparent temperature drops by about 2 °C. This same pattern is observed when the relative humidity is high. While higher wind speeds do help in keeping the apparent temperature lower than the ambient temperature, the benefits are lost due to the much higher air drag force.

9.2.10 Riding Alone

Even though a group ride may start off with 50+ cyclists, you may find yourself riding alone after 5-10 kilometers. This can happen if you are not part of a sub-group within the larger group of riders or you are not able to keep pace with a main group. You may also decide to ride alone, since you can ride at your own pace, schedule, and set the route.

The lone rider does not have a group of cyclists to provide motivation and encouragement and must be self-reliant to persevere through a long ride. Since you will be alone, a pre-bike checkup is critical. A flat tyre is the most likely failure that will stop your ride. You will need to carry tools and a spare tube to fix a flat tyre. There is nothing more deflating than being stuck far away from home on the side of a road with a flat tyre. The other components that can fail include the brakes and chain.

The following is a minimal list of items that you should carry - a portable pump, tyre levers, mini-toolkit, spare tubes, cash, identification, cell phone, bicycle lights (if you will be riding in the dark), water bottles, some food, helmet, and glasses to protect your eyes. With your cell phone and GPS, there are a few apps that can track your ride and relay your current location to friends and family.

If you prefer to train alone, then you can set your own schedule and ride as much or as little as necessary. When you are tired and want to stop, you can do so without forcing your fellow riders who may feel obliged to stop as well. Since you will be riding alone, you should have a good idea of your route and the resources you will need to return home. Getting lost is less likely with a GPS device, but is a possibility that cannot be ruled out. Finally, you should save enough energy to make it back home. The incremental strategy to train for longer distances in the following section is a guideline that you can use to ride longer and further.

9.2.11 Schedule for a 100 km. Ride

Beyond riding faster on short distances, a long ride is a common goal for many cyclists. A long ride can range from 50 to a thousand or more kilometers. Riding a metric century or 100 kilometers may seem daunting at first, but is within the reach of the vast majority of cyclists. Riding clubs and even bicycle shops conduct events for a 100 kilometer ride and riding with a large group makes the ride enjoyable as well as an achievement.

The schedule below from *Bicycling* magazine [5] was originally written for a century in miles which is 60% longer than a century in kilometers. The duration of the training is ten weeks with specific distances for each day. While it maybe hard to ride exactly the same distances mentioned in the schedule, reasonably close distances should be sufficient to finish 100 km. on the last day of the schedule.

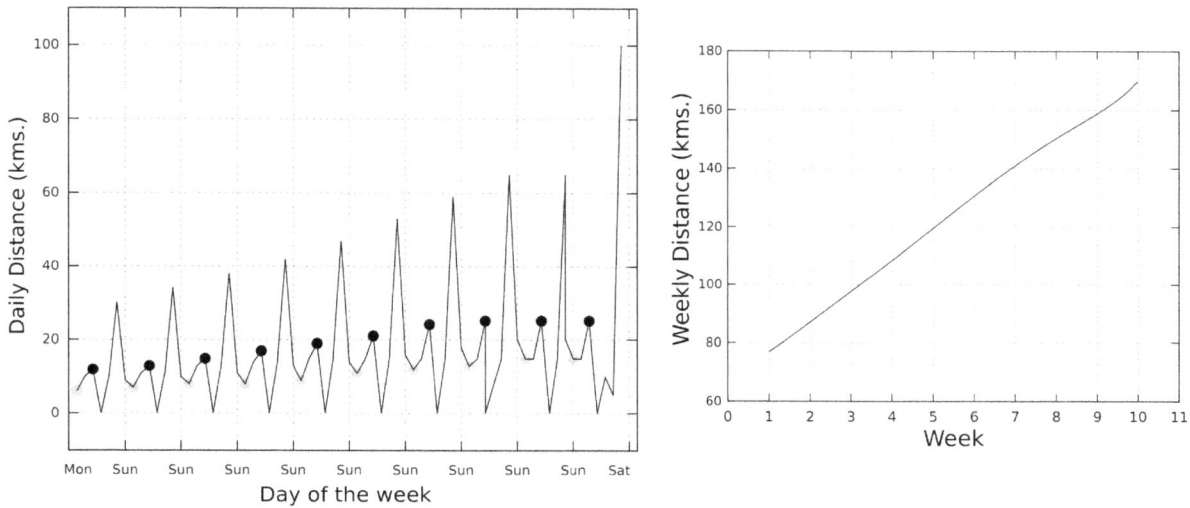

Figure 9.20: Training Schedule for a 100 Km. Ride

The schedule begins on a Monday and ends 10 weeks later on a Saturday (see Figure 9.20). The lightly shaded circles and darkly shaded circles represent training at an easy and brisk pace respectively. On a Monday, the pace is easy while on a Wednesday, the pace is brisk. An easy pace is when your heart rate is in zone 1, while in a brisk pace, your heart rate is in the upper range of zone 2. The remaining rides are all at the speed you expect to ride on the last day.

Every Thursday is a rest day and every Saturday is a long ride representing the valleys and peaks in Figure 9.20 respectively. The peaks or long rides get progressively longer by about 10% starting from 30 kms. in week 1 to 65 kms. in weeks 8 and 9. In the right side of the figure, the total distance covered per week shows a steady increase from 77 kms. on week 1 to 155 kms. on week 9. The total distance covered increases by about 12% per week in weeks 1-3, by about 10% per week in weeks 4-7, and about 3% in the final weeks 8-9. While following a schedule with fixed distances may seem regimented, a tested schedule gives you a skeleton plan that you can adapt to build up endurance.

9.2.12 Schedule for a 200 km. Ride

If you have completed a 100 km. ride, then a 200 km. ride may appear to be more of the same, except longer. Several issues to consider in a ride of 8 hours or more include riding in the night, extra batteries, a USB charger, warm clothes, food, and fluids for a longer ride. The schedule (from Bicycling magazine) for the 200 km. ride is very similar to the schedule for the 100 km. ride, except that the duration is increased from 10 weeks to 16 weeks and the distances covered per week and day are also higher (see Figure 9.21).

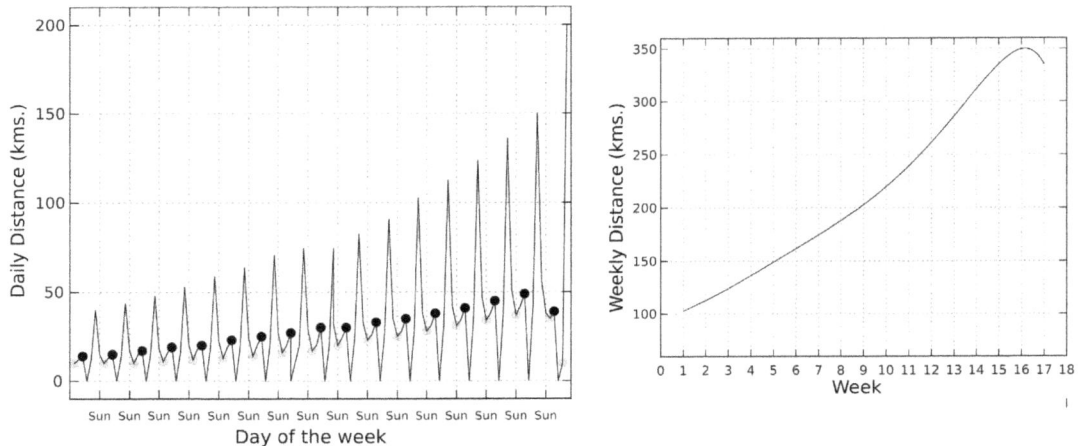

Figure 9.21: Training Schedule for a 200 km. Ride

A small injury or pain on a short ride can be ignored till the end of the ride. In a longer ride, pain can become more intense and it will not be possible to ignore it for too long. Posture on the bicycle and ensuring that the bicycle fits you are critical in long rides. A rough guideline for the time to complete a 200 km. ride is double the time for a 100 km. ride, plus two hours.

9.2.13 300 km. and Beyond

Riding 300 kilometers or more may seem to be beyond the average rider's ability and also require an enormous amount of training time. It maybe surprising that annually many cyclists ride 300 kms. in less than 20 hours [35]. Most of these cyclists are not professionals and do have day jobs.

However, you wouldn't want to attempt a 300 kms. ride without at least riding 100 kms. or more. Each ride becomes a training ride for a successive longer ride. If you can ride 60 kms. with a single water bottle in about two hours, then you cannot extrapolate the same pace for a much longer ride and the same level of hydration. As you ride further, you begin to learn more about your body's ability to cope with long hours on the saddle and the pace that you can sustain for several hundred kilometers.

You do not need a very expensive bicycle to ride 300 kms. A simple road bicycle with an aluminium frame and brazes for bottles is sufficient. While a lighter bicycle will make you a little faster, you do not need a carbon fiber bicycle – a simple, reliable, and comfortable bicycle is adequate.

Surprisingly long distances can be covered at even around 20 kmph., if you maintain a steady pace for the duration. Near the end of the ride, your pace may slow down. When you are tired, making progress can seem sluggish. The odometer on your cyclocomputer is not ticking along as it did at the beginning of the ride. But with a little patience, you can complete the ride, albeit at a slightly slower pace. You will ride further if you ride at a steady pace instead of alternating between riding fast and then coasting. When you ride fast, air drag becomes a major drag force and the extra power that you generate will be spent overcoming this resistance. When you coast, you will slow down and you will have to accelerate again.

If you ride distances of 300 kms. or more, then you can expect to ride in the night. This type of ride may not be appealing to all cyclists. The challenges include drunk drivers, less visibility of the road, and encounters with stray dogs. Since the road ahead is not clearly visible, you cannot ride as fast as you would in the daytime, unless the road is very familiar. A reflective vest and a prominent rear light is essential in the night. Still, a tired motorist may veer into your path or there may not be much of a shoulder on the road increasing your risk of making contact with a larger and heavier vehicle. Tiredness is a factor that you have to deal with and knowing when to stop riding is important to avoid accidents. Often, a brief rest period is enough to resume riding again.

Although you will be sore for a few days after a long ride, rarely will you suffer any permanent harm from an endurance ride. Much like a marathon run, your body will recover after the event with a little rest. The memory of completing a long ride that lingers in your mind is usually the experience of overcoming the difficulties from tiredness, weather conditions, and terrain plus the added happy experience of finishing the ride.

References

[1] Robert Wood,Topend Sports Website. Resting heart rate chart, 2008. URL https://bit.ly/2QotnuV.

[2] Bjarne Martens Nes, Imre Janszky, Ulrik Wisloff, Asbjorn Stoylen, and Trine Karlsen. Age-predicted maximal heart rate in healthy subjects: The Hunt fitness study. *Scandinavian journal of medicine and science in sports*, 23 6:697–704, 2013.

[3] Bicycling.com website. 5 max heart rate training myths, busted, 2018. URL https://bit.ly/2Fa2qtC.

[4] Tim Collins, Ivan Miguel Pires, Salome Oniani, and Sandra Woolley, theconversation.com Website. How reliable is your wearable heart-rate monitor?, 2018. URL https://bit.ly/2Fjt2sx.

[5] Edmund R. Burke and Ed Pavelka. *The Complete Book of Long Distance Cycling*. Rodale Books, 2000.

[6] Dr. Andrew Coggan, Trainingpeaks.com Website. Power profiling, 2008. URL `https://bit.ly/2RXJe45`.

[7] Bikeradar.com Website. What is FTP for cycling?, 2018. URL `https://bit.ly/2OK06ZZ`.

[8] Hunter Allen and Andrew Coggan. *Training and Racing with a Power Meter*. 2010.

[9] Wesley Doyle , Cycling.co.uk Website. Start your engine: aerobic vs. anaerobic, 2016. URL `https://bit.ly/2A9mc2V`.

[10] Wikipedia Website. VO2 Max. URL `https://bit.ly/2zfI6Sj`.

[11] Manu Konchady. GPS Trip Analyzer Android app, 2020. URL `http://bit.ly/2MZBsr1`.

[12] Andrew Heffernan, Experiencelife.com website. Steady-state cardio vs. high-intensity interval training, 2014. URL `https://bit.ly/1A6Vr5a`.

[13] John Ford. 15 minute HIIT cycling workout, 2016. URL `https://bit.ly/2AmeVNn`.

[14] Bicycling.com Website. This is how the dreaded interval boosts your speed – and answers to other cycling training questions, 2018. URL `https://bit.ly/2Eh5QcM`.

[15] James L. Devin, Michelle M. Hill, Marina Mourtzakis, Joe Quadrilatero, David G. Jenkins, and Tina L. Skinner. Acute high intensity interval exercise reduces colon cancer cell growth. *The Journal of Physiology*, 2019.

[16] Chris Carmichael, Trainright.com Website. How long should your longest training ride be?, 2018. URL `https://bit.ly/2SafP6D`.

[17] CyclingWeekly.com website. How to tell whether you are overtraining, and how to avoid it, 2018. URL `https://bit.ly/2GNwZGj`.

[18] Selene Yeager , Bicycling.com Website. Do these 3 rides every week if you want to get faster, 2018. URL `https://bit.ly/2Ks3x79`.

[19] Wikipedia Website. VAM (bicycling), 2018. URL `https://bit.ly/2EwkQ6Z`.

[20] Active.com Website. Why fast pedaling makes cyclists more efficient, 2019. URL `https://bit.ly/2Cmp3YE`.

[21] Takaishi Tetsuo, Yasuda Yoshifumi, Ono Takashi, and Moritani Toshio. Optimal pedaling rate estimated from neuromuscular fatigue for cyclists. *Medicine & Science in Sports & Exercise:*, 28 (12):1492–1497, 1996.

[22] Asker Jeukendrup. Fat burning, using body fat instead of carbohydrates as fuel. URL `https://bit.ly/2HMRO5A`.

[23] Jeff S. Volek, Timothy Noakes, and Stephen D. Phinney. Rethinking fat as a fuel for endurance exercise. *European Journal of Sport Science*, 2015.

[24] John Hughes. Cycling aches and pains, part 1 - a pain in the butt, 2019. URL `https://bit.ly/2BhPj5s`.

[25] Cyclingabout.com Alee Denham. Saddle comfort for cyclists: The best bicycle touring seats, 2015. URL `https://bit.ly/2uRFOWZ`.

[26] Health.com Website. 20 ways to do a plank, 2019. URL `https://bit.ly/2B9UaG7`.

[27] Cycling inform.com Website. What causes muscle leg cramps and spasms when cycling, 2019. URL `https://bit.ly/2FxzgEd`.

[28] Max Glaskin. Ideal cycling cadence: Why amateurs should not try to pedal like Chris Froome, 2018. URL `https://bit.ly/2DqUwtb`.

[29] Selene Yeager. Is high cadence cycling actually slowing you down?, 2019. URL `https://bit.ly/2DK13Rc`.

[30] Cycling Weekly Website. Power transfer, all in your feet, 2013. URL `https://bit.ly/2DW8fIt`.

[31] Hunter Allen and Stephen S. Cheung. *Cutting-Edge Cycling*. Human Kinetics, 2012. URL `https://bit.ly/2I1cXsy`.

[32] Danielle Zickl. 7 ways to overcome the mental blocks ruining your workouts, 2018. URL `https://bit.ly/2tkq6CS`.

[33] Thijs Defraeye, Bert Blocken, Erwin Koninckx, Peter Hespel, and Jan Carmeliet. CFD analysis of drag and convective heat transfer of individual body segments for different cyclist positions. *Journal of Biomechanics*, 2011.

[34] Vcalc.com Website. Australian Apparent Temperature (AT), 2019. URL `https://bit.ly/2IsCEma`.

[35] Jan Heine. 10 common misconceptions about randonneuring, 2015. URL `https://bit.ly/2IwSlbZ`.

A Appendix

A.1 Inseam Length and Frame Sizes

Riding a bicycle that is the right size based on your inseam length (measured from the uppermost inner part of your thigh to the sole of your foot) and height can make a large difference in comfort and riding efficiently. The tables below are reproduced from the book titled - "Biomechanics of Cycling" by Rodrigo R. Bini and Felipe P. Carpes, published by Adis-Springer in 2014. Road and mountain bicycle sizes are typically described in centimeters and inches respectively. Unfortunately all manufacturers do not follow a standard for the remaining parts of a frame of the same size, such as the lengths of the head tube, seat tube, and top tube. You can also use your height to find a suitable frame size.

Inseam Leg Length (inches, cms.)	Road Bicycle (cms.)	Mountain Bicycle (inches)
28.3",72	48	14.5
29.1",74	49	15
29.9",76	50	15.5
30.7",78	51	16
31.5",80	53	16.5
32.3",82	54	17
33.0",84	55	17.5
33.8",86	57	18
34.6",88	58	18.5
35.4",90	59	19
36.2",92	61	19.5
94.4",94	62	20

Height (inches, meters)	Road Bicycle (cms.)	Mountain Bicycle (inches)
4'11"-5'3",1.5-1.6	48	15
5'3"-5'7",1.6-1.7	50,52,54	16-17
5'7"-5'11",1.7-1.8	54,55,56	18-20
5'11"-6'3",1.8-1.9	57,58	21-22
6'3"+,1.9+	60,62	22+

A.2 Drag Force Calculations

The calculation of drag forces on a bicycle is complex for several reasons. There are three types of drag forces - aerodynamic drag, slope resistance, and rolling resistance. Aerodynamic drag calculation is usually not as accurate since it is dependent on the square of the velocity including the head wind velocity. The velocity of wind and direction is often not constant and using an average value of wind speed and direction is an approximation. The formula for calculating the air drag force is $\frac{1}{2}\rho \times A \times C_d \times v^2$ where ρ is air density, A is front area, C_d is the coefficient of drag, and v is the velocity of wind relative to the bicycle.

Front Area   Air drag is also directly proportional to front area A. A single value for the front area is an approximation. The value of front area will change depending on the rider who may change from a racer position to a relaxed position at any given time. One equation (DuBois formula) for calculating the total surface area of a person is $0.164 \times h^{0.422} \times w^{0.515}$ where h and w are in meters and kgs. respectively. So for a person of height 1.78 meters and weight 70 kgs., the total surface area is $1.87\ m^2$. However, for a cyclist only $\frac{1}{3}$ or less of this area is considered in the calculation of drag force. Still, this estimate is higher than commonly used values which range from $0.3 - 0.6\ m^2$.

Another empirical way of calculating your individual front area is to take a photo of yourself on the bicycle from the front and capture the silhoutte (see FigureA.1). You can calculate the area of the silhoutte in a tool like Photoshop or Gimp. The area of the silhoutte is given by the number of pixels (186,612) in the cutout. To convert from pixels to meter squared, capture another image from exactly the same distance as the cyclist holding a meter scale or some other object with a known length.

Figure A.1: Capturing Front Area Image (courtesy Trek)

Now, you can find the length of the meter scale in pixels. This value will be in the form of x pixels per meter. If the value of x is 610 pixels per meter, then number of pixels in one square meter is 610^2 or 372,100 pixels. Therefore 186,612 pixels corresponds to $\frac{186612}{372100}$ or $0.5\ m^2$. In addition to your front area, you will need to include the front area of the bicycle as well to get an estimate of the total front area.

Coefficient of Drag   The coefficient of drag C_d is a factor in the air drag formula and changing the value of C_d can make substantial changes in the calculation of power. The problem with calculating C_d is that you first need to know the drag force, which in turn depends on C_d. Therefore, estimates of C_d are used. One formula for calculating C_d (from Heil D.P.) is $4.45 \times m_b^{-0.45}$ where

m_b is the mass of the bicyclist. So, the value for C_d for a bicyclist weighing 70 kg. is 0.66. This formula gives lower values for C_d with higher masses. The C_d for a bicyclist weighing 100 kg. is 0.56 compared to 0.76 for a 50 kg. bicyclist. None of the other components of the drag force - air density (ρ), front area (A), and velocity (v) depend on mass.

Air Density There are several ways of measuring air density and this is one method to calculate air density based on the temperature, relative humidity, and altitude. Air density is a factor when calculating aerodynamic drag. The density of dry air at sea level at a temperature of 15°C is approximately equal to $1.225\ kg/m^3$. If the humidity of air increases, then the density decreases. Similarly if elevation or temperature is higher, density decreases as well. You can calculate the air pressure P_{loc} for a given temperature and altitude using the hypsometric formula.

$$P_{loc} = \frac{P_{sea}}{(1.0 + \frac{h \times 0.0065}{T})^{5.257}}$$

where h is the altitude in meters, T is the temperature in kelvin, and P_{sea} is the pressure at sea level (1013.25 hPa). To account for relative humidity R_h, the saturated pressure is first calculated using Teton's formula

$$P_s = 0.61078 \times e^{(\frac{17.27 \times T}{T + 237.3})}$$

Then the air density ρ is calculated as

$$\rho = \frac{0.0034848}{T + 273.15} \times (P - 0.0037960 \times R_h \times P_s)$$

While a higher relative humidity appears to make the air more dense for a bicycle rider, air density is actually lower with higher humidity. This is due to the lighter weight of water molecules compared to oxygen and nitrogen molecules. A cubic meter of air has some fixed number of molecules and higher humidity means lighter water molecules displacing other air molecules leading to lower air density. Riding in humid air may take more effort because it is harder to breathe since the number of oxygen molecules is fewer and also stay cool at the same time.

A.2.1 Calculating Velocity from Power

The formula for calculating power P (see Chapter 4) given all drag forces is $P = v_b \times (K_a \times v_t \times v_t + F_r + F_s)$ where v_b is the velocity of the bicycle, K_a is the aerodynamic drag factor, v_w is the velocity of the wind, $v_t = v_b + v_w$, F_r is the rolling resistance drag force, and F_s is the slope resistance drag force. To find velocity v_b from a given power P, the function $f(v_i)$, the difference between an estimated value of v_b i.e. v_i and the power P is minimized (η is an efficiency factor, typically 0.95).

$$f(v_i) = v_i \times (K_a \times (v_i + v_w) \times (v_i + v_w) + F_r + F_s) - \eta \times P$$

or

$$f(v_i) = K_a v_i^3 + K_a v_i v_w^2 + 2K_a v_i^2 v_w + F_r v_i + F_s v_i - \eta \times P$$

If v_i is the estimated value of v_b, then the solution of $f(v_i) = 0$ will return a close value for v_b. If v_0 is the first estimate of v_b, then the next estimate v_1 of v_b is defined as

$$v_1 = v_0 - \frac{f(v_0)}{f'(v_0)}$$

The function $f(v_i)$ is differentiated with respect to v_i, and $f'(v_i)$ can be defined as

$$f'(v_i) = 3K_a v_i^2 + K_a v_w^2 + 4K_a v_i v_w + F_r + F_s$$

After a few iterations, the difference between v_{i+1} and v_i should be close to zero, signalling a convergence to a solution. You can also include F_a the force needed to accelerate in the equation. However F_a is the product of mass and $\frac{dv}{dt}$ the acceleration, which makes the resulting equation a differential equation. Numerical methods can be used to solve this equation to find the value of v. With a GPS device the empirical acceleration can be easily calculated, since the previous velocity, current velocity, previous timestamp, and current timestamp are available.

A.2.2   Calculating Altitude from Pressure

The hypsometric formula allows for conversion from altitude to pressure and vice versa for a given temperature. The pressure P_{sea} at sea level is assumed to be 1013.25 mbar, the pressure P_{loc} is the pressure at a given location, the temperature T is defined in Kelvin, and the altitude h in meters at the location is defined as -

$$h = \frac{((\frac{P_{sea}}{P_{loc}})^{\frac{1.0}{5.257}} - 1.0) \times T}{0.0065}$$

A.3   Calories

While you are cycling, you are performing work to move the bicycle and yourself some distance x meters. The amount of work w is calculated as the product of the applied force f and the distance x. Some energy is consumed to generate the applied force. The quantity of work and energy is measured in the same units (joules). Therefore the energy required to move a bicycle and rider a distance of x meters on a flat road with no wind is the product of the applied force and distance x divided by the efficiency of the rider. There are at least three ways to calculate the energy consumed on a ride.

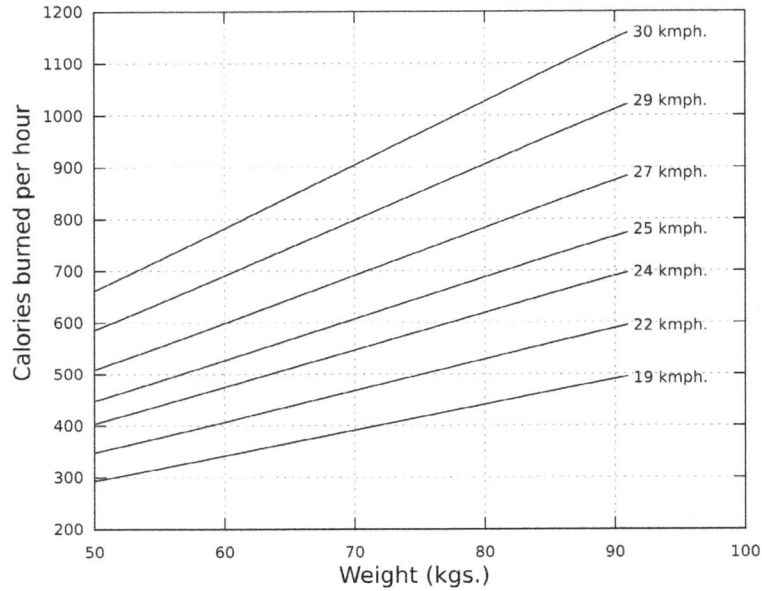

Figure A.2: Calories burned per hour vs. weight in kgs. for different speeds

Method 1 The figure above shows the calories consumed at different speeds and cyclist weights (from a table published in Bicycling, May 1989). The table assumes that there is no wind and the ride is on flat terrain in an upright position for a period of one hour. As you would expect, a heavier rider consumes more calories at the same speed than a lighter rider. The increase in calories consumption is roughly linear with the increase in weight and has a slightly steeper slope at higher speeds.

The actual calories burned will often be higher than the values shown in the Figure. It is very rare that a cyclist will ride on a perfectly flat terrain with no wind for an hour outdoors. The next method takes into account the power generated by a cyclist, which depends on the wind direction and gradient, to calculate the calories consumed.

Method 2 The faster you ride, the greater the applied force and correspondingly your generated power is also higher. Instead of calculating the product of force and distance to estimate the energy consumed, it is easier to use the average power and duration of the ride. The formula for energy consumed is $\frac{t \times p}{\eta \times 1000}$ where t is the time in seconds, p is the power in watts, and η is the cyclist's efficiency in converting energy to force (the result is converted from joules to kilojoules). With an average of 150 watts for an hour, a rider and bicycle weighing a total of 80 kgs. would cover a distance of about 28 kms. At an efficiency of 25%,the energy used is $\frac{3600 \times 150}{0.25 \times 1000} = 2160$ kilojoules or $\frac{2160}{4.2} = 514$ calories. This is the formula used in `bikecalculator.com`.

The number of calories burned is largely dependent on the efficiency of the rider. If efficiency is reduced from 25% to 15%, then the number of calories consumed increases from 514 to 857 calories. A highly efficient rider with 30% efficiency would consume half the number of calories

344

of a rider with 15% efficiency. You would expect professional cyclists to have higher efficiencies than the efficiencies of amateur cyclists.

Method 3   The third method to calculate calories is based on the metabolic equivalent or METs measurement for an activity (see `caloriesburnedhq.com/calories-burned-biking/`). The METs for a given ride was estimated using the closest speed from a table. The Compendium of Physical Activities also publishes the number of METs for cyclists at different speeds.

Speed (kmph.)	METs
8.8	3.5
15.0	5.8
16.0	6.8
19.2	8.0
22.4	10.0
25.6	12.0
32.0	15.8

This method uses the weight of the rider, the duration of the ride in minutes, and average speed to calculate the number of calories. The formula to calculate calories is $\frac{m \times 3.5 \times w \times t}{200}$ where m is the number of METs, w is the weight of the rider in kgs., and t is the time in minutes. The number of METs m is calculated from the table using the METs associated with the closest speed in the table. For cyclists riding at 22 and 25 kmph., m is 10 and 12 METs respectively. A rider weighing 80 kgs. who rides for 60 minutes at an average of 22 kmph. would consume 840 calories. Increasing the average speed from 22 kmph. to 25 kmph. would imply an increase of 168 calories to 1008 calories.

Summary   The first method is relatively hard coded for certain conditions and a fixed time of one hour. The second method is more accurate, since it takes into account the power generated during a ride. Changes in generated power accurately reflect the changes in gradient and wind direction and therefore using the product of the average power and duration of the ride does give a more accurate estimate of the energy consumed. None of the methods described here take into the account the bicycle type (a MTB or a road bicycle), weight of the bicycle or the type of road.

Method 2 is more likely to be accurate than other methods, since power generated does depend on the weight and aerodynamics of the bicycle. A method based on METs will be less accurate since the relationship between speed and METs does not accurately measure effort. Riding downhill at 40 kmph. you are consuming much less energy than a ride at the same speed on a flat road.

A.4   GPS Trip Analyzer Menu

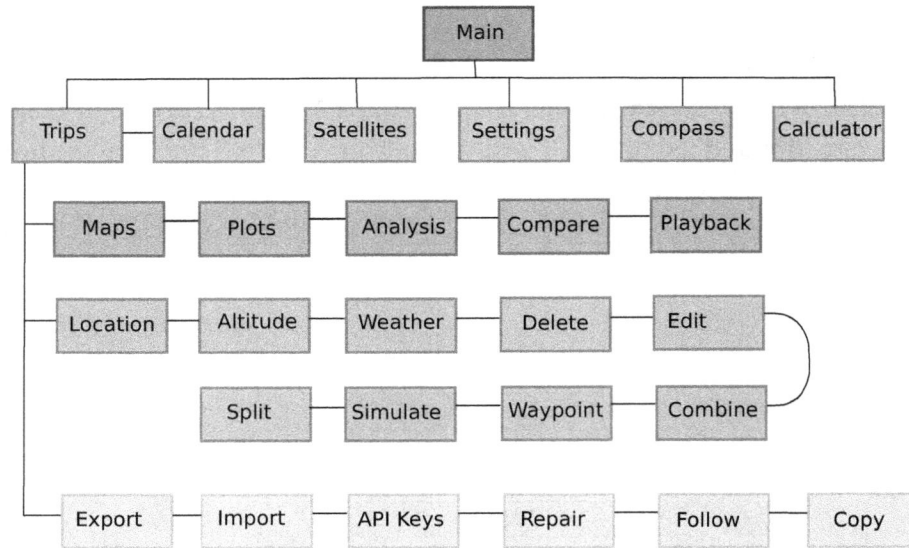

Figure A.3: GPS Trip Analyzer Menu Structure and Functions

A.5   Notation

Symbol	Description
A	Front Area
c_d	Coefficient of Drag
c_r	Coefficient of Rolling Resistance
F_{air}	Drag force due to air resistance
F_{roll}	Drag force due to rolling resistance
F_{slope}	Drag force due to slope or gradient
F_N	Normal force
g	Acceleration due to gravity and also used to measure braking force in terms of gs
I_{wheel}	Rotational mass of a wheel used to calculate rotational inertia or angular mass ω
$PM_{2.5}$	Particulate matter of diameter 2.5 μm
PM_{10}	Particulate matter of diameter 10 μm
K_a	Aerodynamic drag factor
W_a	Wind factor
α	Rate of change of angular velocity ω
μ_s	Coefficient of static friction, used to calculate frictional and normal forces
μ_k	Coefficient of kinetic friction, used to calculate braking forces
ρ	Air density, used to calculate air drag
ω	Angular velocity, used to calculate rotational kinetic energy

Glossary

Aerodynamic Drag: is the air resistance which acts in a direction opposite to the movement of the cyclist. Reducing aerodynamic drag is key to ride at high speeds.

Apparent Wind: is the wind generated by the combined effect of the true wind and the movement of the cyclist.

Attack: is the sudden acceleration of one or more riders who have been riding in a pack to get ahead of the group.

Bicycle Path / Greenway / Bikeway: is a path exclusively for non-motorized vehicles, often shared with pedestrians.

Bonking: occurs when you don't have anymore energy to keep riding. In most cases, the stored carbohydrates or glycogen in your body has been completely depleted and there is no fuel source for muscular energy. In other endurance activities, the phrase, "hitting the wall", is also used instead of bonking.

Breakaway: in a race is a small group of riders who are ahead of the peloton or main group.

Buzzed: is a motorized vehicle moving at a relatively high speed a few centimeters from your bicycle. It maybe interpreted as a sign of impatience with a slow cyclist.

Cadence: is measured in pedal revolutions per minute (below 60 rpm is low and above 100 rpm is high)

Calorie: is a measure of energy (kilojoules). The number of calories consumed in a ride depends on the power you generated, duration of the ride, and your body's efficiency.

Cassette: is a set of cogs kept in place by a lock ring on the rear hub. The ratcheting mechanism is part of hub unlike a freewheel.

Carbon Footprint: measures the amount of CO_2 released as a result of an activity.

Center of Mass / Center of Gravity: is a central location in a body from which the mass is roughly equally distributed in all directions.

Cleat / Clipless Pedals: is a projection on the sole of the shoe that fits into the pedal. It keeps your foot locked into the pedal until you twist or pull your foot out. Prior to cleats, toe clips were used to keep a shoe firmly locked into the pedal. A pedal for cleats does not have toe clips and is called a clipless pedal.

Clincher Tyre: is the most common type of tyre that has an inner tube and a bead on the circumference of the tyre that sits in a groove on the wheel rim.

Cog: See Sprocket

Compact Crank / Crankset : has fewer teeth, say 50 and 34, on the front crank compared to a normal crank with 54 and 39 teeth. A crankset will have 2-3 chain rings.

Contact Area: is the area the tyre makes when in contact with the road. Rolling resistance depends on the shape and size of the contact area.

Cycle Track / Lane: is a riding lane exclusively for cyclists on a shared road, separated from motor traffic.

Derailleur: is the mechanism to move the chain to a higher / lower cog or chain ring. The derailleur is usually controlled by levers mounted on the handlebar.

Digital Elevation Model: is the altitude at any given location defined by a latitude or longitude. The altitude or gradient is used to calculate the slope resistance.

Doored: is when your bicycle hits a car door that a parked motorist unexpectedly opens.

Drafting: is riding closely behind another rider to lower air drag (see Chapter 3: How much speed will you gain by slip streaming or drafting?).

Drag Force: is the combination of the aerodynamic drag, slope resistance, and rolling resistance.

Drive Train: is the set of components including the pedals, crank, chainrings, chain, cogs and derailleur used to transmit power from the pedals to the rear hub.

End Over / Falling Over / Head over handlebar: is a crash over the handle bars, often accompanied with one or more painful injuries. This accident occurs when the bicycle comes to an abrupt stop and the rider continues to move forward over the handlebars.

Fixie or Fixed Gear: is a bicycle where the drive train is fully controlled by pedalling. You cannot coast on a fixie. Pedalling backwards can substitute for a rear brake, which by itself is not sufficient to stop within a safe distance.

Frame Geometry: is the dimensions of different parts of the bicycle frame (See Chapter 2 to find a bicycle that fits your dimensions).

Freewheel: is similar to a cassette with upto 7 cogs, but is threaded onto the hub and includes a ratcheting mechanism

Frontal Area: is the two dimensional area made by the shape of the rider and bicycle in the direction of motion. It is directly proportional to the aerodynamic drag.

Gain Ratio: is the ratio of the distance travelled by the pedal to the distance travelled by the rear wheel or the ratio of the radius of the rear wheel to the crank length times the gear ratio.

Geared Bicycle: is a bicycle with gears and a derailleur in the front and/or rear wheels. A geared bicycle is easier to ride uphill and can be faster downhill.

Gear Ratio: is the number of turns of the rear wheel for a single turn of the crank or the ratio of the number of teeth in the chain ring to the number of teeth in the rear cog.

Groupset: refers to the parts that make up the brakes, gears, and drivetrain. Typically, the parts from a single manufacturer are used together to form a set.

HIIT is high intensity interval training, a repeated schedule of intense activity for a short period followed by a period of recovery.

Intervals: in training represents a schedule of alternating periods of intense hard effort and easy riding.

Lactate Threshold: is a threshold for the intensity of exercise at which lactate builds up in the body. If lactate accumulates at a faster rate than at which it can be removed, you will be forced to stop the activity due to nausea.

MET: is a unit to measure the amount of oxygen consumed. One Metabolic Equivalent (of Task) or MET is the amount of oxygen consumed while at rest.

Pace Line: is a group of riders in a line such that the rider at the front encounters the most air drag while the other riders in the rear follow in the slip stream. Riders take turns to be the lead rider to share the burden of generating power to overcome the head wind at the head of the line.

Panniers: are large bags used in touring bicycles that can hang off a rack on both sides of the front or rear wheel.

Pinch Flat: is a type of flat with a pair of holes in the tube which occurs when a sharp edge on the road pushes the tyre and tube to the rim with sufficient force to perforate the tube.

Peloton: is the main group of riders in a race who ride together to save energy.

Power Meter: is a device attached to the crank or hub to measure the power generated by a rider.

Presta: is a type of valve that is typically longer and more slender than the Schrader valve.

PSI: is a measure of the pressure in an inner tube or type. It represents the number of pounds of force per square inch.

Right or Left Hook: is when a motorist overtakes a cyclist on the left or right and then crosses the bike path forcing the cyclist to brake or risk a collision.

Rolling: occurs when the distance the bicycle moves is exactly the same as the distance the tyre rotates.

Rolling Resistance: is the resistance due to the bicycle tyre rolling on a surface.

Schrader: is a type of valve that is found in car and bicycle tyres. It is usually shorter and wider than a Presta valve.

Single Speed Bicycle is a bicycle with a single gear that also allows you to coast when not pedalling.

Skidding: occurs when the bicycle moves more than the distance that the tyre rotates.

Slick Tyre: is a type of tyre with a smooth surface. It has a better grip on a smooth road due to the larger contact area.

Slipping: occurs when the tyre rotates more than the distance the bicycle moves.

Slipstream: see Drafting

Slope Resistance: is one of the drag forces and represents the additional force due to gravity that you have to overcome when you ride uphill.

Spoke Tension: is the force on a spoke that keeps it stretched. A spoke temporarily loses some tension when the wheel is in contact with the road.

Sprocket / Rear Cog: is a metal wheel with teeth to keep a chain in place.

Stopping Distance: is the braking distance for a bicycle and depends on the type of tyre, road surface, and braking force.

Toe Clip is a metal attachment to the front of the pedal to keep the foot in place

Toe Overlap is when your toe hits the back of the front tyre when steering

Tubeless Tyres unlike clincher tyres do not have an inner tube but does need a specific type of wheel rim

VAM is a measure of fitness and speed. It tracks the average ascent speed in meters per hour

VO_2Max **or Aerobic Capacity** measures how much oxygen your body can consume during an intense ride. An individual's VO_2Max capacity is determined largely by genetics and with some improvement through training.

Wheelset is a type of wheel that is sold in pairs by a particular manufacturer.

Index